Deutsche
Forschungsgemeinschaft

Thermal Processing of Food: Potential Health Benefits and Risks

Symposium

Deutsche
Forschungsgemeinschaft

Thermal Processing of Food: Potential Health Benefits and Risks

Symposium

Editors
Gerhard Eisenbrand (Chairman)
Karl-Heinz Engel, Werner Grunow, Andrea Hartwig,
Dietrich Knorr, Ib Knudsen, Josef Schlatter, Peter Schreier,
Pablo Steinberg, Stefan Vieths

Scientific Secretariat
Sabine Guth, Michael Habermeyer, Monika Kemény
and Doris Wolf

Senate Commission on Food Safety (SKLM)

BICENTENNIAL 1807 WILEY 2007 BICENTENNIAL

WILEY-VCH Verlag GmbH & Co. KGaA

DFG

Deutsche Forschungsgemeinschaft
German Research Foundation
Kennedyallee 40, D-53175 Bonn, Federal Republic of Germany
Postal address: D-53170 Bonn
Phone: ++49/228/885-1
Telefax: ++49/228/885-2777
E-mail: postmaster@dfg.de
Internet: www.dfg.de

British Library Cataloguing-in-Publication Data
A catalogue record for this book is available from the British Library.

Bibliographic information published by the Deutsche Nationalbibliothek
The Deutsche Nationalbibliothek lists this publication in the Deutsche Nationalbibliografie; detailed bibliographic data are available in the Internet at http://dnb.d-nb.de.

ISBN 978-3-527-31909-1

Printed on acid-free paper

Cover Design and Typography: Dieter Hüsken
Wiley Bicentennial Logo: Richard J. Pacifico
Composition: K+V Fotosatz, Beerfelden
Printing and bookbinding: buch bücher dd ag, Birkach

Printed in the Federal Republic of Germany

Inhalt/Contents

Thermal Processing of Food: Potential Health Benefits and Risks
DFG, Deutsche Forschungsgemeinschaft

ISBN: 978-3-527-31909-1

Vorwort

Das DFG-Symposium „Thermal Processing of Food: Potential Health Benefits and Risks" der Senatskommission zur Beurteilung der gesundheitlichen Unbedenklichkeit von Lebensmitteln (SKLM) fand vom 25. bis 27. September 2005 in Kaiserslautern statt. Ziel war eine kritische Bestandsaufnahme und Bewertung des gegenwärtigen Erkenntnisstands. Die SKLM hat im Sinne ihres Beratungsauftrags Schlussfolgerungen und Empfehlungen zu weiterem Forschungsbedarf erarbeitet, die gemeinsam mit den Einzelbeiträgen der Redner und den Posterbeiträgen in diesem Band veröffentlicht werden.

Die SKLM dankt der Deutschen Forschungsgemeinschaft (DFG) für die Möglichkeit im Rahmen dieser Symposienreihe Themen von besonderer Bedeutung für ihre wissenschaftliche Beratungsarbeit in Diskussion mit den besten international auf dem Themengebiet ausgewiesenen Wissenschaftlern zu diskutieren und die Ergebnisse als Symposiumsband zu veröffentlichen. Der Dank des Vorsitzenden gilt den Teilnehmern des Symposiums für ihre wissenschaftlichen Beiträge sowie den Mitgliedern und Gästen der Senatskommission für ihre Mithilfe bei der Abfassung der vorliegenden Veröffentlichung. Ich danke auch den Mitgliedern des Redaktionskomitees, vor allem den Vorsitzenden und Rapporteuren, Prof. Knudsen, Prof. Steinberg, Prof. Knorr, Prof. Engel, Prof. Schreier, Prof. Grunow, Prof. Vieths und Herrn Dr. Schlatter für ihre wertvolle Unterstützung bei der Formulierung der Schlussfolgerungen und Empfehlungen. Das wissenschaftliche Sekretariat der SKLM mit Dr. S. Guth, Dr. M. Habermeyer, Dr. M. Kemény und Dr. D. Wolf hat wesentlich zum Zustandekommen dieses Bandes beigetragen. Ihnen gilt mein herzlicher Dank, ebenso dem Lektorat des Verlages. Für stete, nachhaltige Unterstützung der Arbeit der SKLM sei der Leiterin des Fachreferates bei der DFG Frau Dr. H. Strelen besonders gedankt.

Prof. Dr. Gerhard Eisenbrand
Vorsitzender der DFG-Senatskommission zur Beurteilung
der gesundheitlichen Unbedenklichkeit von Lebensmitteln

Thermal Processing of Food: Potential Health Benefits and Risks
DFG, Deutsche Forschungsgemeinschaft

ISBN: 978-3-527-31909-1

Preface

The DFG Symposium "Thermal Processing of Food: Potential Health Benefits and Risks", organized by the DFG Senate Commission on Food Safety (SKLM), took place from 25 to 27 September 2005 in Kaiserslautern, Germany. The objective of the symposium was a critical appraisal of the situation and an evaluation of the present level of knowledge. In its advisory capacity, the SKLM has drawn conclusions and proposed recommendations for further research, which are published in this volume together with the contributions of the individual speakers and the poster contributions.

The SKLM is grateful to the Deutsche Forschungsgemeinschaft (DFG) for providing the opportunity, during this series of symposia, to discuss topics of particular importance to its scientific advisory work with an international group of recognised scientific experts working in this field, and to publish the results as a symposium volume. I would like to thank those who participated in the symposium for their scientific contributions, as well as the members and guests of the Senate Commission for their assistance in preparing this publication. I am also grateful to the members of the editorial committee, particularly its chair and the rapporteurs, Prof. Knudsen, Prof. Steinberg, Prof. Knorr, Prof. Engel, Prof. Schreier, Prof. Grunow, Prof. Vieths and Dr. Schlatter, for their valuable support in the formulation of the conclusions and recommendations. Members of the Scientific Office of the SKLM, Dr. S. Guth, Dr. M. Habermeyer, Dr. M. Kemény and Dr. D. Wolf, have made a vital contribution to the preparation of this volume, and I would like to express my thanks to them and to the publisher's editorial office. I am indebted to Dr. H. Strelen, Head of the Life Sciences Division 1 of the DFG, for her unstinting support of the SKLM's work.

Prof. Dr. Gerhard Eisenbrand
Chair of the DFG Senate Commission on Food Safety

Thermal Processing of Food: Potential Health Benefits and Risks
DFG, Deutsche Forschungsgemeinschaft

ISBN: 978-3-527-31909-1

I Bericht und Schlussfolgerungen

1 Einleitung

Die Senatskommission der Deutschen Forschungsgemeinschaft (DFG) zur Beurteilung der gesundheitlichen Unbedenklichkeit von Lebensmitteln (SKLM) hat vom 25. bis 27. September 2005 unter Beteiligung von Experten aus dem In- und Ausland ein Symposium zum Thema „Thermal Processing of Food: Potential Health Benefits and Risks“ durchgeführt. Zur Diskussion standen gesundheitlich positive und Risikoaspekte der thermischen Behandlung von Lebensmitteln, insbesondere auch unter dem Blickwinkel der hiermit verbundenen Potentiale innovativer Lebensmitteltechnologie. Die SKLM hat hieraus in Wahrnehmung ihrer bewertenden und beratenden Funktion Schlussfolgerungen formuliert und Wissenslücken aufgezeigt.

2 Allgemeine und lebensmitteltechnologische Aspekte

Das Erhitzen ist eines der ältesten, bewährtesten und das heute gebräuchlichste Verfahren der Lebensmittelbehandlung, das entscheidend zur nachhaltigen Sicherung der Ernährung beigetragen hat. Die thermische Behandlung von Lebensmitteln, wie beispielsweise Kochen, Trocknen, Rösten, Braten, Grillen und Frittieren verändert diese auf vielfältige Art und Weise. Beispielsweise werden Geschmack und Aroma sowie Textur verändert, und die Haltbarkeit kann verbessert werden. Ein Großteil dieser im erhitzten Lebensmittel stattfindenden Prozesse beruht auf der sog. nicht enzymatischen Bräunung, der Maillard-Reaktion.

Die thermische Behandlung von Lebensmitteln lässt sich mit anderen konventionellen Verfahren oder neuartigen Technologien kombinieren, wodurch eine produktschonende und energieeffiziente Verarbeitung eines Lebensmittels ermöglicht wird. So lassen sich beispielsweise durch den Synergieeffekt von Hochdruckbehandlung und schonender Erhitzung effizient Mikroorganismen

Thermal Processing of Food: Potential Health Benefits and Risks
DFG, Deutsche Forschungsgemeinschaft

ISBN: 978-3-527-31909-1

abtöten oder Enzyme inaktivieren, während wertgebende Verbindungen, wie Vitamine, Farb- und Aromastoffe, weitgehend unbeeinflusst bleiben.

Generell ist beim Einsatz neuartiger technologischer Behandlungs- und Verarbeitungsverfahren sicherzustellen, dass sich hieraus keine zusätzlichen mikrobiellen, toxikologischen oder allergenen Risiken als Folge ergeben. Bei der Anwendung solcher Verfahren hat der Gesetzgeber mit der Verordnung EG Nr. 258/97 über neuartige Lebensmittel und neuartige Lebensmittelzutaten ein Instrument zur Gewährleistung der Sicherheit entsprechend behandelter Lebensmittel geschaffen.

3 Gesundheitliche Aspekte thermischer Behandlung von Lebensmitteln

Bei der Beurteilung der thermischen Behandlung von Lebensmitteln sind sowohl positive als auch nachteilige Aspekte zu berücksichtigen, die exemplarisch im Rahmen des Symposiums diskutiert wurden.

3.1 Vorteilhafte Aspekte

Aspekte der Ernährung und Akzeptanz

Das Erhitzen von Lebensmitteln schafft vielfach erst die Voraussetzung für den bestimmungsgemäßen Verzehr. Häufig entsteht erst durch den thermischen Prozess ein Lebensmittel, z. B. im Falle von Brot oder Backwaren. Andere Lebensmittel, wie gewisse Rohgemüse, Kartoffeln, Reis und Teigwaren, werden erst nach dem Kochen für den Menschen genießbar. Die thermische Behandlung von Lebensmitteln ist also ein Prozess, der zur Genießbarkeit und Bekömmlichkeit von Lebensmitteln beiträgt und eine nachhaltige und ausgewogene Ernährung ermöglicht. Auch die Entstehung von Aroma und Geschmack und die damit verbundene geschmackliche Aufwertung sind weitere wichtige Funktionen der Lebensmittelerhitzung. Geruch, Geschmack, Farbe, Textur und das äußere Erscheinungsbild werden durch die Erhitzungsvorgänge erheblich beeinflusst und tragen so maßgeblich zur Akzeptanz des Lebensmittels bei.

Mikrobielle Aspekte

Erfahrungsgemäß kann die Erhitzung von Lebensmitteln die mikrobiologische Sicherheit und Haltbarkeit verbessern. Über die hitzeinduzierte Inaktivierung von pathogenen Organismen/Mikroorganismen, Toxinen oder Enzymen hinaus können zusätzlich antimikrobiell wirksame Substanzen bzw. enzyminhibitorische Verbindungen entstehen, die sich ebenfalls günstig auf die Haltbarkeit auswirken können.

Allergene

In der Regel scheinen thermische Verfahren eher eine Inaktivierung von Allergenen in Lebensmitteln zu begünstigen als das allergene Potential von Lebensmitteln zu erhöhen. So wurde eine schnelle Abnahme der Allergenität schon nach milder thermischer Behandlung z. B. für bestimmte Allergene in Früchten beobachtet. In Studien zum Einfluss des Röstens von Erdnüssen auf das allergene Potential wurde vereinzelt über einen Anstieg berichtet.

Der Nachweis einer Veränderung der Allergenität nach thermischer Behandlung erweist sich meist als schwierig, da u. a. die Löslichkeit der Proteine nach thermischer Behandlung häufig abnimmt. Derzeit weist aber wenig darauf hin, dass während der thermischen Prozessierung von Lebensmitteln neue Epitope gebildet werden. Es gibt zwar einige Beispiele dafür, dass die Maillard-Reaktion in Einzelfällen zu einer Verstärkung der IgE-Antikörperreaktivität von Lebensmittelproteinen führen kann, jedoch konnte das Auftreten von „Maillard-Epitopen", die im Rohzustand nicht vorhanden sind, bisher nicht gezeigt werden.

Weitere Aspekte

Die Aufklärung von Mechanismen bzw. Verfahren, die die Ausprägung gesundheitlich positiver Effekte thermischer Lebensmittelbehandlung begünstigen, ist ein Schwerpunkt des gegenwärtigen wissenschaftlichen Interesses. Mechanistische Aspekte betreffen beispielsweise die Bildung resistenter Stärke und deren Einfluss auf die Darmgesundheit, die Generierung von Substanzen mit antioxidativem oder anderweitigem chemopräventivem Potential oder die Steigerung der Bioverfügbarkeit wertgebender Lebensmittelinhaltsstoffe, wie z. B. von Lycopin in Tomaten bzw. Tomatenprodukten. Gezielte Sortenauswahl und/oder entsprechende Prozesssteuerung kann ebenfalls dazu beitragen, gesundheitlich positive Effekte zu verstärken bzw. nachteilige zu vermindern.

3.2 Nachteilige Aspekte

Thermische Behandlung von Lebensmitteln kann aber auch unerwünschte Folgen haben, z. B. den Verlust an Nährstoffen wie Vitaminen, essentiellen Aminosäuren und ungesättigten Fettsäuren. Ebenso ist eine thermisch induzierte Bildung von gesundheitlich bedenklichen Stoffen möglich. Dies stellte einen weiteren wesentlichen Teil der wissenschaftlichen Diskussion dar, wobei Acrylamid, heterozyklische aromatische Amine, Furan und 3-Monochlorpropan-1,2-diol vertieft behandelt wurden.

Acrylamid

Acrylamid kann beim Backen, Rösten und Frittieren von Lebensmitteln entstehen, besonders in Kartoffelprodukten (z. B. Pommes Frites, Bratkartoffeln, Kartoffelchips), Getreideprodukten (z. B. geröstete Cerealien, Brot, Backwaren), Kaffee und Kakao. Die Bildung erfolgt beim Erhitzen von Lebensmitteln über 120 °C im Zuge der Maillard-Reaktion, im Wesentlichen durch Kondensation der Aminosäure Asparagin mit reduzierenden Zuckern wie Fructose und Glucose.

Acrylamid gilt als wahrscheinliches Kanzerogen für den Menschen (IARC Gruppe 2A, MAK-Kommission Krebserzeugend Kategorie 2), wobei sein Metabolit Epoxypropanamid (Glycidamid) als eigentlich genotoxische Wirkform anzusehen ist. Glycidamid-DNA-Addukte wurden im Tierversuch an Nagern bei entsprechend hoher Dosierung in allen untersuchten Geweben gefunden. Darüber hinaus wirkt Acrylamid im Tierversuch toxisch auf das Nervensystem und die Reproduktionsorgane.

Der Abstand der im Tierversuch kanzerogenen Dosis zur höchsten geschätzten mittleren Acrylamidaufnahme beim Menschen ist verhältnismäßig gering. Die bisherigen Abschätzungen des Tumorrisikos für den Menschen durch verschiedene wissenschaftliche Gremien unterscheiden sich allerdings nicht unerheblich. Bislang vorliegende epidemiologische Studien geben keinen Hinweis auf eine Korrelation zwischen Acrylamidaufnahme und Krebsinzidenz, sind aber in ihrer Aussagekraft limitiert.

Heterozyclische aromatische Amine

Heterozyclische aromatische Amine (HAA) werden beim Braten oder Grillen bzw. vergleichbaren Erhitzungsverfahren von Fleisch und Fisch in Abhängigkeit von Proteintyp, Temperatur und Erhitzungsdauer in unterschiedlichen Konzentrationen gebildet. Die Bildung von HAAs lässt sich bei Einhaltung entsprechender Bedingungen aber weitgehend vermeiden.

Es werden zwei maßgebliche Klassen von HAAs unterschieden, die je nach Temperaturbedingungen und Vorläuferverbindungen in unterschiedlichem Maße entstehen: Isochinolin- (sog. IQ-Verbindungen) und Carbolin-Derivate.

Viele HAAs zeigen mutagenes Potential in bakteriellen und säugerzellbasierten Testsystemen und sind kanzerogen im Tierversuch. Die Bioaktivierung der HAAs zu krebsauslösenden Metaboliten unterliegt beim Menschen starken interindividuellen Unterschieden. Zwischen einzelnen Spezies können ebenfalls große Unterschiede bestehen.

Stark durchgebratenes bzw. übererhitztes rotes Fleisch kann zusätzlich zu HAAs auch weitere kanzerogene Substanzen in vergleichbaren Konzentrationen enthalten. Nicht eindeutig geklärt ist bisher, ob nur HAAs oder eher ein komplexes Gemisch thermisch induzierter genotoxischer Verbindungen als Krebsrisikofaktoren anzusehen sind.

Furan

Furan ist eine flüchtige Verbindung, die in einer Vielzahl erhitzter Lebensmittel wie Kaffee, Brot sowie Gemüse- und Fleischkonserven und in Gläschennahrung für Kleinkinder nachgewiesen wurde. Das Vorkommen in den unterschiedlichsten Lebensmitteln lässt verschiedene Bildungswege vermuten. Die Datenlage zum Vorkommen in verschiedenen Lebensmittelkategorien ist begrenzt, und die meisten Daten wurden nicht im verzehrsfertigen Lebensmittel bestimmt. Die Exposition kann bei dieser begrenzten Datenlage noch nicht zuverlässig abgeschätzt werden.

Nach oraler Gabe von Furan wurden in Ratten und Mäusen unterschiedliche Tumormuster gefunden. In Ratten traten dosisabhängig hepatozelluläre Karzinome, ein hoher Anteil cholangiozellulärer Karzinome und Leukämien auf, in Mäusen hepatozelluläre Adenome und Karzinome. Es wird vermutet, dass die kanzerogene Wirkung zumindest anteilig über einen genotoxischen Mechanismus ausgelöst wird, wobei vor allem der Metabolit *cis*-2-Buten-1,4-dial als genotoxisch wirkendes Mutagen infrage kommt. Auf der Basis der bisher vorliegenden begrenzten Daten zu Toxizität, Stoffwechsel und Exposition ist eine adäquate Sicherheitsbewertung derzeit nicht möglich.

3-Monochlor-1,2-propandiol (3-MCPD)

3-Monochlor-1,2-propandiol (3-MCPD) gilt als Leitsubstanz für eine Reihe unerwünschter Verbindungen, die sog. Chlorpropanole. 3-MCPD kann im µg/kg-(ppb-)Bereich bei der Herstellung von Würzsoßen aus Pflanzenproteinhydrolysat, aber auch beim Backen oder Rösten von Brot bzw. Toast, Getreide und Kaffeebohnen gebildet werden. Die Entstehung von 3-MCPD setzt bei Temperaturen ein, die deutlich über 100 °C liegen, wobei verschiedene Bildungsmechanismen diskutiert werden. Neuere Studien zeigen, dass neben Chlorpropanolen auch deren Ester in Mengen vorkommen können, die in manchen Lebensmitteln um das 50–150-fache höher liegen als jene an 3-MCPD.

3-MCPD hat im Tierversuch bei höheren Dosierungen zur Bildung von Tumoren geführt. Da jedoch bei Wertung aller Daten ein genotoxischer Wirkmechanismus nicht anzunehmen ist, wurde sowohl vom SCF[1] als auch vom JECFA[2] eine tolerable tägliche Aufnahmemenge von 2 µg/kg Körpergewicht für den Menschen abgeleitet. Die Bildung von 3-MCPD in Würzen lässt sich durch Variation der Prozessbedingungen steuern und minimieren. Der Mechanismus der Bildung von 3-MCPD wurde bisher nur am Beispiel säurehydrolysierter Pflanzenproteine untersucht.

4 Fazit

Die thermische Behandlung von Lebensmitteln ist notwendig, um Lebensmittel verdaubar und bekömmlich zu machen, um die mikrobiologische Sicherheit zu gewährleisten und um einen spezifischen Geschmack, Aroma und Textur zu erzeugen. Dies kann zu weitreichenden Veränderungen im Lebensmittel führen. Die Prozesse bei der thermischen Behandlung von Lebensmitteln sind noch längst nicht vollständig verstanden, und eine Vielzahl durch thermische Reaktionen gebildeter Verbindungen ist noch nicht charakterisiert. Der Aufklärung des Einflusses der Zusammensetzung von Rohstoffen und deren Veränderung, z. B. bei Wachstums- und Lagerbedingungen kommt besondere Bedeutung zu. Dies gilt noch mehr für das Verständnis der Mechanismen, die zur Bildung von Farb-, Aroma-, Geschmackskomponenten und von Inhaltsstoffen sowohl mit positiver als auch mit gesundheitlich nachteiliger biologischer Wirkung führen. Diese Informationen erlauben eine gezielte Einflussnahme auf Bildung und Gehalte solcher Verbindungen im Lebensmittel, beispielsweise durch Rohstoffauswahl, Prozesssteuerung bzw. Kombination thermischer Prozessierung mit anderen Verfahren. Die Bewertung muss jeweils als Einzelfall unter Berücksichtigung von Nutzen und Risiko des jeweiligen thermischen Prozesses durchgeführt werden.

5 Forschungsbedarf

Die SKLM sieht Forschungsbedarf sowohl bezüglich positiver und negativer Effekte der thermischen Lebensmittelbehandlung im Allgemeinen als auch bezüglich der vertieft behandelten toxischen Stoffe/Stoffgruppen im Einzelnen.

---

[1] Scientific Committee on Food (2001): Opinion on 3-Monochloro-Propane-1,2-Diol.

[2] Joint Expert Committee on Food Additives (2002): 3-Chloro-1,2-Propanediol, WHO Food Add. Ser. 48.

Gesundheitlich positive Auswirkungen thermischer Lebensmittelbehandlung

Gesundheitlich positive Effekte thermischer Behandlung, die in experimentellen Systemen *in vitro* (Zellen) oder *in vivo* (Tierversuch) erhoben worden sind, müssen in ihrer Bedeutung für den Menschen wissenschaftlich belegt werden. Gleichzeitig muss die gesundheitliche Unbedenklichkeit gewährleistet sein. Die SKLM empfiehlt die weitere Erforschung von Mechanismen, auch unter technologischen Aspekten, die gesundheitlich positive Effekte thermischer Lebensmittelbehandlung verstärken. Dies gilt in erster Linie für die Aufklärung von Effekten auf Matrix- und Inhaltsstoffe und ihre Auswirkung auf Verdaulichkeit, Nährwert und Darmgesundheit. Besondere Beachtung verdienen dabei:

- Auswirkungen auf die Bioverfügbarkeit wertgebender Inhaltsstoffe;
- Förderung der Bildung von Stoffen mit antioxidativem oder anderweitigem chemopräventivem Potential;
- Entwicklung sensitiver Biomarker für Exposition und Wirkung;
- Verminderung der Bioverfügbarkeit von gesundheitlich nachteiligen Stoffen, z. B. durch Bindung an die Matrix.

Gesundheitlich nachteilige Auswirkungen thermischer Lebensmittelbehandlung

Die Entstehung gesundheitlich bedenklicher Stoffe durch eine Behandlung von Lebensmitteln ist generell unerwünscht und sollte durch geeignete Verfahren minimiert werden. Die SKLM sieht Forschungsbedarf betreffend der Auswirkung solcher Minimierungsmaßnahmen auf andere Lebensmittelinhaltsstoffe. Dieser Forschungsbedarf besteht insbesondere hinsichtlich:

- Verbesserung der Datenbasis zur Exposition mit thermisch generierten, gesundheitlich bedenklichen Stoffen;
- Entwicklung sensitiver Biomarker für Exposition und Wirkung;
- Verbesserung der Datenbasis zu toxikologisch relevanten Wirkungen vergleichend in Versuchstier und Mensch (soweit ethisch vertretbar), möglichst unter Einschluss von Kinetik, Metabolismus, Bioverfügbarkeit und molekularen Mechanismen;
- Fortentwicklung von Methoden der molekularen Epidemiologie zum Studium individueller Sensibilität bzw. zur Identifizierung von Risikogruppen;

- Verbesserung der Erfassungsmöglichkeiten der Allergenität nach thermischer Behandlung in Einzelfällen (wo sie geboten erscheint).

Forschungsbedarf zu speziell beim Symposium behandelten Stoffen

Acrylamid:

- Entwicklung/Optimierung von Biomarkern für Exposition und Wirkung, insbesondere verbesserte Methoden zur Erfassung von DNA-Schäden, Adduktbildung, Mutationen und Reparatur der Addukte sowie zur Frage einer Korrelation von Glutathion/Mercaptursäure/Hämoglobin-Addukten. Vordringlich ist der quantitative Vergleich dieser Prozesse zwischen Versuchstier und Mensch.
- Vertiefende Einsichten in den Mechanismus der Tumorentstehung: möglicher Einfluss von Acrylamid auf das Hormonsystem.

HAAs:

- Vertiefende Studien an definierten Risikogruppen mit Enzympolymorphismen und Erfassung von Zusammenhängen zwischen Exposition und möglichem Krebsrisiko;
- Klärung einer möglichen endogenen Bildung von HAAs bei hoher Creatininaufnahme.

Furan:

- Verbesserung der Datenlage zu Entstehung und Vorkommen von Furan in Lebensmitteln;
- Verbesserung der Datenbasis zur Toxikologie unter Einschluss von Entwicklungs- und Reproduktionstoxikologie;
- Vergleichende Untersuchungen an substituierten Furanen um deren toxikologische Bedeutung zu klären.

3-MCPD:

- Verbesserung der Datenbasis zu Entstehung und Vorkommen von 3-MCPD in Lebensmitteln sowie zu weiteren Monochlorpropandiolen und Dichlorpropanolen. Der Beitrag von veresterten Chlorpropanolen an der Gesamtbelastung ist ebenfalls bislang nur unzureichend untersucht. Die Daten zu Bioverfügbarkeit, Toxikokinetik und Toxizität von Chlorpropanolen und Estern sind zu ergänzen.

II Report and Conclusions

1 Foreword

The Senate Commission on Food Safety (SKLM) of the Deutsche Forschungsgemeinschaft (DFG) hosted a symposium entitled "Thermal Processing of Food: Potential Health Benefits and Risks" from 25th to 27th September 2005. Participants included experts from Germany and abroad. The main topics discussed were positive health effects and risk aspects of thermal food processing, also encompassing associated opportunities for innovative food technology. Following its mandate for evaluation and advice concerning human health the SKLM has elaborated conclusions and has highlighted areas that require further research.

2 General and Food Technological Aspects

Heating is one of the oldest and most reliable methods of food treatment that has substantially contributed to ensuring sustainable supply of food. Thermal treatment of foods, such as boiling, drying, roasting, baking, grilling and frying, affects food in many ways. For example, it can alter the taste, aroma and texture, and can also lengthen the shelf life. A large number of the processes taking place in foods during heating are based on non-enzymatic browning, i.e. the Maillard reaction.

Thermal processing of foods can often be combined with other conventional methods or novel technologies in order to ensure a product and energy saving process. The synergy effect resulting for example from combining high-pressure treatment and gentle heating can efficiently kill microorganisms or inactivate enzymes while desirable compounds, such as vitamins, colourants and flavourings, remain largely unaffected.

The application of novel technological treatment and processing methods in general presupposes that this does not lead to any additional microbial, toxicolog-

Thermal Processing of Food: Potential Health Benefits and Risks
DFG, Deutsche Forschungsgemeinschaft

ISBN: 978-3-527-31909-1

ical or allergenic risks. With respect to the use of such methods, the legislation has created an instrument to ensure the safety of correspondingly processed foods within the EU Directive 258/97 on novel foods and novel food ingredients.

3 Health Aspects of Thermal Food Processing

Evaluation of thermal processing of foods needs to take into account both, positive and negative aspects. These were discussed with examples during the symposium.

3.1 Advantageous Aspects

Nutritional aspects

Heating of foods often is a condition for intended consumption. In many cases, it is only the thermal process that produces an edible food, as is the case for bread or bakery products. Other foods, such as some raw vegetables, potatoes, rice and pasta, need to be cooked to render them palatable. The thermal treatment of foods is thus a process contributing to palatability, and ensures a sustainable and balanced diet. Heating of foods also plays an important role in the generation of aroma and taste thus improving palatability. Aroma, taste, colour, texture and appearance are greatly influenced by the heating process thus contributing decisively to acceptance of foods.

Microbial aspects

Heating of foods improves microbiological safety and shelf life. In addition to heat-induced inactivation of pathogenic organisms/microorganisms, toxins or enzymes, additional antimicrobial substances or enzyme inhibitors may be produced that also have a favourable effect on the shelf life.

Allergens

As a rule, thermal processes tend to inactivate food allergens rather than to increase the allergenic potential of food. A rapid decrease in the allergenicity of certain fruit allergens was observed after mild thermal processing. However, sporadic studies, concerning the influence of roasting on peanut allergens, have reported an increase in allergenicity.

A change in the allergenicity after thermal processing is usually difficult to detect since e.g. the solubility of the proteins frequently decreases after such treatment. At present, there is little evidence for the formation of new epitopes during the thermal processing of foods. Although there are examples of the Maillard reaction being able to enhance the IgE antibody reactivity of food proteins in individual cases, there are as yet no structural data confirming the presence of "Maillard epitopes" that are not present in the raw state.

Other aspects

The elucidation of mechanisms or processes that promote the development of positive health effects during thermal food processing is within the focus of current scientific interest. For example, mechanistic aspects are important with respect to the formation of resistant starches and their influence on gastrointestinal health, the formation of substances with antioxidative or other chemopreventive properties or enhanced bioavailability of nutritional constituents, such as lycopene in tomatoes or tomato products. Intelligent selection of varieties and/or corresponding processing controls can also contribute to favour positive health effects or to reduce negative ones.

3.2 Disadvantageous Aspects

Thermal processing of foods may also have undesirable consequences, e.g. the loss of nutrients such as vitamins, essential amino acids and unsaturated fatty acids. Likewise, thermally induced formation of harmful substances is also possible. The latter constituted another main subject of the scientific discussions, addressing in greater detail acrylamide, heterocyclic aromatic amines, furan and 3-monochloro-1,2-propanediol.

Acrylamide

Acrylamide may be formed by baking, roasting and frying of foods, particularly of potato products (e.g. chips, fried potatoes, crisps), cereal products (e.g. roasted cereals, bread, bakery products), coffee and cocoa. It is formed on heating of foods above 120°C by the Maillard reaction, preferentially by condensation of the amino acid asparagine with reducing sugars, such as fructose and glucose.

Acrylamide is considered a probable human carcinogen (IARC Group 2A, MAK Commission Carcinogen Category 2), implying that its metabolite epoxypropanamide (glycidamide) is the actual active genotoxic agent. In animal experiments with rodents, glycidamide-DNA adducts were found in all tissues investigated at high dosage. Furthermore, animal experiments have shown that acrylamide exhibits toxic effects to the nervous system and the reproductive organs.

There is a relatively small margin between the carcinogenic dose found in animal experiments and the highest estimated mean acrylamide intake in humans. However, hitherto quantitative estimates of the carcinogenic risk for humans by various scientific committees differ quite considerably. As yet, available epidemiological studies do not provide convincing evidence of a correlation between the acrylamide intake and tumour incidence; however, their ability to detect such an association is limited.

Heterocyclic aromatic amines

Heterocyclic aromatic amines (HAA) are formed in varying concentrations by frying or grilling or by comparable methods of heating meat and fish, depending on the type of protein, the temperature and the heating time. However, the formation of HAAs can be largely avoided by following certain precautions.

Two main types of HAAs are formed in varying amounts, depending on the temperature and precursor compounds: isoquinoline (so-called IQ compounds) and carboline derivatives.

Many HAAs show a mutagenic potential in bacterial and mammalian cell-based test systems and are carcinogenic in animal experiments. In humans, bioactivation of HAAs to carcinogenic metabolites varies greatly between individuals. There are also significant differences between species.

In addition to HAAs, thoroughly fried or overcooked red meat may contain other carcinogenic substances in comparable concentrations. It has not yet been unambiguously determined whether only HAAs or rather a complex mixture of thermally induced genotoxic compounds should be regarded as carcinogenic risk factors.

Furan

Furan is a volatile compound that has been detected in a number of heated foods, such as coffee, bread, vegetable and meat preserves, as well as in baby food. Its presence in widely different types of foods suggests that it is formed by different routes. There is only a limited database on its occurrence in different food categories and most data has not been measured in food ready for consumption. It is therefore not possible as yet to provide a reliable estimate on exposure levels from this limited data.

Oral application of furan to rats and mice led to different tumour patterns. Rats exhibited dose-dependent hepatocellular carcinomas and a high percentage of cholangiocellular carcinomas and leukemias; mice exhibited hepatocellular adenomas and carcinomas. It is suspected that the carcinogenic activity, at least in part, is based on a genotoxic mechanism. In particular the metabolite cis-2-butene-1,4-dial is suspect to be the genotoxic mutagen. On the basis of the limited data available on toxicity, metabolism and exposure, an adequate risk assessment is not possible at present.

3-Monochloro-1,2-propanediol (3-MCPD)

3-Monochloro-1,2-propanediol (3-MCPD) is regarded as a reference substance for a series of undesirable compounds known as chloropropanols. 3-MCPD may be produced in the µg/kg (ppb) range during manufacture of seasoning sauces from hydrolysed vegetable protein as well as by baking or toasting of bread, cereals and coffee beans. 3-MCPD is formed at temperatures significantly above 100 °C; various formation mechanisms are discussed. More recent studies show that certain foods may contain not only chloropropanols but also their esters in quantities that can be 50 to 150 times higher than those of 3-MCPD.

3-MCPD has caused tumours in animal experiments when applied at high dosage. However, based on the evaluation of all data a genotoxic mechanism is not considered likely. Both SCF[1] as well as JECFA[2] have therefore derived a tolerable daily intake of 2 µg/kg body weight for humans. The formation of 3-MCPD in seasonings can be controlled and minimised by varying the processing conditions. So far, the formation mechanism of 3-MCPD has only been investigated with acid-hydrolysed vegetable proteins.

4 Conclusions

Thermal processing of foods is necessary to make food digestible and palatable, to ensure microbiological safety and to produce a distinct taste, aroma and texture. This may lead to extensive changes in the foods. The processes occurring during thermal treatment of foods are by far not completely understood and a multitude of compounds formed by thermal reactions are not characterised.

1 Scientific Committee on Food (2001): Opinion on 3-Monochloro-Propane-1,2-Diol.

2 Joint Expert Committee on Food Additives (2002): 3-Chloro-1,2-Propanediol, WHO Food Add. Ser. 48.

Elucidating the influence of composition of raw materials (e.g. choice of variety, growth and storage conditions) is of prime importance. This applies even more to the mechanisms that lead to the formation of colour, taste and flavourings, and to constituents with positive or adverse biological effects. Knowledge of these factors allows to specifically influence the formation and concentrations of such compounds in foods, for example, by appropriate choice of raw materials, control of the processing conditions or by combining thermal processing with other methods. A case-by-case evaluation is needed for each individual thermal process taking into account the associated benefits and the risks.

5 Research Needs

The SKLM sees a need for further research on both the positive and negative effects of thermal food processing in general, as well as on details of individual toxic substances/groups of substances.

Positive health effects of thermal food processing

The relevance for humans of positive health effects from thermal food processing observed in experimental systems *in vitro* (cells) or *in vivo* (animal) have to be proven. At the same time, wholesomeness of thermally treated foods must be guaranteed. The SKLM recommends further research to be carried out on mechanisms and processing aspects that enhance the positive effects of thermal food treatment. This applies primarily to the elucidation of effects on matrix materials and constituents and how they affect digestibility, nutritional value and gastrointestinal health. Particular attention should be paid to the following:

- to study effects on bioavailability of nutritional constituents
- to increase the formation of substances with antioxidative or other chemopreventative properties
- to develop sensitive biomarkers for exposure and effect
- to reduce the bioavailability of substances with adverse health effects, e.g. by binding to the matrix.

Adverse effects on human health of thermal food processing

The formation of harmful substances as a consequence of the treatment of food is generally undesirable and should be minimised by appropriate methods. The

SKLM sees a need for further research on the impact of such minimisation measures on other food constituents. Further research is needed, in particular, on the following:

- more data on exposure to thermally generated substances with adverse health effects
- development of sensitive biomarkers for exposure and effect
- more data on toxicologically relevant effects comparing animals and humans (as far as ethically justifiable) and including, as far as possible, kinetics, metabolism, bioavailability and molecular mechanisms
- continued development of improved methods in molecular epidemiology to investigate individual sensitivities or to identify risk groups
- improved methods to determine allergenicity after thermal treatment in individual cases (where this appears necessary).

Further research needs, relating to specific substances discussed during the symposium

Acrylamide:

- Development/optimization of biomarkers of exposure and effect, particularly improved methods to measure DNA damage, adduct formation, mutations and repair of adducts, and investigations on a correlation of glutathione/ mercapturic acid/haemoglobin adducts. Quantitative comparisons of these processes in animals and humans are of particular interest
- improved understanding of mechanism of tumour formation: Possible influence of acrylamide on the hormonal system

HAAs:

- more in depth studies on defined risk groups carrying enzyme polymorphisms and determination of relationships between exposure and incidence of cancer
- clarification of possible endogenous formation of HAAs in conjunction with a high creatinine intake

Furan:

- more data on the formation and occurrence of furan in foods
- improvement of the toxicological database, including developmental and reproductive toxicology
- investigation on substituted furans in comparative studies, to elucidate the toxicological relevance of these compounds

3-MCPD:

- more data on the formation and occurrence of 3-MCPD and other monochloropropane diols and dichloropropanols in foods. The contribution of esterified chloropropanols to the total content has not been studied adequately so far. More data are required on the bioavailability, toxicokinetics and toxicity of chloropropanols and their esters.

III Contributions

1 Thermal Processing of Foods: Technological Aspects

Dietrich Knorr *, Volker Heinz, and Cornelius Luscher

1.1 Introduction

The aim of this contribution is to discuss some of the basic technological aspects of food processing. It will focus on thermal preservation processes including their desired impact and undesirable side effects on food materials.

Foods are usually processed for two reasons: preservation of the food to extend its shelf life and modification of the raw material to affect physicochemical and sensory properties or to simply make it edible. In traditional technologies both effects are working hand in hand to create the product. Sometimes the combination is desired, such as in baking of bread, where the thermal processing steps make the dough consumable. In some other processes, the modification of texture, taste, color and flavor was originally an undesired necessity of the preservation process, such as in the case of dried fruit. However, meanwhile dried fruit is a typical traditional food with accepted quality, although other and gentler technologies for fruit preservation exist. Preservation processes, like freezing, where any modification of the freshness of the raw material is considered undesirable, is a relatively new approach which was not accessible in human history on a large scale until recently [1].

This current way of thinking implies that traditional thermal processing steps, besides the production of established products like bread, are sometimes seen as detrimental to the quality of the product. However, a number of benefi-

* Department of Food Biotechnology and Food Process Engineering, Berlin University, Königin-Luise-Str. 22, 14195 Berlin, Germany.

Thermal Processing of Food: Potential Health Benefits and Risks
DFG, Deutsche Forschungsgemeinschaft

ISBN: 978-3-527-31909-1

Table 1.1: Selected thermal effects on foodstuff.

Appearance	Nutrition	Function	Safety	Quality
Color	Digestibility	Structure	Microorganisms	Enzymes
Flavor	Availability	Phase transition	Toxins	Sensory attributes
Texture	Palatability	Density	Allergens	Shelf life
Sound	Delivery	Volume	Viruses	Convenience
		Water	Anti-nutrients	Degassing
		Mobility		
		Weight		
		Solubility		
		Concentration		
		Assembly		
		Coating		

cial effects can be attributed to the thermal treatment of food (Tab. 1.1). The selection given may be incomplete, but it should be indicated that thermal treatments often increase the digestibility of products; they can destroy allergens and toxins, improve the handling of foodstuff, modify the flavor and, most importantly, kill harmful bacteria which may be present in non-processed food.

1.2 Basics of Food Preservation

The key to the understanding of the preservation of food is to learn that preservation is generally a combination of "hurdles" for spoilage microorganisms, high enough so that they cannot overcome them. This means that a number of attributes of the environment for microorganisms, which in this specific case is the food matrix, can be modified to make it more difficult for them to propagate. Microorganisms might grow at a certain level of reduced water activity, if all other conditions are favorable. However, if nutrients and oxygen are restricted and the pH is reduced, it is not very likely that the same microorganisms still grow at the same level of product dryness. The microorganisms might survive the initial hurdles, however, only some can survive the following ones and it is not very likely that many of them overcome all subsequent hurdles, if they are sufficiently high.

To understand the role of heat within this concept, it must be stressed that heat is not a internal matrix factor of the food like the ones mentioned above, which are constantly present throughout the storage of a product. Heat is usually applied during the manufacturing of food. It reduces the number of microorganisms at the beginning of the process. To assure extended shelf life, storage of the product must be carried out in the appropriate way. Otherwise, one or more hurdles disappear.

It should not be neglected that food spoilage is a lot more complex than the spoilage attributed to microorganisms, especially considering the effects of enzyme activity. However, microbial activities are the biggest threat to product safety and thus to the consumer. An additional positive aspect of conventional heat treatment is the reduction of enzyme activity as well as inactivation of microorganisms.

1.3 Concepts of Thermal Processing

Traditional thermal food processing relies on some traditional concepts and indices, which are working well, even though current food research has raised some objections against these concepts.

The traditional concept of the *D*-value and of the z-value is illustrated in Figure 1.1. To determine a *D*-value of a particular strain of microorganisms, samples of these microorganisms are exposed to an elevated temperature for a number of time steps. The residual number of microorganisms is plotted on a log-scale versus treatment time and the data points (should) align linearly. The *D*-value is defined as the time necessary to reduce the number of microorganisms at that respective temperature by one log cycle. The z-value describes the dependence of the *D*-value on the temperature. It is defined as the temperature increment necessary to reduce the *D*-value by one log cycle [2].

These values are commonly used to describe the heat inactivation of microorganisms. However, all thermal decay reactions, including the thermal destruction of heat sensitive food constituents, obey these laws. Figure 1.2 shows a scheme explaining a common concept of thermal food processing, the high temperature-short time treatment (HTST). According to the *D*-value and z-val-

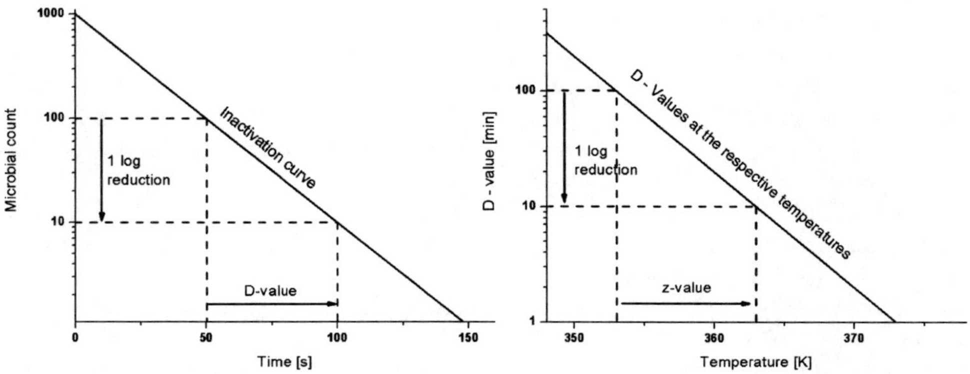

Figure 1.1: Concept diagram of the *D*-value and z-value.

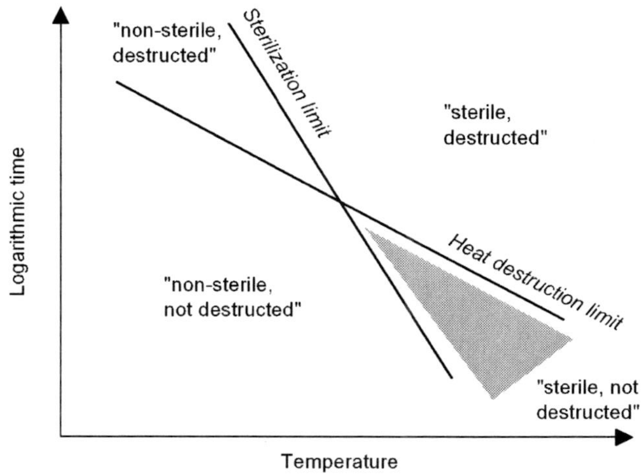

Figure 1.2: Basic concept of high temperature-short time (HTST) treatments in food processing. At high temperatures sterilization may be achieved without influencing the product with excessive thermal load.

ue, different time-temperature combinations can be found, which are sufficient to achieve the same effect of pasteurization or sterilization. These combinations are represented by the sterilization limit in the diagram. This means all time-temperature combinations to the lower left of the line represent treatments which are not sufficient for food safety reasons. Thermal decay reactions, which are detrimental to the quality of a product, are summarized in the second line, which is the heat destruction limit. All points to the lower left of the line represent treatments which result in acceptable products. Temperature induced chemical decay reactions are accelerated by a factor of about 3, if the temperature is raised by 10 K, whereas the factor is 10 for microorganisms [2]. This relation leads to different slopes of the described limits in the diagram, hence an interesting window opens, in which combinations of high temperature und short time can be found, which are adequate to achieve pasteurization/sterilization without severely damaging heat sensitive foods.

This concept represents a rule of thumb, because it depends to a large extent on indicator reactions which are used to determine the heat destruction limit and the sterilization limit. For example, if a decay reaction of a very sensitive compound is used, the processing window might shift to treatment parameters which are not feasible anymore.

Another intrinsic technical problem is the uniform distribution of heat within the product. In addition to the fact that foods are often not uniform with respect to their heat transfer properties (pores, bones etc.) the heat has to be transferred to the food. It is generally a problem to transfer the heat to the center of a piece or package. Figure 1.3 shows some arbitrary heating curves for some typical heating problems in food processing [3]. The sterilization of a container (glass or can) after packaging is the biggest problem, resulting in high

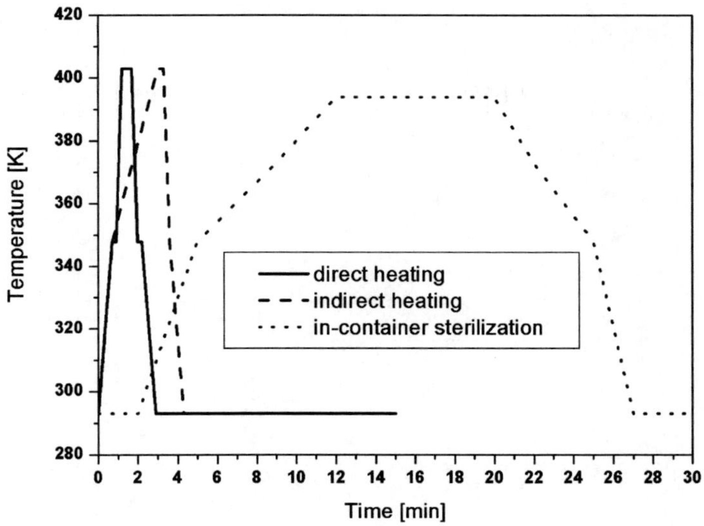

Figure 1.3: Schematic plots of product temperature during direct heating and indirect heating processes in comparison with traditional in-container sterilization (core temperature).

thermal loads due to the necessity to transfer the heat to the center of the container via its surface. If aseptic packaging is an option, heat can be transferred to liquids by passing thin layers of product between hot surfaces in a plate heat exchanger. In this way, it is avoided that major process inhomogeneities at the periphery of the product occur, which receive a thermal overload, while the center of a package is heated. However, this is still an indirect way of heating the food. A more beneficial way of heating is to bring the heat directly into the food, which can be accomplished by direct steam injection into liquids or via microwave or radio frequency heating [3].

1.4 Challenges to Traditional Thermal Processing

In addition to the general limits of thermal processing already described, additional points should be raised which represent challenges for thermal food processing. One point which was already suggested during the presentation of the *D*-value is that some basic assumptions which are underlying all conventional thermal processes are not true in all cases. Most of all, the assumption that microbial death obeys linear first-order kinetics is often a simplification of a very complex issue. Figure 1.4 shows an early diagram which presents one common non-linear effect called tailing [4]. It describes the fact that during sterilization

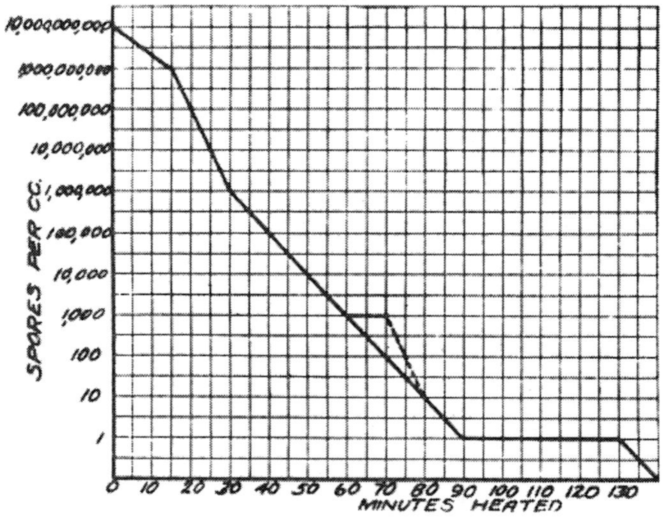

Figure 1.4: Non-log-linear inactivation kinetics found for *Clostridium botulinum* in 1922 [4].

processes it is getting increasingly difficult to reduce the residual microbes within an element of volume. The reasons for this behavior are still under discussion including the single hit theory suggesting that the probability of inactivation decreases with decreasing numbers of microorganisms. A speculation of additional reasons including the diversity of a microbial population would exceed the approach of this paper. However, it has to be taken into account that this raises a safety problem, when designing thermal processes for foods. Already a small number of microorganisms can proliferate within a short time, if they find optimum conditions. It should be noted that the concept of Stumbo [5] was based on the data of Esty and Meyer [4] by just utilizing the log-linear portion of the inactivation curve obtained by the authors. The *D*-concept seems successful mainly because the food industry usually uses safety margins to overcome this problem, which in turn leads to increased thermal loads and subsequent quality losses of the processed foodstuff.

Another point beside microbial and enzyme inactivation is the problem of BSE (*bovine spongiform encephalopathy*), the so called mad cow disease. The associated problem is that the agent which causes the disease (virus or prion) is not sufficiently inactivated by conventional thermal treatments which are usually applied in food or feed processing. As a consequence, risk assessment of all raw materials and processes in food industry has to be reconsidered since it is not true anymore that a proper sterilization process at the end of the food processing chain eliminates the risk of transmitting diseases by food.

Health risks associated with food are not necessarily only microbial or other infectious diseases either. A severe problem results from toxic substances which are generated during processing. It is known that some thermal processes such as smoking produce chemical compounds, which are generally regarded to be a health risk [6]. However, substances like acryl amide which have not been considered until recently were found in processed foods produced at temperatures above 175 °C, such as in frying or baking [7].

1.5 Possibilities of Non-thermal Processing

As an alternative to classical thermal processing a number of processes have been developed which are summarized under the label "non-thermal" processes. This term already suggests the main intention of these processes, which is the preservation of food with minimum thermal effects. These non-thermal processing concepts include especially irradiation, pulsed electric field treatment and high pressure processing. All these processes have a common goal of aiming at the destruction of complex living organisms without using an amount of heat energy which is detrimental to all food components. Irradiation is not widely used in Europe due to consumer's reservations. Pulsed electric field treatment aims at the destruction of cellular membranes. It destroys vegetative microorganisms without the use of heat. Cells are exposed to high electrical field strengths (about 20 kV/cm) which induce pores in the electrically non-conductive phosphor-lipid layers of biological membranes. If used for pasteurization processes, it has to be considered that a considerable amount of electrical energy dissipated to the product is leading to an increase in temperature. Consequently, it makes sense to design treatments which make use of the combined efficiency of electrical fields and mild heat, even though the original concept of pure non-thermal processing is partly abandoned [8, 9].

During high pressure processing food (in flexible containers or in non-packaged form) is exposed to hydrostatic pressure in vessels filled with pressure transferring liquids, usually water. In industrial applications, pressure of about 600 MPa is generated which is almost uniformly transmitted to the food through the pressure transferring medium. Most microorganisms do not survive this pressure and the food matrix is not exposed to excessive heat during the process. However, it has to be considered that during pressure generation the compression causes heating which is quasi-adiabatic, as the heat transfer through a pressure vessel is low. Consequently in this case it also makes sense to integrate mild heating in the processing concept, especially since the combination of pressure and temperature is additive. Figure 1.5 shows some pressure–temperature combinations at which certain process targets could be achieved at reasonable treatment times [8, 10].

The main intention of these non-thermal processes (with or without combination with mild heat) is to reduce the heat load of the product and to achieve fresh-like preserved foods. However, it should be stressed that pressure acts on the food in a unique previously unknown way and with completely different in-

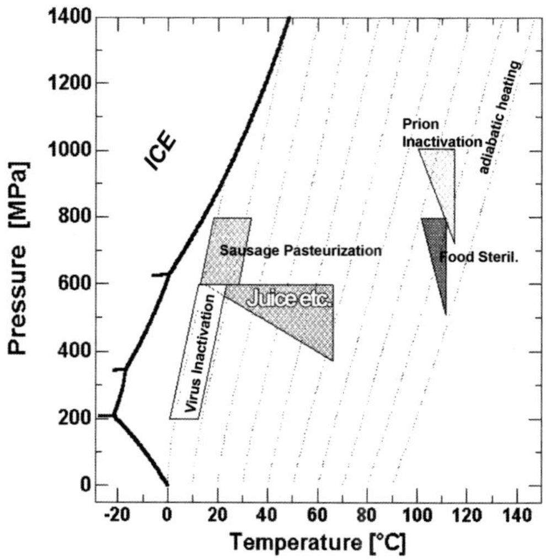

Figure 1.5: Potential processing parameters for combined pressure-temperature processing of food.

activation mechanisms of microorganisms or enzymes than heat [11]. These pressure effects can also be used to create completely new matrices for food such as gelatinization of proteins or polysaccharides with new and unique physico-chemical properties [12, 13].

1.6 Summary

The application of heat is one of the main concepts of conventional food processing. However, heat is not only used for pasteurization or sterilization purposes at the end of processing of foods, but it often is also a part of the assembly of a product itself. It would not be possible to digest many foods without thermal processing. Thermal energy generates color, structure and taste and reduces anti-nutrients, allergens and toxins. However, it can also destroy nutrients like vitamins and generate toxins. It is not possible and not desirable to replace thermal processing, but it is necessary to be aware of limits and problems of thermal processing, as well as to consider alternatives. High pressure processing is clearly an attractive alternative currently entering the food market with commercial products and processes on a world wide scale with over 90 industrial high pressure vessels currently in use [14].

References

1. Welch, R.W. and P.C. Mitchell (2000) Food processing: a century of change. *British Medical Bulletin* **56**(1): 1–17.
2. Holdsworth, S.D. (1985) Optimisation of Thermal Processing – A Review. *Journal of Food Engineering* **4**: 89–116.
3. Edmond, S.P. (2001) Continuous heat processing. In: Thermal technologies in food processing. P. Richardson, Ed. Woodhead Publishing, Cambridge.
4. Esty, J.R. and K.F. Meyer (1922) The heat resistance of the spore of Bacillus botulinus and allied anaerobes. *Journal of Infectious Diseases* **31**: 650–663.
5. Stumbo, C.R. (1948) Bacteriological considerations relating to process evaluation. *Food Technology* **2**(2): 115–132.
6. Tritscher, A.M. (2004) Human health risk assessment of processing-related compounds in food. *Toxicology Letters* **149**(1–3): 177–186.
7. Tareke, E., P. Rydberg, P. Karlsson, S. Eriksson and M. Törnqvist (2002) Analysis of acrylamide, a carcinogen formed in heated foodstuffs. *Journal of Agricultural and Food Chemistry* **50**(17): 4998–5006.
8. Knorr, D. and V. Heinz (2001) Development of nonthermal methods for microbial control. In: Disinfection, Sterilization and Preservation. S.S., Block. Ed. Lippincott, Williams & Wilkins, Philadelphia: 853–877.
9. Heinz, V., S. Toepfl and D. Knorr (2003) Impact of temperature on lethality and energy efficiency of apple juice pasteurization by pulsed electric fields treatment. *Innovative Food Science & Emerging Technologies* **4**(2): 167–175.
10. Ludikhuyze, L., A. Van Loey, Indrawati and M. Hendrickx (2001) Combined high pressure thermal treatment of foods. In: Thermal technologies in food processing. P. Richardson. Ed. Woodhead Publishing, Cambridge.
11. Ananta, E., V. Heinz and D. Knorr (2004) Assessment of high pressure induced damage on *Lactobacillus rhamnosus GG* by flow cytometry. *Food Microbiology* **21**: 567–577.
12. Hendrickx, M. and D. Knorr (2002) Ultra High Pressure Treatments of Foods. Kluwer Academic/Plenum Publishers, New York.
13. Bauer, B. A. and D. Knorr (2005) The impact of pressure, temperature and treatment time on starches: pressure-induced starch gelatinisation as pressure time temperature indicator for high hydrostatic pressure processing. *Journal of Food Engineering* **68**: 329–334.
14. Tonello, C., personal communication.

2 Thermal Processing: More than Extending the Shelf Life of Foods

Thomas Hofmann * and Peter Schieberle * *

Abstract

Thermal treatment of foods, such as baking, toasting, frying or charring, initiates a number of chemical reactions, thereby generating new food constituents from precursors in the raw materials. One of the most prominent thermally induced reactions is the so-called Maillard reaction (MR) of free amino acids or protein side chains with reducing carbohydrates. In former decades, the scientific focus was mainly on the identification of the chemical structure of major Maillard reaction products formed in processed foods or model systems, respectively, and a tremendous number of compounds has been identified in the past. More recent studies have, however, tried to develop methods for the identification of biologically active compounds among the numerous reaction products generated. In this contribution, recent advances in the identification of biologically active compounds showing impact on the human physiology will be highlighted.

2.1 Purpose and Motivation of the Lecture

Today, scientists date back the use of fire by mankind to about 1 million years, and it seems probable that a flash during a thunderstorm brought the fire to earth. Following, it was only a matter of time until our progenitors became aware of the benefits of a thermal treatment of foods. Thus, until today, heat processing is used in the food industry to inactivate enzymes or microorganisms as well as to improve the palatability and the texture of foods like bread or meat.

In addition, already our progenitors have for sure been aware of the fact that roasting or baking of raw materials generates a unique aroma and taste and, thus, enhances the acceptance of such foods. However, if one talks to a

* Institute for Food Chemistry, University of Muenster, Corrensstr. 45, 48149 Muenster, Germany.

* * Chair of Food Chemistry, Technical University of Munich, Lichtenbergstr. 4, 85748 Garching, Germany.

Thermal Processing of Food: Potential Health Benefits and Risks
DFG, Deutsche Forschungsgemeinschaft

ISBN: 978-3-527-31909-1

consumer of nowadays, food processing like frying, grilling or baking has a bad image. The reasons for this are quite obvious: It is well-known that the thermal treatment of foods leads, at least, partially to the degradation of essential micronutrients, such as vitamins, unsaturated fatty acids or essential amino acids. Furthermore, numerous compounds may be formed, which are at least suspected to be carcinogenic, such as the polycyclic aromatic hydrocarbons, the heterocyclic amines, the chloropropandiols or, most recently, acrylamide. Because results of research studies focused on such topics are often immediately released to the public, sometimes even in non-scientific journals and in an imbalanced way, consumers have often concerns about the "healthiness" of their diet.

Thus, from a molecular point of view, thermal processing of foods generates "desirable" chemicals, which should be present in the food we eat, and others, which we might assign as "physiologically detrimental" chemicals, which should be avoided in our diet. This clearly corroborates the scientific view that the quality of foods is determined by their chemical composition (Fig. 2.1).

So, in simple words it is clearly a challenge in the application of thermal food processing (i) to minimize the formation of physiologically detrimental compounds, and (ii) to improve the formation of desirable compounds showing a positive effect on the human mood and physiology. However, because "frying in a pan" generates such compounds from the same precursors, e.g. carbohydrates, the knowledge on formation pathways and transient intermediates has to be extended.

Besides the peroxidation of unsaturated lipids, one of the most important reactions leading to the formation of new foods constituents during thermal processing of raw materials is the Maillard reaction (MR), named after the French chemist Louis Camille Maillard. About 95 years ago, he performed a simple reaction between glucose and the amino acid glycine. After only seconds, a brown colour was formed accompanied by generation of carbon dioxide

Food safety	⇒	compounds showing toxic effects on humans
Nutritional safety	⇒	compounds with essential function for human life
Hedonic value	⇒	compounds stimulating human odour or taste receptors
Technofunctional value	⇒	compounds influencing the macromolecular appearance

⇒ Quality attributes of foods are determined by their chemical composition

Figure 2.1: The molecular view of food quality.

and an intense odour [1, 2]. Today, the complex reaction cascade behind the Maillard reaction is known as source of hundreds or even thousands of compounds [3]. Although some of these reaction products are already well-known for their "bioactivity" with respect to their influence on human behaviour or health [4–7], it can be assumed that still many compounds remain to be discovered.

Another challenge is that the same compound might be regarded as positive in one food, but as negative in another. For instance, to some extent bitter tasting compounds might be desirable in roasted coffee, but are perceived as tasting unpleasant in roasted meat or bread crust. As another example, brown coloured compounds are definitely linked in consumer minds with as a high-quality indicator of bread crust, but should be avoided during manufacturing of, e.g., milk powder.

2.2 The Need for Screening Methods

Because the processing conditions have an extreme influence on the yields of Maillard-type food constituents, it is a main goal in food chemistry research to get a deeper insight into single pathways of this complex reaction cascade. But due to the enormous complexity of the MR reaction and the tremendous number of reaction compounds already known to be formed by Maillard-type reactions, it has to be realized that there is a need of activity-guided screening methods to select, in particular, those compounds responsible for certain "bioactivities" from the bulk of physiologically "inactive" components.

In general, regarding their physiological activity compounds formed during Maillard-type reactions can be grouped into at least six classes (Fig. 2.2): colorants, odorants, tastants, and antioxidants which may be desired in a food, and food-borne toxicants as well as post-translationally modified amino acid residues in proteins, such as carboxymethyllysine, which should be diminished during food processing.

But, how is it possible to identify unknown food constituents showing a certain bioactivity in a complex mixture of thousands of components occurring in an extremely broad range of concentrations from 10^{-6} to 10^{-14} grams per kilogram? In general, the methods used by our groups are based on the principle of Paracelsus implying that the bioactivity of a chemical is clearly a function of its concentration. So, if the concentration of each bioactive constituent is lowered by dilution, finally, the compound will reach its threshold of activity, i.e. does not show any influence on the "bioassay" used (Fig. 2.3). In case of colorants, odorants or tastants, the human eyes, the nose or the tongue can be used in such bioassay. In case of antioxidants, the screening can be based on *in vitro* systems or animal trials. However, for food borne toxicants, only *in vitro* methods can be used to determine the threshold, the no observable effect level. This

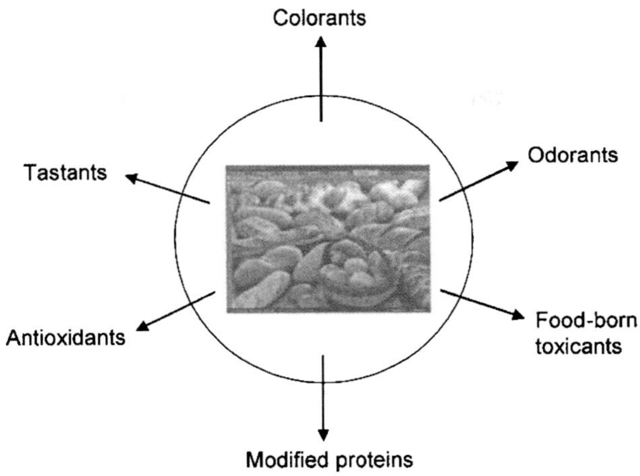

Figure 2.2: Classes of "bioactive" compounds generated by Maillard-type reactions.

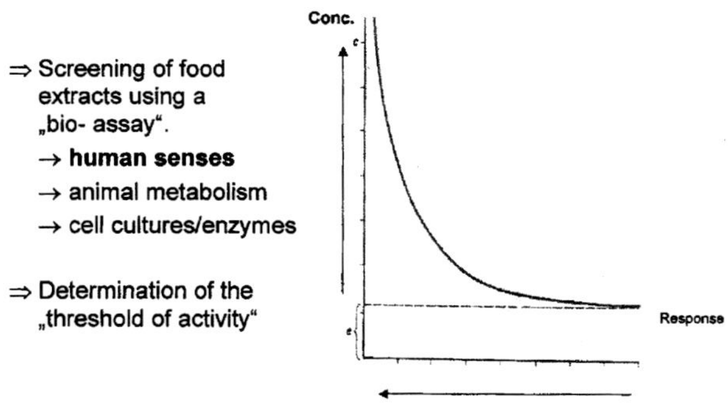

Figure 2.3: How to identify unknown "bioactive" food constituents?

principle of the dilution to activity threshold is visualized for a coloured solution in Figure 2.4. In this example a solution containing coloured products generated by a MR was diluted in 1:1 steps, thereby indicating the colouring efficacy of the colorants.

In the following, results of our groups obtained by applying activity-guided approaches will be presented focusing on chemicals occurring in processed foods, which are responsible for the respective aroma, taste, colour, and antioxidant activity.

However, as an introduction and to highlight the complex reactions occurring in the Maillard reaction, the so-called "primary reaction pathways" of the MR will be briefly discussed.

Figure 2.4: The colour dilution factor: a useful tool to determine the efficacy of a colourant.

2.3 Primary Pathways of the Maillard Reaction

In the first reaction step, an imine between the aldehyde function of the reducing carbohydrate and an amine is formed, giving rise to the so-called Amadori product (Fig. 2.5). From the Amadori product, three different deoxyosones can then be formed by eliminating one molecule of water at either carbon atom 1, 3 and 4 of the carbohydrate (Fig. 2.6). In case of the 3- and the 4-deoxyosone, the amino acid is either kept in the molecule or is liberated by hydrolysis. If no hydrolysis occurs and the reaction has taken place with, e.g., the ε-amino group of lysine in proteins, the so-called glycated proteins are formed (R=protein-bound lysine; Fig. 2.6). The deoxyosones are rather unstable transient reaction intermediates and, in consequence, so called "advanced" Maillard reaction products are generated by secondary reactions upon further heating.

In general, two prominent reactions can be differentiated at this point: First, the deoxyosones formed may loose further molecules of water leading to, e.g., furans like 5-hydroxymethylfurfural (HMF) from the 3-deoxyosone or 2-

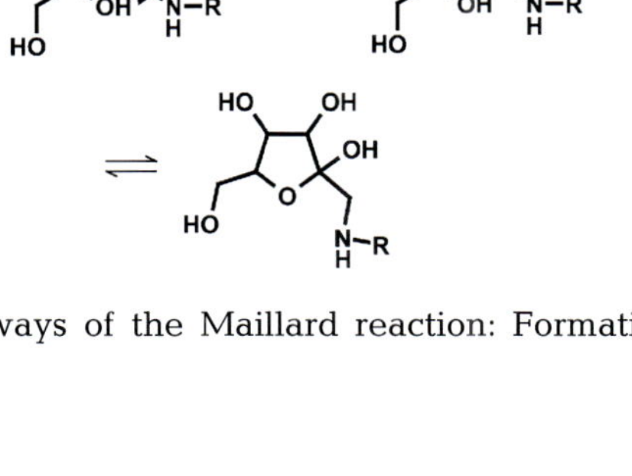

Figure 2.5: Primary pathways of the Maillard reaction: Formation of the Amadori compound.

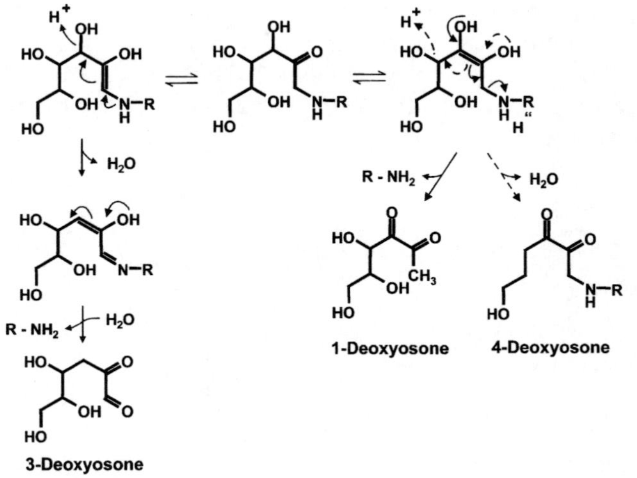

Figure 2.6: Primary pathways of the Maillard reaction: Formation of deoxyosones.

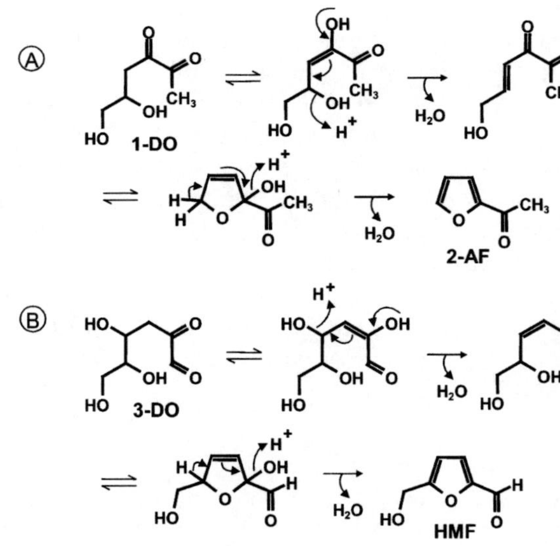

Figure 2.7: Formation of (A) 2-acetylfuran (2-AF) from the 1-deoxyosone, and (B) 5-hydroxymethylfurfural (HMF) from the 3-deoxyosone.

acetylfuran (2-AF) from the 1-deoxyosone (Fig. 2.7). Depending on their structure, such dehydration products may already be odour-active, but can also undergo condensation reactions with other transient MR intermediates to yield coloured compounds and pre-melanoidins, respectively [5, 6].

Secondly, the carbohydrate skeleton may be cleaved leading to very reactive short-chain dicarbonyls, such as 2-oxopropanal or glyoxal. These intermediates can then either undergo a reaction (i) with the amino acid side chains of proteins to form the so-called advanced glycation end products (AGE's) or (ii) take part in the Strecker degradation of free amino acids (cf. Fig. 2.8), leading to very potent aroma-active volatiles [7]. These types of chemical reactions are the common basis for the formation of numerous flavour compounds generated upon thermal treatment of nearly all foods.

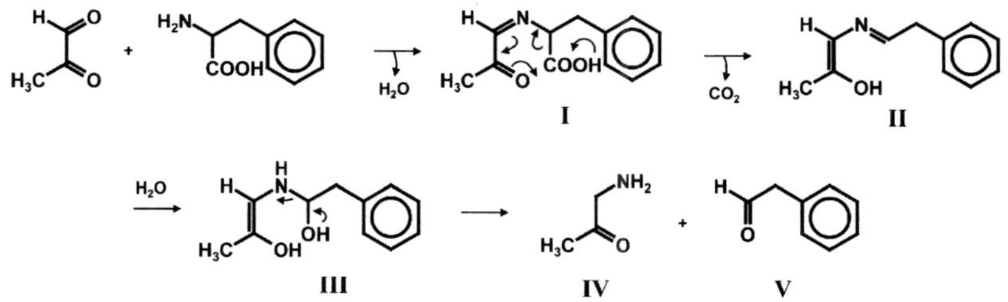

Figure 2.8: Strecker degradation of phenylalanine to phenylacetaldehyde (V) initiated by 2-oxopropanal.

As the aroma and taste perception is caused by volatile and non-volatile compounds, respectively, coming into contact with G-protein coupled receptors in the human nose or the taste buds on the surface of the tongue, such compounds can be regarded as "bio-active". Because these olfactory and gustatory receptors are very selective, food volatiles clearly differ in their "flavour potential", which can be expressed, e.g., by the odour threshold. As indicated in Table 2.1 large differences can be observed among selected compounds known to be formed by the MR. Based on these differences in human perception, it is quite obvious that the most important odorants in foods can only be detected by means of activity-directed screening techniques. In the following, the application of "bioactivity"-based concepts will be demonstrated first on aroma and taste components as examples.

Table 2.1: Odour thresholds of selected volatiles known as Maillard reaction products.

Compound	Odour threshold [µg/L water]
2-Methyl-3-hydroxy-pyran-4-one (Maltol)	35000
2-Furanaldehyde	3000
4-Hydroxy-2,5-dimethyl-3(2H)-furanone	60
2,3-Butandione	4
2-Acetyl-1-pyrroline	0.1
2-Furfurylthiol	0.005
2-Methyl-3-furanthiol	0.0004

2.4 Key Aroma Compounds in Cereal Malt

Aimed at identifying the aroma compounds in cereal malt, we followed the analytical concept given in Figure 2.9. Using this approach, which has been developed by our group and used in the identification of a large number of aromas of thermally processed foods, like bread crust, roasted coffee, or cooked meat, each analytical step is "controlled", by a human biosensor, the "nose, with respect to the aroma activity of the compounds identified. The application of the concept will be explained using malt as the example.

Malts are prepared by germination of different cereals followed by thermal processing to generate the typical caramel-like aroma. Besides being used in the production of whisky and beer, special malts are widely used to enhance the aroma of baked goods, sweets, and breakfast cereals. In a flavour extract from Caramalt we could previously identify the 12 odour compounds given in Figure 2.10 based on the application of a dilution to odour threshold approach, the aroma extract dilution analysis (AEDA) [8]. However, the flavour dilution (FD) factors obtained by this "sniffing" procedure (cf. Fig. 2.9) do not give a direct idea of the contribution of a single compound to a food aroma, because, amongst other factors, the odour evaluation during high-resolution gaschromatography/olfactometry (HRGC/O) is based on the odour thresholds in air. To

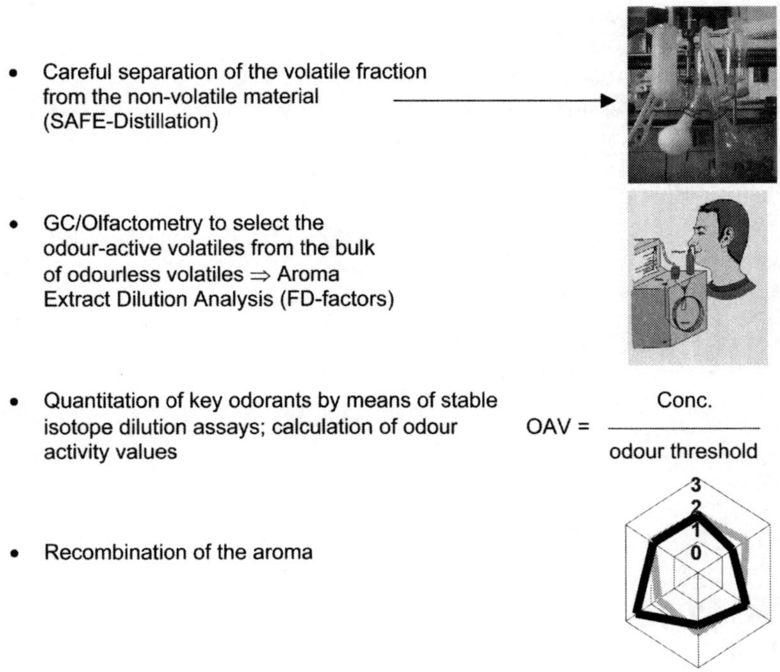

Figure 2.9: Analytical concept to identify key aroma compounds in foods.

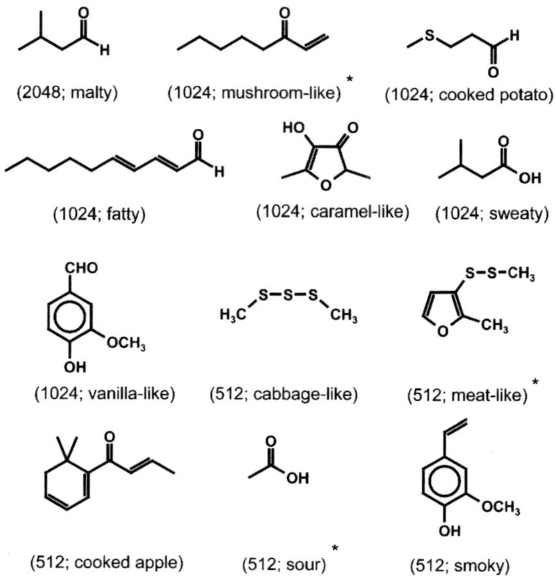

Figure 2.10: Most odour-active compounds identified in Caramalt (FD-factor, odour quality). Compounds marked with an asterisk have been identified for the first time in Caramalt [8].

link the data with the food itself, it is necessary to determine the odour thresholds of each compound identified using a matrix, which is very close to the matrix of the food under investigation, e.g. in starch for Caramalt. Then, the concentrations of each compound have to be determined and, finally, the aroma contribution is evaluated by calculation of Odour Activity Value (OAV; ratio of concentration to odour threshold) [9]. By applying this concept to Caramalt, we could show that sixteen of the thirty-nine odour-active compounds identified in Caramalt showed odour activity values >10 (Tab. 2.2), or in other words, ex-

Table 2.2: Odour activity values of the 16 most odour-active compounds (OAVs>10) in Caramalt (Fickert and Schieberle, unpublished results).

Odorant	OAV	Odorant	OAV
3-Methylbutanal	235	1-Octen-3-one	45
2-Methylbutanal	130	Hexanal	40
Methional	130	2-Methylbutanoic acid	30
Dimethylsulfide	85	Vanillin	24
Dimethyltrisulfide	80	2,3-Butandione	20
2-Methylpropanal	70	(E)-β-Damascenone	15
4-Hydroxy-2,5-dimethyl-3(*2H*)-furanone	60	3-Methylbutanoic acid	15
(E,E)-2,4-Decadienal	55	3-Hydroxy-4,5-dimethyl-2(*5H*)-furanone (Sotolon)	13

ceeded their odour threshold by the factors given in the table. The results showed that, in particular, 3-methylbutanal and 2-methylbutanal eliciting malty odours, but also (E,E)-2,4-decadienal with a deep-fat fried aroma or 1-octen-3-one with a mushroom-like odour quality are significantly involved in the overall aroma of the cereal malt.

Based on such "untypical" odour qualities it is not easy to understand that these odorants are necessary in eliciting the overall malt aroma. However, this was proven by reconstitution of the overall malt flavour by mixing the 19 odorants in their "natural" concentrations. As can be seen from Table 2.3, honey-,

Table 2.3: Comparative odour profile analysis of Caramalt and the aroma recombinate (Rec) in water or starch[a].

Odour quality	Intensity in	
	Caramalt	Rec (starch)[b]
Malty	2.6	2.4
Fatty	0.4	0.5
Caramel-like	1.5	1.4
Honey-like	1.8	1.5
Seasoning-like	0.8	1.0

[a] The recombinate consisted of a mixture of 19 odorants in the "natural" concentrations.
[b] 4 of 10 panellists were able to detect the recombinate in a blind test.

Table 2.4: Important aroma generating processes during malt production.

Maillard Reaction	
→ Carbohydrate degradation:	4-Hydroxy-2,5-dimethyl-3(2H)-furanone 3-Hydroxy-4,5-dimethyl-2(5H)-furanone 2,3-Butandione
→ Strecker reaction:	2- and 3-Methylbutanal 2-Methylpropanal Methional Phenylacetaldehyde 2- and 3-Methylbutanoic acid
Lipid oxidation	
→ Linoleic acid degradation:	(E,E)-2,4-Decadienal 1-Octen-3-one Hexanal
Miscellaneous reactions	
→ Methionine degradation:	Dimethylsulfide Dimethyltrisulfide
→ Carotenoid degradation:	(E)-β-Damascenone
→ Phenolic acid degradation:	Vanillin

caramel- and malt-like aroma notes prevail and undoubtedly mask e.g. the fatty or mushroom-like odour notes.

When the odorants identified are arranged according to the underlying formation mechanisms, it becomes obvious that the MR is not the only flavour generating process in malt: three compounds, namely 4-hydroxy-2,5-dimethyl-3(2H)-furanone, 3-hydroxy-4,5-dimethyl-2(5H)-furanone, and 2,3-butandione undoubtedly stem from carbohydrate degradation, while most of the other aroma compounds are derived from a Strecker reaction or lipid oxidation (Tab. 2.4).

Compared to aroma-active volatiles, relatively little attention has been paid to tongue responses induced by non-volatiles which are not present in the foods *per se*, but are generated during food processing by Maillard-type reactions from carbohydrates and amino acids. One reason for that lack of information is that most studies focused primarily on the quantitatively predominating compounds, rather than selecting the target compounds to be identified with regard to taste-activity. In the following, an activity-directed concept will be demonstrated for the elucidation of key taste compounds in beef bouillon as an example.

2.5 Taste Enhancer in Beef Bouillon

Beef bouillon is frequently used as a tasty base for savoury dishes, processed food compositions, or convenience foods. As preliminary experiments undoubtedly revealed that the taste profile and, in particular, the intensity of the sweet taste quality of beef broth could not be sufficiently covered by the taste compounds known so far, it has to be concluded that the attractive and unique taste of beef broth is due to yet unknown taste enhancing compounds which might be formed from tasteless precursors during thermal treatment.

In order to locate hydrophilic, taste enhancing compounds formed during cooking of meat, a freshly prepared beef bouillon was de-fatted by solvent extraction and de-odourized by means of high-vacuum distillation as given in Figure 2.11. Aimed at mapping of compounds contributing to sweet taste, taste-inactive proteins and high molecular weight melanoidins were separated from the taste-active low molecular weight compounds by means of gel permeation chromatography (GPC) on Sephadex G-15 [10]. As each analytical step is "controlled" by the tongue used as a human biosensor with respect to the taste activity of the compounds, the individual fractions obtained are evaluated in their taste activity. Using this approach, GPC fraction III was identified to contain a sweetness enhancing compound. This fraction was further separated by means of RP-HPLC and the fractions collected were evaluated for their sweetness enhancing activity by means of an HPLC/degustation technique, the so-called taste dilution analysis (TDA) (Fig. 2.11).

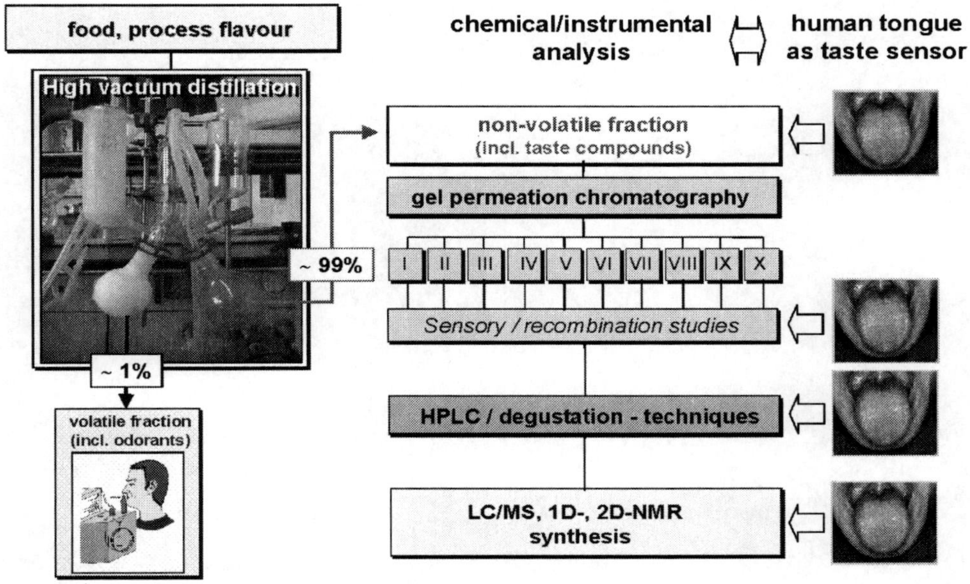

Figure 2.11: Activity-guided identification of non-volatile taste compounds and taste modifiers in foods.

Similar to the AEDA used in aroma research, the TDA opens the possibility to resolve complex fractions isolated from foods into individual components which are then evaluated based on a dilution to taste threshold approach. To achieve this, each HPLC fraction (Fig. 2.12) isolated from GPC fraction III was lyophilized, dissolved in a two-fold hyperthreshold sucrose solution, and was then stepwise 1+1 diluted with water until the sweetness threshold was reached. HPLC fraction 3 was evaluated with a high taste dilution (TD) factor of eight for sweetness, whereas all the other fractions did not lower the sweetness threshold of sucrose (Fig. 2.12). To gain further insight into the chemical structure of the compound enhancing the sweet taste of sucrose, fraction 3 was isolated and analyzed by HPLC-MS/MS and HPLC-DAD. On the basis of identical spectroscopic, chromatographic, and sensory data, the sweetness enhancer could be successfully identified as *N*-(1-carboxyethyl)-6-hydroxymethyl-pyridinium-3-ol inner salt (Fig. 2.12) and, finally, confirmed by synthesis [10, 11]. The identification of this compound, named alapyridaine, verified for the first time the natural occurrence of taste enhancing MR products in thermally processed foods.

After quantification of alapyridaine in beef bouillon, sensory analysis of synthetic beef taste reconstitutes containing "natural" concentrations of amino acids, sugars, 5′-nucleotides, organic acids, and minerals in the absence or presence of "natural" amounts of synthetic alapyridaine (419 µg/L) revealed a significant increase in sweetness and umami character of the taste imitate when the alapyridaine was present (A and B in Fig. 2.13) [10]. Comparison of the taste profile of the taste imitate containing alapyridain to the taste quality

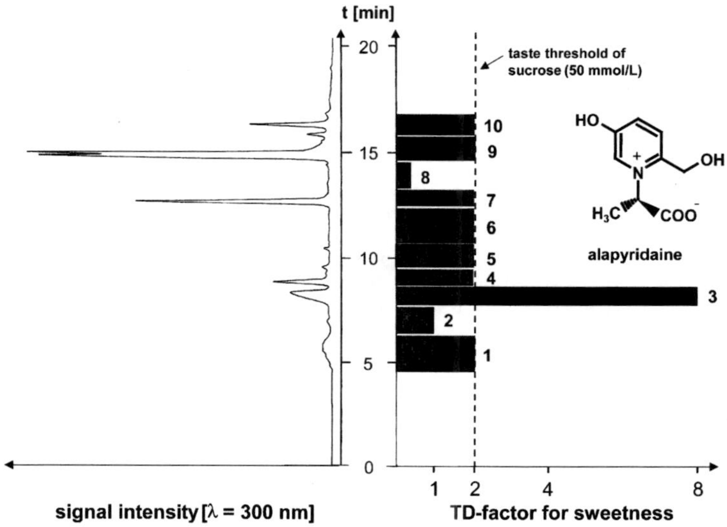

Figure 2.12: RP-HPLC chromatogram of GPC fraction III (left hand side) and taste dilution chromatogram (right hand side).

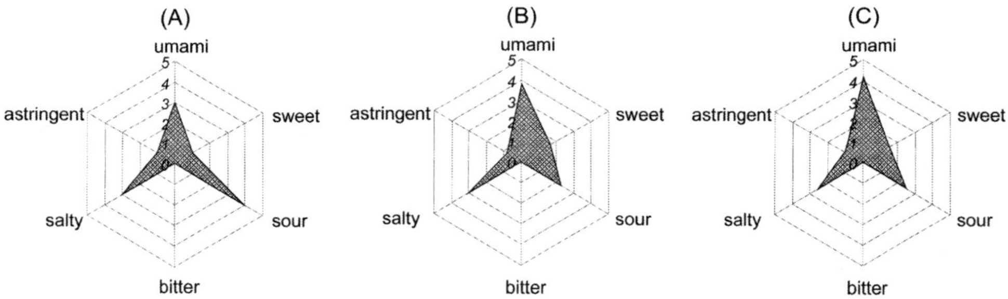

Figure 2.13: Taste profiles of (A) beef taste recombinate, (B) beef taste recombinate plus "natural" concentration of alapyridaine (419 µg/L), and (C) authentic beef broth.

of an authentic beef bouillon clearly demonstrated the similarity of the taste profiles (Fig. 2.13). These data demonstrate for the first time that, in "natural" concentrations, the alapyridaine exhibited a pronounced effect on the overall taste quality of beef broth, in particular, on the sweet and umami character.

Furthermore, investigations on the influence of alapyridaine on the taste intensity of aqueous solutions of sucrose, monosodium glutamate (MSG), and sodium chloride (NaCl), caffeine, and citric acid, respectively, revealed that the novel MR compound is able to enhance not only the sweetness perception, but also the intensity of umami as well as salt perception (Fig. 2.14) [12]. On the basis of these data it might be concluded that alapyridaine is a multivalent taste enhancer increasing the human sensitivity for sweetness, saltiness, and umami taste, whereas bitter and sour taste perception was unaffected by alapyridaine [12].

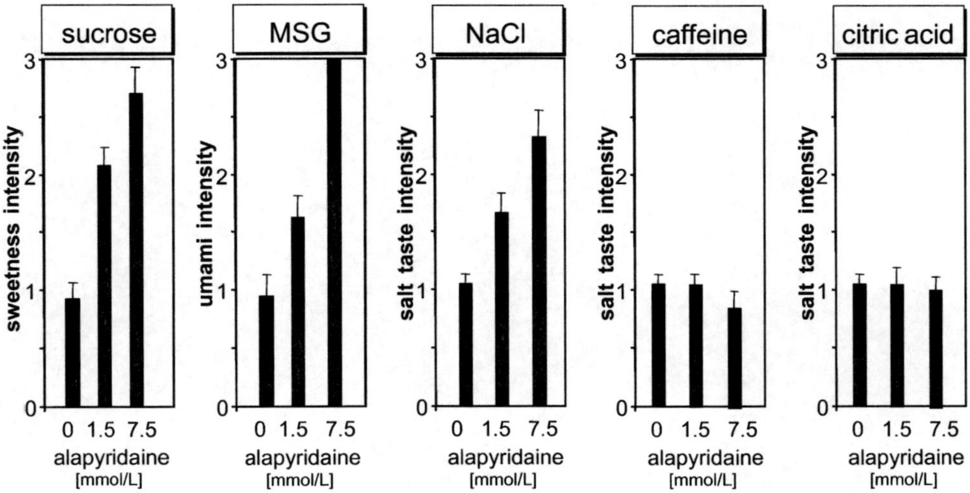

Figure 2.14: Influence of alapyridaine on the human perception of sweet (sucrose, 12.5 mmol/L), umami (monosodium glutamate, MSG, 1.5 mmol/L), salty (NaCl, 10.0 mmol/L), bitter (caffeine, 2.0 mmol/L), and sour taste (citric acid, 3.0 mmol/L).

2.6 Browning Compounds Formed by MR

Besides the unique aroma and taste, the typical brown colour developing during thermal food processing mainly originates from the Maillard reaction [12]. For example in roasted coffee, breakfast cereals, bread crust, roasted meat or kiln-dried malt, this browning is highly desirable and is intimately associated in consumers minds with a delicious, high-grade product. This browning is mainly due to the so-called melanoidins, which are assumed to be water-soluble, brown coloured, nitrogen-containing, polyfunctional macromolecules with masses up to 100 000 daltons [12].

Model studies on carbohydrates and amino acid side chains of food proteins as well as ESR spectroscopic experiments gave strong evidence for the involvement of radicals in thermal melanoidin genesis during thermal food processing [14, 15]. ESR studies, LC-MS/MS spectroscopy as well as chemical synthesis then led to the identification of the previously unknown protein-bound 1,4-bis-(5-amino-5-carboxy-1-pentyl)pyrazinium radical cation, named Crosspy [14, 16], formed as transient reaction intermediate during carbohydrate-induced protein oligomerization. As outlined in Figure 2.15, the carbohydrate degradation product glyoxal reacts with the ε-amino group of protein-bound lysine to form the corresponding glyoxal imine (I). Upon redox reaction with reductones formed by carbohydrate degradation, these imines are then rapidly reduced into the protein-bound *N*-(5-amino-5-carboxy-1-pentyl)aminoacetaldehyde (II), which, upon dimerization, forms the 1,4-bis-(5-amino-5-carboxy-1-pentyl)-1,4-

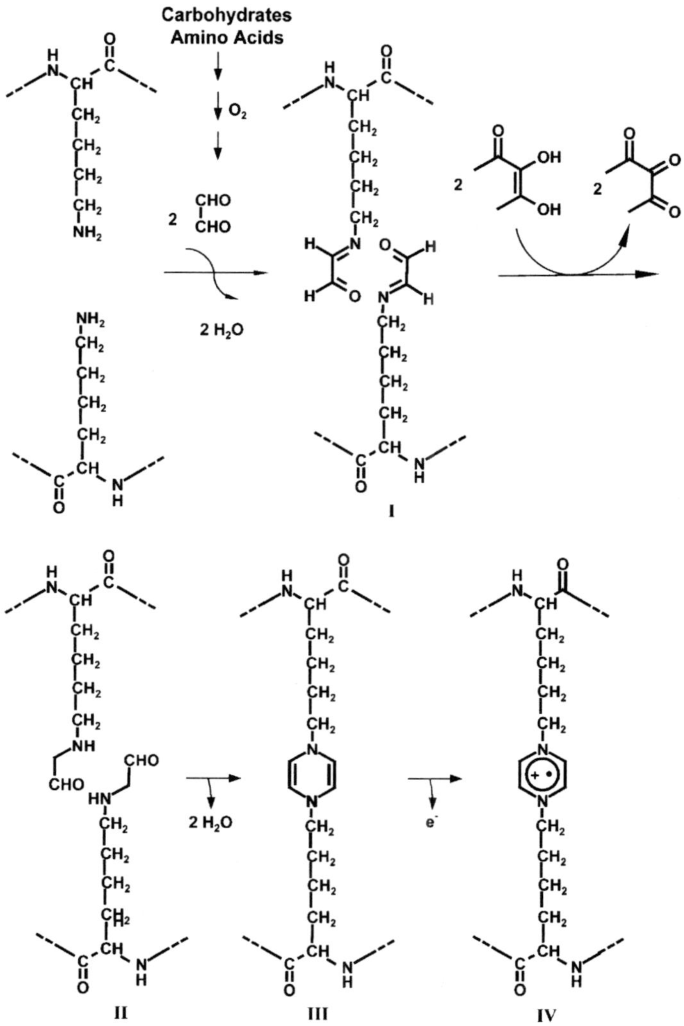

Figure 2.15: Reaction pathway leading to protein crosslinking by reductone-induced Crosspy formation from lysine side chains and glyoxal.

dihydropyrazine (III). Such dihydropyrazines are well-accepted in the literature to undergo subsequent oxidation, thereby giving rise to the radical cation IV.

Based on the observation that Crosspy containing solutions undergo rapid browning, we assumed that these radical cations might be important intermediates in browning development and melanoidin formation. Model studies on heated mixtures of glucose and N_α-acetyl-L-lysine, used as a model substance for protein-bound lysine residues, revealed that after the concentration of Crosspy went through a maximum, the radical concentration decreased again, whereas the formation of browning products was strongly accelerated (Fig. 2.16). On the

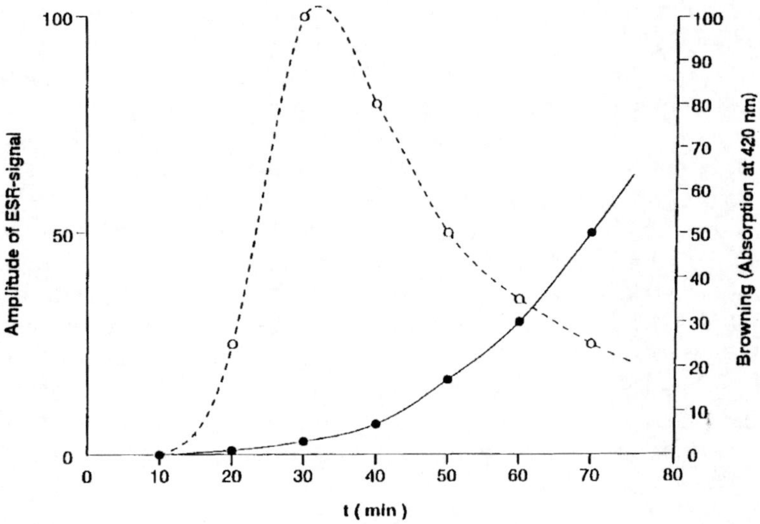

Figure 2.16: Time course of browning development and Crosspy formation in a thermally treated mixture of glucose and N_{α}-acetyl-L-lysine.

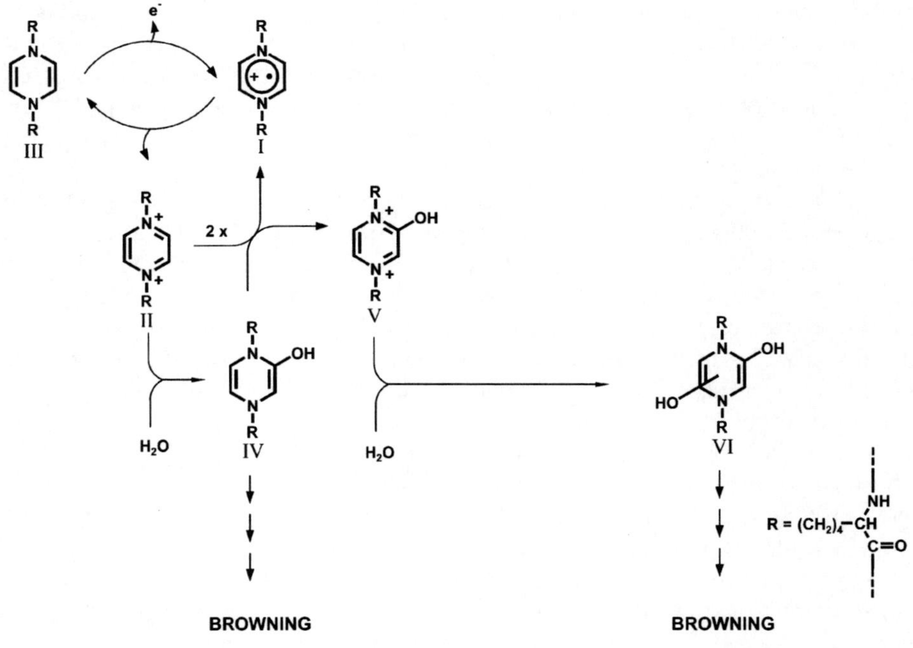

Figure 2.17: Reaction pathways leading to non-enzymatic browning via Crosspy (I) and hydroxylated dihydropyrazines (IV, VI) as the key intermediates.

basis of the tight relationship observed between Crosspy degradation and browning development, Crosspy was assumed as an important intermediate in melanoidin genesis [16]. Based on the results obtained from an array of model experiments, a mechanism for Crosspy degradation was proposed in Figure 2.17 showing the formation of browning compounds via Crosspy as a radical intermediate. Crosspy (I) generates its diquarternary pyrazinium salt (II) and the 1,4-dihydropyrazine (III) upon disproportionation. Whereas the dihydropyrazine rapidly regenerates the Crosspy by oxidation, the diquats were shown to subsequently form 2-hydroxy-1,4-dihydropyrazines (IV) upon hydratization as the primary reaction product. Redox reaction between the diquat (II) and the hydroxydihydropyrazine (IV) was recently shown to regenerate Crosspy and to form bishydroxylated dihydropyrazine (VI), most likely upon hydratization of an intermediate hydroxylated diquat (V). Both the hydroxy- (IV) and the bishydroxydihydropyrazines (VI) possess strong nucleophilic character and, as proven by synthetic experiments, are penultimate precursors of browning compounds [16].

2.7 Antioxidative Effects Caused by MRPs *In Vitro*

In foods, oxidative processes are responsible for the degradation of, in particular, unsaturated fatty acids, and it is well-known that compounds generated in the MR may act as antioxidants, thereby inhibiting the oxidation processes. Although endiol-α-oxo compounds having a reductone structure as found in ascorbic acid are suggested as the active principle, this has never been proven by reliable experiments using pure compounds.

Nearly 50 years ago Ivainsky and Franzke [17] showed that addition of a melanoidin fraction (MRPs) to the dough of baked goods was able to inhibit lipid peroxidation significantly. Numerous studies have later on addressed this topic by studying the antioxidative effect of extracts prepared from processed amino acid/carbohydrate model solutions [18, 19] or from processed foods [20–22]. However, in most cases no attempts were made to correlate the antioxidative effects with either monomeric compounds or structural elements present in the high molecular weight melanoidins.

Very recently, we have been able to characterize 2,4-dihydroxy-2,5-dimethyl-1-(5-acetamino-5-methoxycarbonyl-pentyl)-3-oxo-2H-pyrrol, formed as a very potent antioxidant in the MR of N_{α}-acetyl-carboxymethyl-L-lysine and the carbohydrate degradation product acetylformoin [4]. Bound to lysine residues of cereal proteins, this compound, named pronyl-L-lysine, was made responsible for the significant antioxidative properties of extracts from rye bread crusts. The systematic approach used by us to correlate the antioxidative properties with a molecule will be highlighted in the following:

In a first experiment [4], rye breads were prepared based on a sourdough procedure and with increasing baking time. The crusts and the crumbs were separated and extracted with water (fraction I; Fig. 2.18) followed by 60% aqueous ethanol (fraction II; Fig. 2.18) and 50% aqueous isopropanol (fraction III; Fig. 2.18). Compared to the extracts obtained from the flour and the crumb, the crust extracts showed the highest antioxidative potential. The most pronounced effect was elicited by fraction II. Separation of fraction II by molecular weight indicated that the antioxidative activity was present in the low molecular weight subfraction (MW<1000) as well as the high molecular weight fraction (MW≥30000), which, however, showed the highest activity (weight vs. tro-

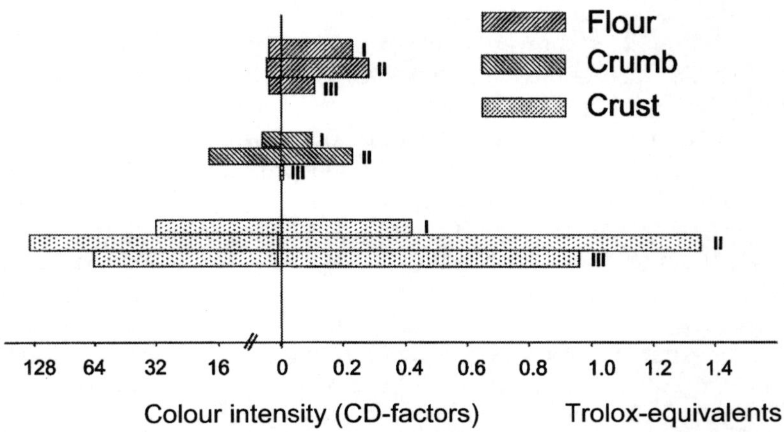

Figure 2.18: Correlation of antioxidative properties in bread crust extracts with colour intensity (fraction I: water; fraction II: 60% ethanol; fraction III: 50% isopropanol).

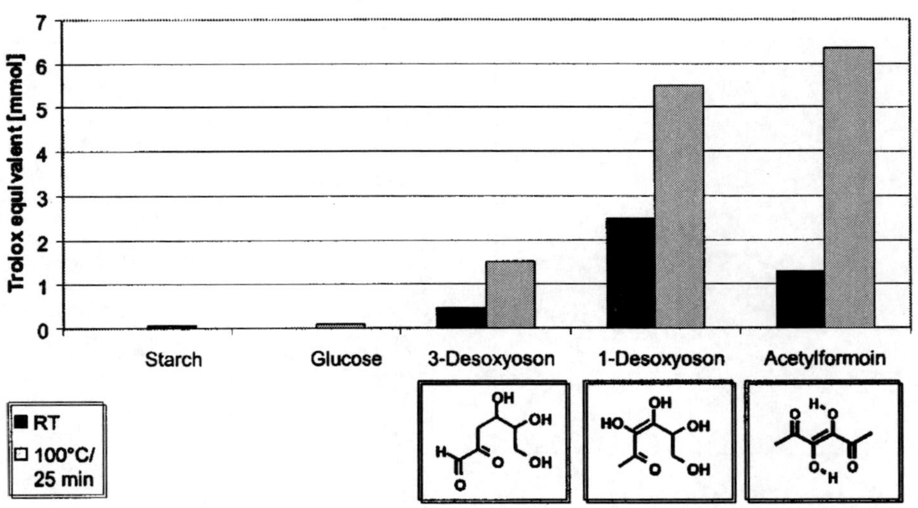

Figure 2.19: Antioxidative potential of N_α-acetyl-carboxymethyl-L-lysine when reacted with carbohydrates or their degradation products, respectively.

lox equivalents). Because all proteinogenic amino acids could be detected after hydrolysis of the HMW fraction, it was concluded that posttranslational modifications of, in particular, the lysine side chains of the cereal proteins are involved in the effective structures [4].

To prove this hypothesis, N_α-acetyllysine-methylester (ALME) was reacted with either glucose, the 1- and the 3-deoxyosone or acetylformoin. The results showed that the highest activity was generated from ALME and acetylformoin (Fig. 2.19). Although already the unprocessed mixtures of the deoxyosones and acetylformoin showed antioxidative activity, this was significantly increased after heating for 25 min at 100 °C. HPLC separation of the processed ALME/acetylformoin mixture and a screening of the antioxidative potential of each of the HPLC subfractions showed one very active area, in which 2,4-dihydroxy-2,5-dimethyl-1-(5-acetamino-5-methoxycarbonyl-pentyl)-3-oxo-2H-pyrrole could be identified (Fig. 2.20). The compound showed an antioxidative potential, which was by a factor of 5 higher than that of ascorbic acid.

Based on the cleavage of the pronylated protein side chains in the cereal proteins using methylhydrazine [4], the amount of pronyl-L-lysine in the protein was finally determined based on the amount of the hydroxypyrazole derivative formed. It was found that in the darker crusts from rye breads baked for a longer time, much higher amounts of pronyl-L-lysine were measured compared to lighter coloured crusts from bread baked for a shorter time. This increase in pronyl-L-lysine concentrations with baking time was in good correlation with the increase in antioxidative efficacy of the respective extracts [4].

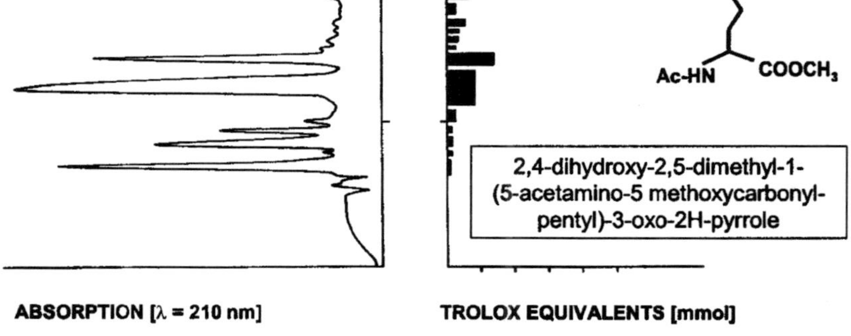

Figure 2.20: Activity guided screening of the reaction products formed from N_α-acetyl-lysine-methylester (ALME) and acetylformoine.

2.8 Health Effects of MRPs

The effects of MRPs in nutrition and toxicology have previously been reviewed [23]. However, most data in the literature up to that time were performed on crude extracts of MRPs.

In particular MRPs with 3(2*H*)-furanone structures have more recently been reported to cause DNA damage [24, 25]. Furthermore, the potential carcinogen acrylamide has been identified very recently to be formed by a carbohydrate assisted degradation of asparagine [26, 27]. On the contrary, other compounds, like 3-hydroxy-4-[(*E*)-(2-furyl)-methylene] methyl-cyclopentene-1,2-dione were found to have anticarcinogenic effects [28]. This compound, also responsible in part for the brown colour of MRPs, significantly inhibited the growth of human tumour cells as established by *in vitro* tests [28].

Besides such contradictory results, another challenge with MRPs arises from the fact that the Maillard reaction also takes place in the human body, in particular of diabetic patients. Interestingly, the same structures as identified in foods or model systems, respectively, have also been identified in physiological samples [29–32]. It should be noted that in the physiological literature such MRPs are assigned as "advanced glycation end products" (AGEs).

As indicated in Figure 2.21, arginine and lysine side chains of proteins are mostly susceptible to a reaction with carbohydrates leading to either posttranslational protein modifications or protein cross-links. Among them carboxy-

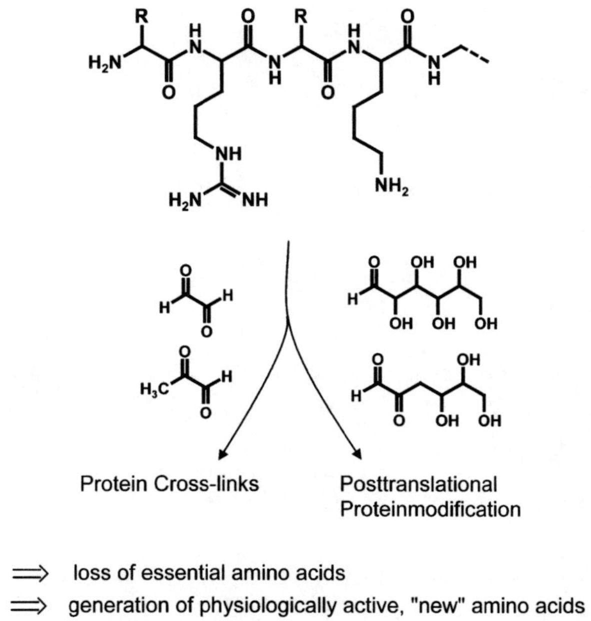

Figure 2.21: Protein modifications initiated by "glycation".

methyl-lysine and carboxyethyl-lysine as well as pentosidine (Fig. 2.22) are known for more than a decade [29–31].

When discussing health effects of MRPs, one of the major questions, which have to be addressed is, whether their resorption from foods takes place in the gut and, finally, whether they reach target organs like the kidney or the liver. This "metabolic transit" of MRPs has been studied previously by Finot and Magnenat [33] feeding radio-labelled MRPs to rats. Their results revealed

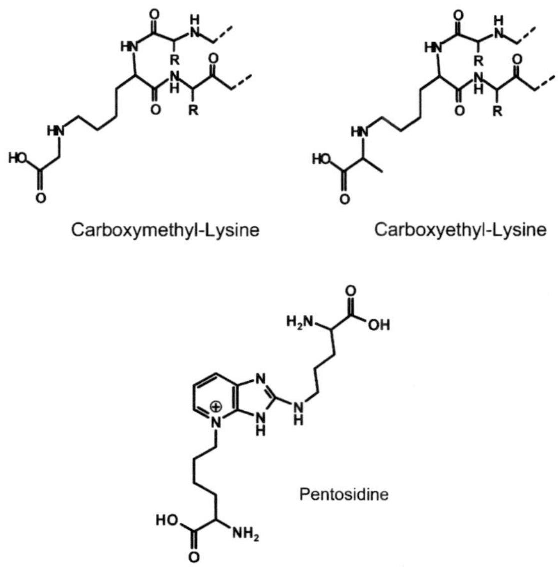

Figure 2.22: AGEs (MRPs) detected in foods and physiological tissues.

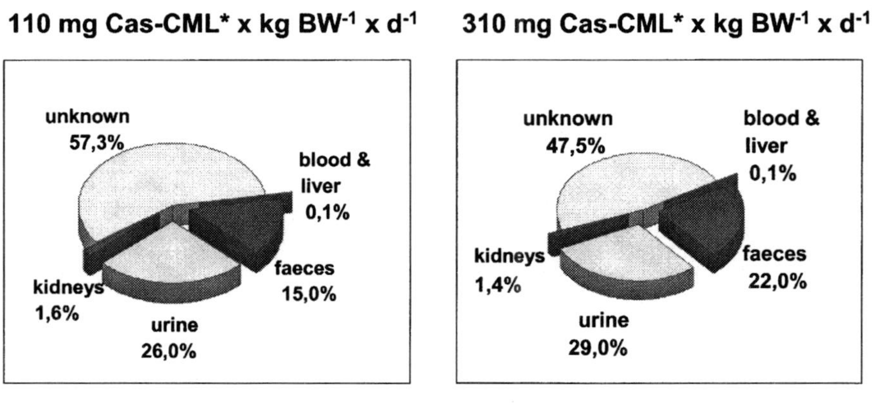

Figure 2.23: Metabolic transit data of casein-linked N_{ε}-carboxymethyl-lysine (CML). Male wistar rats ($n = 10$ per group) were fed a heated casein containing CML for 10 days; urine and faeces samples were collected during the last 5 days of the experiment; tissue samples were collected at the end of experiment [27].

that below 10% of the ingested radioactivity was kept in the bodies, while the major part was excreted via the faeces.

Focusing on the fate of carboxymethyl-lysine in male wistar rats [34] fed with carboxymethyl-lysine (CML) containing casein showed a different picture (Fig. 2.23). While 41% to 51% of the CML was excreted via urine and faeces, depending on the dose, about 50% of CML was not recovered. Compared to liver and blood, the kidney tissue contained much higher amounts indicating that CML had been resorbed from the diet [34].

One approach to study the influence of MRPs on human physiology is the *in vitro* measurement of enzymes involved in the metabolism of xenobiotics, such as the phase I-enzyme cytochrome C reductase (CCR) or the phase II-enzyme glutathione-*S*-transferase (GST). In particular, the induction of GST has been proposed as a promising strategy in cancer prevention [35]. The colon is regarded as the most likely site for the development of different types of dietary induced cancer [35, 36]. Consequently, the intestinal Caco-2 cell line is a promising tool for such vitro studies on the effect of MRPs on phase I/phase II enzymes. Using this approach, extracts of rye bread crust have been offered to Caco-2 cells and the activity of GST and CCR were determined [4]. As shown in Figure 2.24, GST activity was significantly increased by the crust extracts, whereas CCR activity was decreased.

To correlate this effect with pronyl-lysine, bovine serum albumine was reacted with acetylformoin and, also, pronylglycine was synthesized. As shown in Figure 2.25 both compounds clearly enhanced the GST activity and pronylglycine also inhibited CCR activity.

The results discussed so far suggest that the antioxidative effects observed *in vitro* may be correlated with chemopreventive effects *in vivo*. However, it has also been proven by numerous studies that AGEs accumulating in the organisms are clearly "glycotoxins". These have been associated with the progression of chronic inflammation, diabetes mellitus, nephropathy, Alzheimer's disease as well as

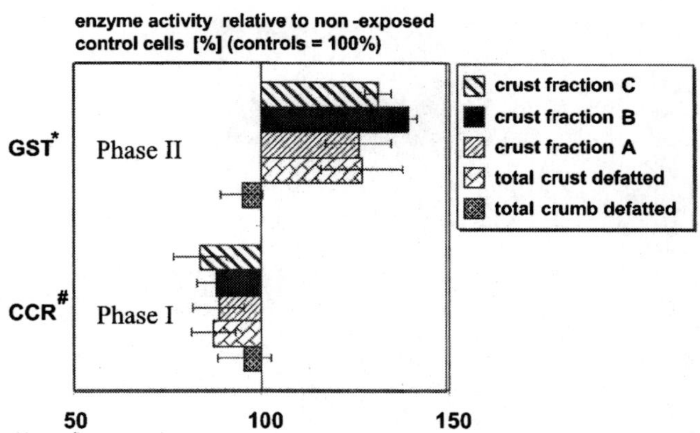

Figure 2.24: Influence of bread crust melanoidins on the activity of the phase I and II enzymes in intestinal epithelium cells (Caco-2).

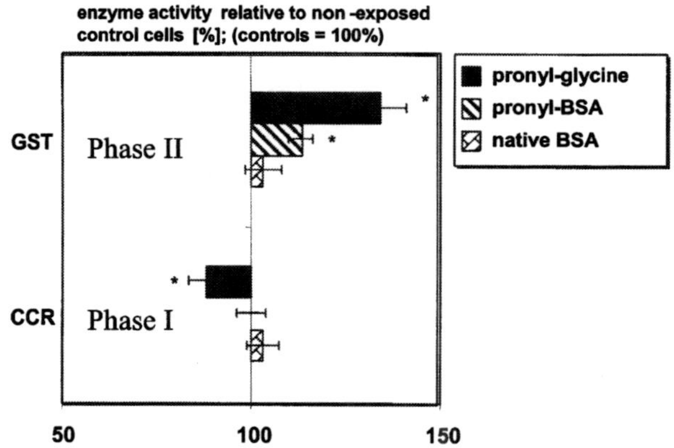

Figure 2.25: Influence of pronylated proteins on the activity of the phase I and II enzymes in intestinal epithelium cells (Caco-2).

atherosclerosis and aging. This part, the "physiological" MR is another interesting topic of the reaction between reducing carbohydrates and amino groups, however, a discussion of these topics would go beyond the scope of this contribution. The interested reader is, therefore, referred to some reviews in this field [37–39].

References

1. L. C. Maillard, C. R. Hebd, *Seances Acad. Sci.*, 1912, **154**, 66.
2. L. C. Maillard, C.R. Hebd, *Seances Mem. Soc. Biol.*, 1912, **72**, 559.
3. F. Ledl, E. Schleicher, *Angewandte Chemie*, 1990, **102**, 597.
4. M. Lindenmeier, V. Faist, T. Hofmann, *J. Agric. Food Chem.*, 2002, **50**, 6997.
5. T. Hofmann, *J. Agric. Food Chem.*, 1998, **46**, 932–940.
6. T. Hofmann, *Helv. Chim. Acta.*, 1997, **80**, 1843–1856.
7. T. Hofmann, P. Schieberle, *J. Agric. Food Chem.*, 2000, **48**, 4301–4305.
8. B. Fickert, P. Schieberle, *Food/Nahrung*, 1999, **53**, 109–136.
9. P. Schieberle, In: Food Analysis: Emerging Methods, A. Goankar (ed)., 1995, Elsevier Science BV, pp. 403–431.
10. H. Ottinger, T. Hofmann, *J. Agric. Food Chem.*, 2003, **51**, 6791.
11. H. Ottinger, T. Soldo, T. Hofmann, *J. Agric. Food Chem.*, 2003, **51**, 1035.
12. T. Soldo, I. Blank, T. Hofmann, *Chem. Senses*, 2003, **28**, 371.
13. T. Hofmann, *J. Agric. Food Chem.*, 1998, **46**, 3891.
14. T. Hofmann, W. Bors, K. Stettmaier, *J. Agric. Food. Chem.*, 1999, **47**, 379.
15. T. Hofmann, W. Bors, K. Stettmaier, *J. Agric. Food. Chem.*, 1999, **47**, 391.
16. T. Hofmann, W. Bors, K. Stettmeier, In: Free Radicals in Food – Chemistry, Nutrition and Health Effects, M.J. Morello, F. Sahidi, C.-T. Ho (eds.), 2002, ACS Symposium Series 807, American Chemical Society, Washington, DC, USA, 49.
17. H. Ivainsky, C. Franzke, *Dtsch. Lebensm. Rundsch.*, 1954, **52**, 129–133.
18. K. Eichner, *Prog. Food Nutr. Sci.*, 1981, **5**, 441–451.
19. H. Lingnert, C. E. Eriksson, G. R. Waller, *J. Agric. Food Chem.*, 1983, **31**, 27–30.

20. M.C. Nicoli, M. Anese, M.T. Parpinel, S. Franceschi, C.R. Lerici, *Cancer Lett.*, 1997, **114**, 71–74.
21. M.C. Nicoli, M. Anese, L. Manzocco, C.R. Lerici, *Lebensm. Wiss. Technol.*, 1997, **30**, 292–298.
22. D. Bright, G.G. Stewart, H. Patino, *J. Am. Soc. Brew. Chem.*, 1999, **57**, 133–137.
23. J. O'Brien, P.A. Morrissey, *Crit. Rev. Fd. Sci. Nutr.*, 1989, **28**, 211–248.
24. K. Hiramoto, R. Aso-o, H. Ni-iyama, S. Hikage, T. Kato, K. Kikugawa, *Mutat. Res.*, 1996, **359**, 17–24.
25. K. Hiramoto, A. Ishihara, N. Sakui, S. Daishima, K. Kikugawa, *Biol. Pharm. Bull.*, 1998, **21**, 102–104.
26. D. Mottram, B.L. Wedzicha, A.T. Dodson, *Nature*, 2002, **419**, 448–449.
27. R.H. Stadler, I. Blank, N. Varga, F. Robert, J. Hau, P. Guy, M.-C. Robert, S. Riediker, *Nature*, 2002, **419**, 449–450.
28. D. Marko, M. Kemeny, E. Bernady, M. Habermeyer, U. Weyand, S. Meiers, O. Frank, T. Hofmann, *Food Chem. Toxicol.*, 2002, **40**, 9–18.
29. D.R. Sell, V. Monnier, *J. Biol. Chem.*, 1989, **264**, 21597–21602.
30. H.F. Erbersdobler, *Bibl. Nutr. Dieta*, 1989, **43**, 140–155.
31. M.U. Ahmed, S.R. Thorpe, J.W. Baynes, *J. Biol. Chem.*, 1986, **261**, 4889–4894.
32. T. Henle, U. Schwarzenbolz, H. Klostermeyer, *Z. Lebensm. Unters. Forsch.*, 1997, **204**, 95–98.
33. P.A. Finot, E. Magnenat, *Prog. Fd. Nutr. Sci.*, 1981, **5**, 193–207.
34. V. Faist, E. Wenzel, H.F. Erbersdobler, *J. Nutr.,* 2002, submitted.
35. H.J. Prochaska, P. Talalay, *Cancer Res.*, 1988, **48**, 4776–4782.
36. W.H.N. Peters, H.M.J. Roelfs, *Biochem. J.*, 1989, **264**, 613–616.
37. R. Singh, A. Barden, T. Mori, L. Beilin, *Diabetologia*, 2001, **44**, 129–146.
38. J.L. Wautier, P.J. Guillausseau, *Diabetes Metab.*, 2001, **27**, 535–542.
39. J.W. Baynes, In: The Maillard reaction. Consequences for the chemical and life sciences, R. Ikan (ed.). Wiley+Sons, Chichester, 1996, pp. 55–72.

3 Nutritional Aspects

Gerhard Rechkemmer *

Abstract

Important nutritional aspects of the thermal treatment of food are the inactivation of enzyme activities and the elimination/reduction and inactivation of microbial contamination and microbial toxins. These changes in general lead to an increase in the ability to store and use food.

Thermal treatment also changes the physico-chemical structure of macronutrients, e.g. starches and proteins, with the generalised effect of a better gastro-intestinal digestion.

Regarding the nutritional quality of heat-treated food one should discriminate between the effects on essential nutrients and non-nutrient bioactive components. While some water-soluble vitamins are heat-sensitive, e.g. vitamins C, B_1, B_2, B_6, and folic acid, the lipid-soluble vitamins are not. A decrease in essential nutrients by heat treatment thus reduces the nutritional value of certain foods.

Thermal treatment also causes changes in plant cell wall structure and thus modifications of the food matrix. These changes may lead to a marked increase in the bioavailability of bioactive food components, e.g. secondary plant metabolites (phytochemicals). Such effects have been demonstrated for carotenoids, e.g. β-carotene, lycopene, and lutein, but also for other substances like the isoflavonoids. These non-nutrient bioactive compounds have various biological functions. They act for example as radical scavengers, antioxidants, antimicrobials, phytoestrogens or have even anticarcinogenic activity and act as signalling molecules. The increase in bioavailability thus is relevant for the biological activity and the nutritional quality of these foods.

By changing the tertiary structure of proteins thermal treatment may also lead to a change in epitopes responsible for allergenicity of certain foods. However, not all allergens can be inactivated by heat treatment.

The examples discussed relate to changes of components already present naturally in food before thermal treatment. However, during thermal treatment new compounds are formed and these compounds may have nutritional relevance. The Maillard reaction products are important for the colour and sensory properties of foods. The biological activities of these compounds are discussed in the following.

* Unit Biofunctionality of Food, Life and Food Sciences Center Weihenstephan, Technical University of Munich, Hochfeldweg 1, 85350 Freising, Germany.

Thermal Processing of Food: Potential Health Benefits and Risks
DFG, Deutsche Forschungsgemeinschaft

ISBN: 978-3-527-31909-1

3.1 Introduction

Using heat to prepare food has been invented by mankind already a long time ago during the evolution of *Homo sapiens*. Technological developments and the use of various industrial as well as home-cooking methods have led to a great number of different ways on how to use thermal treatment to achieve specific food qualities. Surprisingly in the scientific literature only a rather limited number of publications dealing with the influence of thermal treatment on the nutritional quality of food can be found. The majority of publications is concerned with technological and hygienic aspects of thermal treatment. Thermal processing of food may have specific nutritional and health benefits but also certain risks. However, according to the 2002 WHO Health report, thermal treatment or the lack of heating of food, respectively, are not listed as major risk factors for human disease, neither in developed nor in developing countries, but unsafe water, sanitation and hygiene are among the major causes for the global burden of disease (Fig. 3.1) [1].

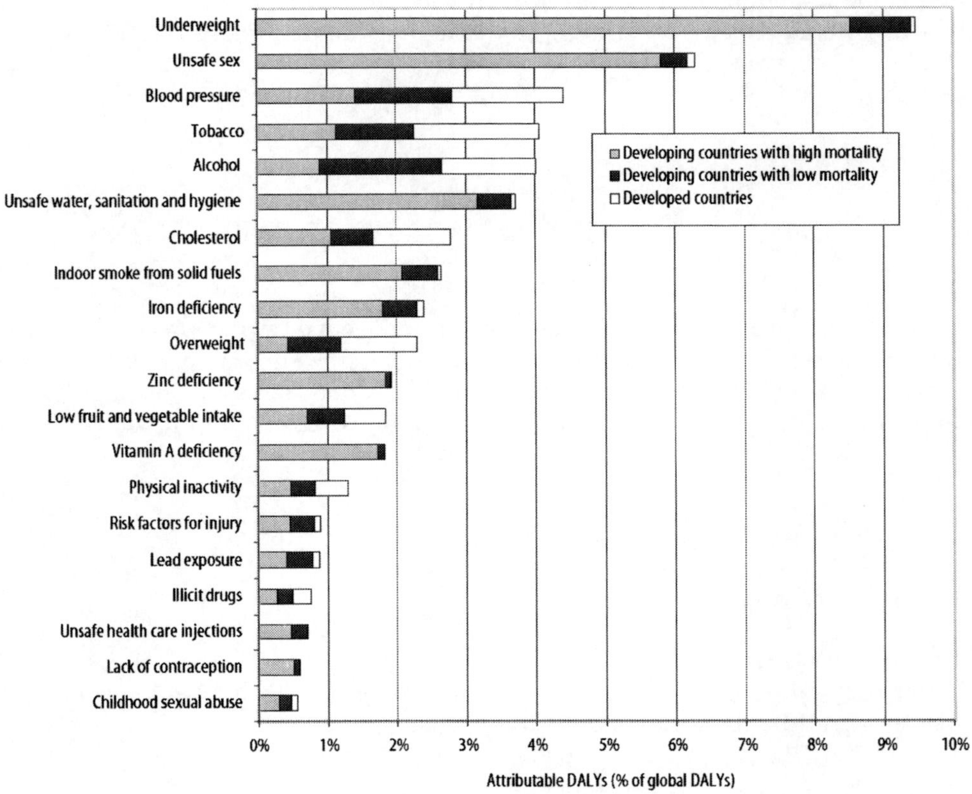

Figure 3.1: The global distribution of burden of disease attributable to 20 leading selected risk factors [1].

The major nutritional benefits of thermal treatment of food are:

- Reduction/Elimination of Harmful Components
  - Microbial Contamination/Parasites
  - Toxins
  - Enzyme Inhibitors
  - Allergens
- Preservation of Food
- Changes in Food Matrix Structure and Texture
  - Improvement of Digestibility
  - Increased Bioavailability
- Generation of Beneficial New Compounds
  - Aroma
  - Antioxidants?

In this chapter the hygienic aspects of microbial or parasitic contamination of foods will not be discussed because the main emphasis should be on the beneficial biological effects of thermally generated compounds. Since the reduction or elimination of potential health risks by thermal treatment of foods may result in a nutritional benefit, toxins, enzyme inhibitors, and allergens are briefly considered. Thermally generated compounds related to aroma, colour, and taste and their functional properties are discussed in Contribution III-2.

3.2 Factors to Consider in Relation to Thermal Treatment of Food

For the nutritional quality of foods and the respective impact on health nutrients and bioactive non-nutrients should be considered separately. For the nutrients thermal effects on macronutrients and micronutrients are of importance and since the effects are rather different they will be discussed individually.

3.2.1 Nutrients

3.2.1.1 Macronutrients

3.2.1.1.1 Proteins

Thermal treatment of proteins leads to denaturation and corresponding changes in three-dimensional structure, which in general leads to a better digestibility by the action of proteolytic enzymes in the human digestive system. Protein denaturation, on the other hand, also imposes a greater risk of rapid subsequent microbial contamination under poor hygienic or storage conditions, respectively, imposing the risk of food poisoning. Immunological properties of food proteins may be altered by heat denaturation as has recently been demonstrated with bovine beta-lactoglobulin. In cell culture model systems of Caco-2 cells and M cells, respectively, it was demonstrated that heat-denatured beta-lactoglobulin was less efficiently transported than the native form [2]. If such findings are confirmed *in vivo*, heat denaturation could have important consequences for the immunological responses in the gut-associated lymphoid tissue.

3.2.1.1.2 Complex Carbohydrates

During thermal treatment in the presence of water starches, e.g. amylose and amylopectin undergo gelatinisation and during subsequent cooling retrogradation, both processes greatly affect the digestibility of these compounds. The degree of gelatinisation depends on temperature, proportion of water, and time of cooking. Gelatinised starches are digested much more rapidly than raw ones [3]. With retrogradation resistant starches are produced which are poorly degraded by α-amylase and thus generally reach the large intestine, where microbial fermentation takes place. Resistant starches (Tab. 3.1) thus may affect the microbial flora composition and the microbial metabolism which may have consequences for gut health by changing for example the short-chain fatty acid patterns [3]. Short-chain fatty acids (SCFA) and particularly butyrate are considered to be important substrates for colonic epithelial cells and relations between butyrate concentration and the risk for colorectal carcinogenesis have been observed. Thus foods which lead to a higher production of butyrate in the colon may have beneficial effects. Therefore the generation of resistant starches by thermal treatment of starchy foods may also be associated with beneficial nutritional and health effects.

The amount of resistant starches in a food influence the glycemic index (GI) of that particular food (Tab. 3.2). The GI is defined as the change in blood glucose concentration during a 2 hour period (area under the curve: AUC) following the ingestion of 50 g of available carbohydrates in the test product as a percentage of the corresponding area following an equivalent amount of carbohydrate from a reference product. Usually as references either glucose ingestion or the consumption of white bread are used, respectively. There is growing evidence that the type of carbohydrate consumed is important in relation to metabolic disease risk, and there is currently particular interest in the role of low-glycemic-index (GI) foods. Observational studies have associated low-GI diets with decreased risk of type 2 diabetes and coronary heart disease (CHD), and improvements in various metabolic risk factors have been seen in some intervention studies [4]. However, findings have been mixed and are inconsistent. High amounts of RS in a food generally are associated with a low-GI and this property may be nutritionally beneficial. Thermal treatment of starchy foods thus favourably changes the nutritional properties of these foods.

3.2.1.1.3 Fats

The physicochemical properties of fats are highly affected by temperature. Unsaturated fatty acids (mono-unsaturated fatty acids: MUFA; poly-unsaturated fatty acids: PUFA) but also cholesterol are rapidly oxidized in the presence of oxygen

Table 3.1: Nutritional classification of resistant starches [3].

Types of Resistant Starch	Examples of Occurrence
RS_1; physically inaccessible	Partly milled grains and seeds
RS_2; resistant granules	Raw potato, green banana, some legumes and high amylose starches
RS_3; retrograded	Cooked and cooled potato, bread, and cornflakes
RS_4; chemically modified	Etherized, esterified, or cross-bonded starches (used in processed foods)

Table 3.2: Main factors affecting the GI of foods and meals [5].

Food Factor	Examples of Influencing Factors	Effect on GI
Structure		
Gross structure	Grinding, *heat treatment*	Higher GI when homogenised
Cellular structure (cell wall integrity)	Ripeness	Higher GI with increased ripeness
Starch		
Granular structure (intact or gelatinised)	*Heat treatment*	Higher GI when gelatinised
Amylose (unbranched)	Genotype of raw material	Lower GI compared to amylopectin
Amylopectin (branched)	Genotype of raw material	Higher GI compared to amylase
Other factors		
Gel-forming types of dietary fibre	Genotype of raw material, added fibres	Lower GI
Organic acids	Fermentation, added acids	Lower GI
Amylase inhibitor	*Heat treatment*	Lower GI
Fructose/glucose-ratio	Genotype of raw material, type of added sugars	Lower GI with increased ratio

and in the absence of food antioxidants, respectively. Furthermore MUFA and PUFA may undergo isomerisation with heat treatment. Fat oxidation compounds in foods are nutritionally undesirable and have not been shown to have any beneficial effects, however these compounds are of toxicological relevance, they are not further discussed since this chapter deals with nutritional benefits.

3.2.1.1.4 Amino Acids and Sugars

Amino acids either in free or in protein-bound form and sugars undergo Maillard-type reactions during thermal treatment and thus change the nutritional and sensory properties of food. Advanced glycation end products (AGE) in food have been demonstrated to show undesired nutritional and health effects, for example they seem to play a role in cellular signal transduction pathways involved in inflammatory processes, cell proliferation, tumour growth and metastasis [6]. (These aspects are dealt with in detail in Contribution III-2).

3.2.1.2 *Essential Micronutrients*

In relation to micronutrients it is known since several decades that particularly the water-soluble vitamins are rapidly degraded by thermal treatment. Thus thermal treatment of foods may cause a nutritionally relevant decrease of heat-labile vitamin content [7]. The magnitude of these thermal processing losses of course depends on the processing variables, e.g. temperature and exposure time. The processing related losses of vitamins are creating nutritional problems only for those vitamins where the recommended dietary intakes in the population or in population subgroups are not achieved, this is of particular importance in specific subgroups with a high requirement, for example folic acid supply in pregnant women. Folic acid is one of the critical vitamins in the industrialized part of the world and in fact some countries require mandatory folic acid fortification of staple foods (for example flour in the USA). Folic acid deficiency is associated with the incidence of neural tube defects and with plasma homocysteine concentrations, an independent risk factor for cardiovascular disease. Folic acid is one of the heat labile vitamins and is rapidly destroyed with increasing temperature [8]. Furthermore the bioavailability of folates from natural sources is highly variable and may depend on the food matrix structure and stabilizing factors in the food to reduce folate degradation for example by thermal treatment [9].

Most nutrition societies promote a high consumption of fresh fruit and vegetables and also encourage the use of frozen vegetables which maintain a high level of vitamins if properly processed and stored.

On the other hand the bioavailability of minerals and trace elements may be enhanced by changes in the food matrix upon heat treatment. A recent human dietary intervention study investigating the bioavailability of selenium from various food sources, however, could not find a difference in apparent absorption or retention of selenium from either cooked or salted fish, respectively [10]. In general, information is lacking on the effects of thermal treatment on the bioavailability of essential micronutrients in human nutrition.

3.2.2 Thermal Effects on Non-nutrients

Non-nutrients are those bioactive compounds in food that do not seem to be essential for avoiding a specific disease or a clinical condition associated with a deficiency for a specific substance, e.g. a nutrient. However, non-nutrients may have important nutritional and biochemical functions by acting for example as antioxidants or having antimicrobial or anticarcinogenic properties, respectively. In the past years it has become increasingly accepted that bioactive compounds may play an important role in the pathogenesis of various chronic diseases, particularly cardiovascular disease and certain cancers [11]. Bioactive non-nutrients, specifically polyphenols, may also be important in the pathophysiology of neuro-degenerative diseases, like Parkinson's and Alzheimer's disease [12, 13], and furthermore with the ageing process, respectively. Thus the effects of thermal treatment on these food components are of importance and highly relevant to human health. There are a number of bioactive compounds in food and in the human diet, probably the largest group being the phytochemicals. Phytochemicals are secondary plant metabolites which are present in small and varying amounts in plants. Phytochemicals belong to different chemical classes among them carotenoids, flavonoids, isoflavonoids, phenolic acids, glucosinolates, monoterpenes, phytosterins, and saponins to name the most important and prominent classes. Especially the dietary phenolics and the carotenoids have attracted great scientific interest in recent years. Large scale epidemiological studies provided evidence for protective effects of largely plant-based diets in the development of cardiovascular disease, certain cancers, obesity, and diabetes type 2. These disease risk reducing effects of diets high in fruit, vegetable and whole grain have been partly attributed to the presence of phytochemicals in these products.

3.2.2.1 Polyphenols

Polyphenols are highly reactive compounds and good substrates for various enzymes, including polyphenoloxidases, peroxidases, glycosidases, and esterases (Figs. 3.2 and 3.3). They undergo numerous enzymatic and chemical reactions during postharvest food storage and processing. Although the occurrence of such reactions and their roles in the development or degradation of food quality are well documented, the structures of the resulting products are still poorly understood and their concentrations in food are usually unknown [14].

The effects of thermal treatment on these compounds so far have not been studied in detail: Particularly the interactions between different bioactive compounds present in the same food during thermal treatment have not been investigated. The interest in the health properties of polyphenols was for a large part stimulated by the so-called "French Paradox", referring to the low incidence of death from coronary heart disease in Southern France despite a high consumption of saturated fats and cholesterol [16]. It was postulated that the low rate of cardiac death was related to the high consumption of red wine in southern France and associated with the high intake of antioxidant polyphenols from this source, respectively [17]. A number of molecular effects on smooth muscle and endothelial cells have been described which may be associated with a protective role of wine polyphenols against atherosclerosis [17], these findings are summarized in Figure 3.4.

Recently the bioavailability and bioefficacy of polyphenols in humans has been reviewed [18, 19]. Data from 97 bioavailability studies [18] and 93 inter-

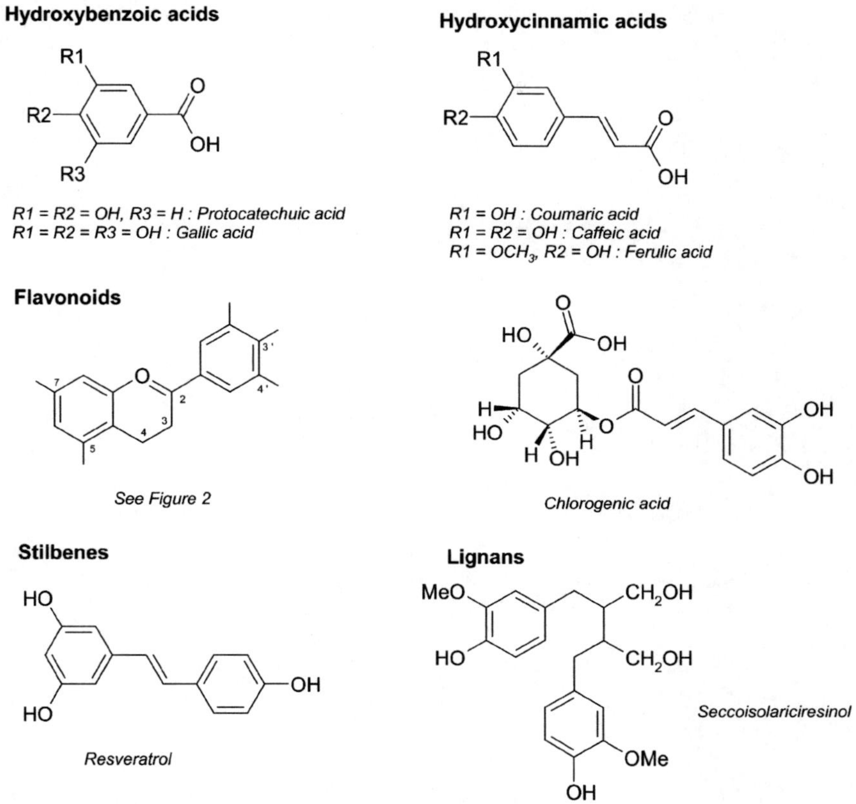

Figure 3.2: Chemical structures of polyphenols [15].

vention studies [19] were presented. It was concluded from these studies that compared with the effects of polyphenols *in vitro*, the effects *in vivo*, although significant, are more limited. The reason for this are 1) lack of validated *in vivo* biomarkers, especially in the area of carcinogenesis; 2) lack of long-term studies; and 3) lack of understanding or consideration of bioavailability in the *in vitro* studies, which are subsequently used for the design of *in vivo* experiments [19]. So far, mainly mechanistic *in vitro* studies with single compounds are published. Studies using whole-foods or combinations of various bioactive compounds are missing. Different polyphenols react differently to thermal treatment as has been shown for example in a study with virgin olive oils [20]. Virgin olive oils contain hydroxytyrosol and tyrosol-like substances and they also contain the lignans 1-acetoxypinoresinol and pinoresinol. The lignans were much less affected by heating to 180 °C for 25 hours, microwave heating for 10 minutes or boiling in a pressure cooker for 30 minutes than hydroxytyrosol and tyrosol-like substances [20]. Thus thermal treatment differentially affects bioactive compounds present in the same food and thus very likely changes the nutritional characteristics of that food considerably.

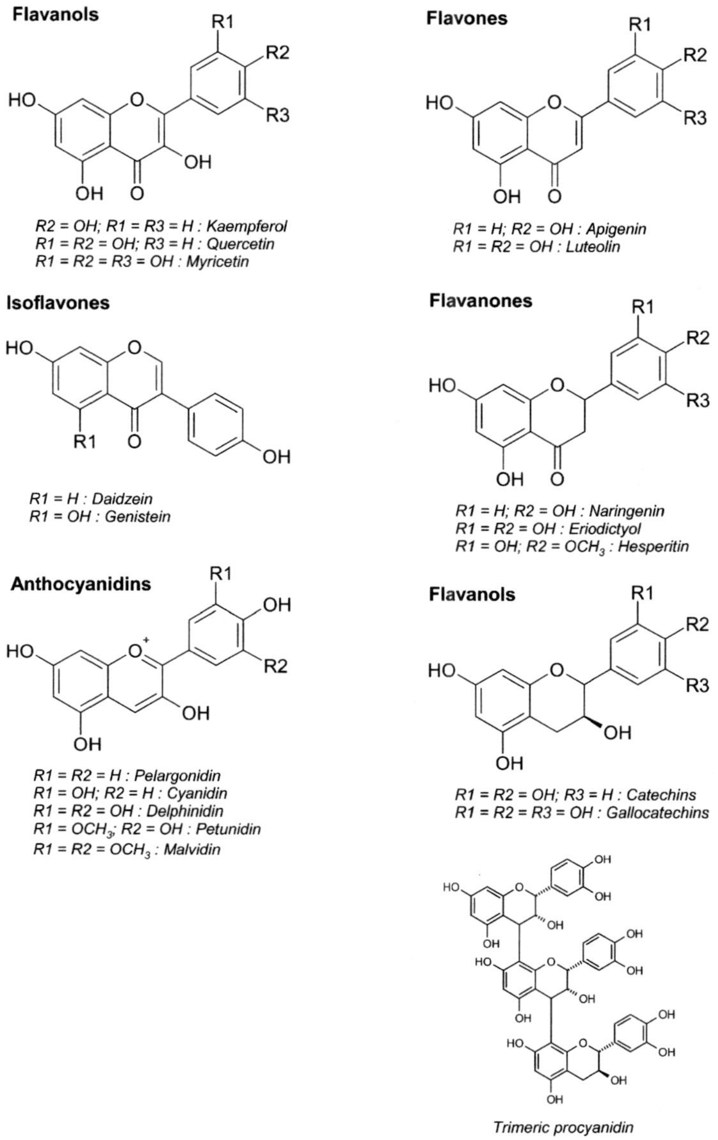

Figure 3.3: Chemical structures of flavonoids [15].

Since it is not possible within the objectives of this chapter to deal with the effects of thermal treatment on all bioactive compounds in foods, a few examples are discussed in more detail.

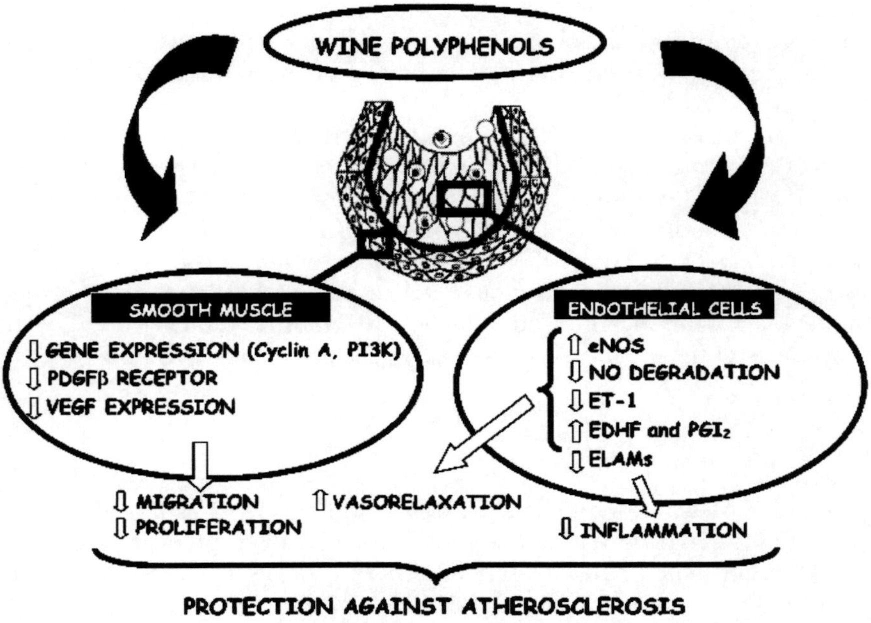

Figure 3.4: Hypothetical model of protective effects of red wine polyphenols on the arterial wall [17].

3.2.2.2 *Anthocyanins*

Anthocyanins are present in different fruits, particularly in berries, and vegetables. They are water-soluble plant pigments and they occur naturally as glycosides. The large variety of different anthocyanins results from the differences in glycosylation patterns. Some anthocyanins occur also in the acylated form, for example in the skin of red radishes [21] or in purple carrots [22]. Anthocyanidins, the aglycon forms of anthocyanins are chemically labile with respect to pH, temperature and oxidation. Anthocyanins are considered to be potent anti-

Table 3.3: Effects of thermal processing and high-pressure treatment of anthocyanins [23].

Compound [mg/L]	Unprocessed	Pasteurisation	High-pressure Treatment
Delphinidin-3,5-diglucoside	207	193	178
Cyanidin-3,5-diglucoside	253	251	242
Petunidin-3,5-diglucoside	205	203	186
Pelargonidin-3,5-diglucoside	15.0	18.9	14.8
Peonidin-3,5-diglucoside	377	408	396
Malvidin-3,5-diglucoside	155	182	176
Total anthocyanins	1210	1260	1190

oxidants, at least *in vitro*. However, several studies have investigated their bioavailability from different foods sources and have generally found a very low bioavailability. The bioavailability of anthocyanins does not seem to be affected by thermal processing [18].

Thermal processing and high-pressure treatment was shown recently to have no effects on the stability of different anthocyanins from grape juices [23], but another study investigating the effects of processing on different strawberry products found a loss of 27–39% after pasteurisation of juice or heating of nectar, respectively (Tab. 3.3) [24]. In a clinical intervention study with raw and cooked purple carrots it was demonstrated that cooking increased the recovery of nonacylated anthocyanins but not acylated anthocyanins [22]. The health consequences of this observation, however, are not clear at present.

3.2.2.3 Carotenoids

Carotenoids are also a large group of plant pigments. It is estimated that in nature about 500–600 different carotenoids exist. However, plants used for human nutrition contain only approximately 50 different carotenoids. The major carotenoids in the human diet are α- and β-carotene, β-cryptoxanthin, lycopene, lutein and zeaxanthin. Chemically carotenoids are terpenoids and they are highly lipophilic molecules. Carotenoids can be divided into two main classes, the carotenes and the xanthophylls. Functionally carotenoids are classified into provitamin A carotenoids (α- and β-carotene, β-cryptoxanthin) and non-provitamin A carotenoids (lycopene, lutein and zeaxanthin).

A general feature of carotenoids is their ability to quench singulett-oxygen and thus act as antioxidants. However, besides those properties they also have direct biological effects for example in cell communication. Carotenoids exert part of their antioxidant activity by incorporation into blood lipids (high-density lipoproteins: HDL; low-density lipoproteins: LDL; very low-density lipoproteins: VLDL). Oxidized LDL particles are regarded as the major culprit in the early phases of the development of atherosclerotic plaques in blood vessels which subsequently may cause heart attacks and stroke. Thus the reduction of LDL oxidation should reduce the risk for developing atherosclerosis and the associated cardiovascular diseases. Sufficient intestinal carotenoid uptake and bioavailability depends on the cellular matrix of foods and on micelle formation during the small intestinal digestive phase and thus requires the presence of small amounts of fat. The main food sources of carotenoids are vegetables (for example: carrots, tomato and tomato products, spinach, broccoli, kale, spinach, lettuce, peppers, pumpkin) and fruits (for example: watermelon, pink grapefruit, papaya, apricots, mango, oranges).

In recent years the most studied carotenoids have been β-carotene, lycopene, and lutein. Particularly lycopene has attracted scientific interest because it is a non-provitamin A carotenoid and has a high singulett-oxygen quenching capacity. By far the most important dietary sources of lycopene are tomato and tomato products. Furthermore tomato products usually undergo intense processing including thermal treatment during their production. During thermal processing isomerisation of lycopene from *all-trans*-lycopene to *cis*-lycopene isomers occurs depending on the temperature and the time of heating [25]. Thermal processing enhances the nutritional value of tomatoes by increasing total antioxidant activity [26]; despite a decrease in vitamin C content with increased duration of processing at 88 °C lycopene content and total antioxidant activity increased, respectively (Tab. 3.4). Similar data have been reported by Gahler et

Table 3.4: Percent changes of vitamin C content, lycopene content, and total antioxidant activity in processed tomatoes in comparison to those in unprocessed tomatoes [26].

	Processing time at 88 °C		
	2 min	15 min	30 min
Vitamin C	–10.2	–15.5	–29.4
Lycopene	54.4	171.1	164.3
Total antioxidant activity	28.1	33.9	62.2

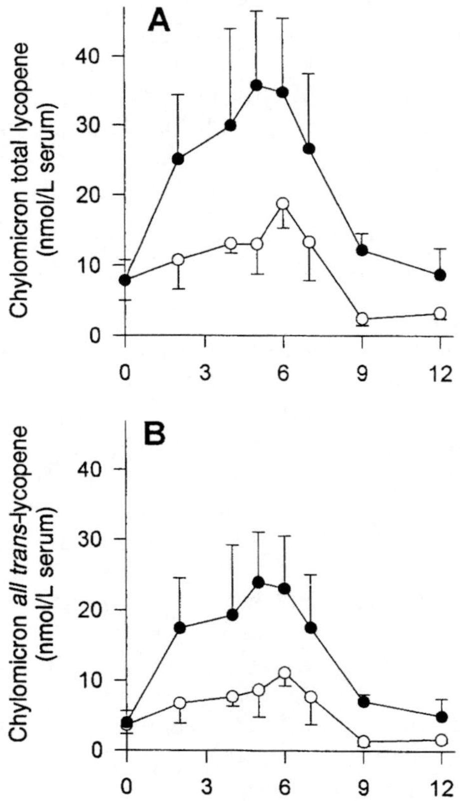

Figure 3.5: Bioavailability of lycopene from fresh tomato or tomato paste [28]. A) Total lycopene, B) all-trans lycopene; ○ fresh tomatoes, ● tomato paste.

al. [27], these authors also found a decrease in vitamin C content and an increase of the total phenolics concentration and the water soluble antioxidant capacity with thermal processing of tomato juice, tomato sauce, and baked tomatoes, with a temperature treatment of up to 220 °C for 45 minutes [27].

The bioavailability of lycopene has been shown to be much higher from processed tomato products (tomato paste) as compared to fresh tomato in a human dietary intervention study (Fig. 3.5) [28].

These initial findings with tomato products have also been confirmed in a human dietary intervention study using lycopene-rich carrot products. It was found that lycopene uptake from carrots treated for 30 minutes at 130 °C was increased 10-fold (Möseneder and Bub, personal communication). A high consumption of tomatoes and tomato products leading to a high supply of lycopene has been shown in several epidemiological studies to be associated with a decreased risk of cancers and in particular of prostate cancer [29, 30]. Prostate cancer has a very high incidence and is now the most prevalent form of cancer in men in the US. Thus increasing the bioavailability of lycopene by thermal processing of tomatoes may reduce the risk of prostate cancer and thus offer a great health benefit.

In another human dietary intervention study where differently processed spinach products were investigated it was observed that the bioavailability of lutein from spinach was higher than that of β-carotene and that enzymatic disruption of the matrix (cell wall structure) enhanced the bioavailability of β-carotene from whole leaf and minced spinach, but had no effect on lutein bioavailability [31].

3.2.2.4 Isoflavonoids

As a last example of bioactive phytochemicals isoflavonoids will be briefly discussed. This group of compounds is often also designated as phytoestrogens because of their structural similarity to human endogenous 17-β-estradiol and their capacity to bind to human estrogen receptors α and β, respectively. The main dietary isoflavonoids are genistein and daidzein and these compounds occur primarily in soy and soy products. As other flavonoids also the isoflavonoids occur in these products primarily in glycosylated form. Soy foods are traditional foods in Asian countries, particularly in Japan and China, however, in the past years they have become more and more accepted also in the Western world, particularly as a good source of vegetable protein as an alternative to animal protein. Soy protein has specific properties and a health claim in relation to cardiovascular health is approved by FDA in the US. The influence of thermal treatment on the stability of soy isoflavones, genistein and daidzein, was studied in model solutions and rapid degradation of genistein and daidzein was observed at pH 7 and thermal treatment at 120 °C for 20 minutes [32]. It should be emphasized, however, that the degradation rates may be different in an intact food matrix as compared to a model system with isolated compounds.

3.2.3 Other Thermal Effects to Be Considered

The allergenicity of certain food allergens may be affected by thermal processing and this aspect is discussed extensively in Contribution III-13 and thus will not be dealt with here in detail. Besides the allergens themselves, however, other modifying factors present in food may play an important role with respect to thermal treatment and heat inactivation of allergens, as has been recently demonstrated in a study on thermal processing on the IgE reactivity of the non-specific lipid transfer protein from apple, Mal d 3 [33]. These compounds are involved in allergy to fresh and processed fruits. In this study it was found that only severe heat treatment caused a significant decrease in the allergenicity of Mal d 3 and that glycation had a protective effect. The presence of sugars in fruits may contribute to the thermostability of the allergenic activity of LTP in heat-processed foods [33].

In some foods, particularly in legumes, natural components are present, such as lectins, amylase, and trypsin inhibitors, that may adversely affect their nutritional properties [34]. Because of their protein nature, enzyme inhibitors and lectins are inactivated under conditions leading to irreversible protein denaturation. Thus heat treatment may lead to the destruction of these antinutritional factors [34] and thus may improve the nutritional quality of the respective foods.

3.3 Conclusions and Future Research

The nutritional effects of thermal treatment of foods are very diverse and so far mainly information of the impact of heat treatment on individual food components was gathered. The effects of thermal treatment can be nutritionally beneficial for example by increasing the bioavailability of bioactive components from an altered food matrix, however, also some of these bioactive components, particularly water-soluble vitamins and some heat-sensitive phytochemicals, will be lost during heat treatment of foods. Although the heating of food has a very long history during the evolution of modern man still the knowledge about the nutritional consequences of chemical and physical changes in food induced by thermal processing are scarce. With the modern analytical methods, like mass-spectrometry and NMR-techniques, the nature and structure of molecules formed during thermal treatment of food can be elucidated in much more detail than in the past. The new technologies now available for nutrition research in the post-genomic era should enable nutritionists to get more detailed insight into the effects of thermal treatment of complex foods and the resulting consequences for health. The sequencing of the human genome has opened the door to the most exciting new era for nutritional science. It is now possible to study the underlying mechanisms for diet–health relationships, and in the near future dietary advice for promoting optimal health could be provided on an individual basis, in relation to genotype and lifestyle [35]. Emphasis in the diet–health relation should be placed on investigations using primarily complex foods to further the understanding of the interactions between the various food components during thermal treatment and the respective consequences for nutrition and health.

References

1. World Health Organization (2002) The World Health report 2002: reducing risks, promoting healthy life. Geneva, Switzerland.
2. Rytkonen J, Valkonen KH, Virtanen V, Foxwell RA, Kyd JM, Cripps AW, Karttunen TJ (2006) Enterocyte and M-cell transport of native and heat-denatured bovine betalactoglobulin: significance of heat denaturation. *J Agric Food Chem* **54**: 1500–1507.
3. Topping D, Clifton PM (2001) Short-chain fatty acids and human colonic function: roles of resistant starch and nonstarch polysaccharides. *Physiological Reviews* **81**: 1031–1064.
4. Aston LM (2006) Glycaemic index and metabolic disease risk. *Proc Nutr Soc* **65**: 125–134.
5. Arvidsson-Lenner R, Asp NG, Axelsen M, Bryngelsson S, Haapa E, Järvi A, Karlström B, Raben A, Sohlström A, Thorsdottir I, Vessby B (2004) Glycaemic index: Relevance for health, dietary recommendations and food labelling. *Scand J Nutr* **48**: 84–94.
6. Zill H, Bek S, Hofmann T, Huber J, Frank O, Lindenmeier M, Weigle B, Erbersdobler HF, Scheidler S, Busch AE, Faist V (2003) RAGE-mediated MAPK activation by food-derived AGE and non-AGE products. *Biochemical and Biophysical Research Communications* **300**: 311–315.
7. Henry CJK, Heppell N (2002) Nutritional losses and gains during processing: future problems and issues. *Proc Nutr Soc* **61**: 145–148.
8. Indrawati, Van Loey A, Hendrickx M (2005) Pressure and temperature stability of 5-methyltetrahydrofolic acid: a kinetic study. *J Agric Food Chem* **53**: 3081–3087.
9. McNulty H, Pentieva K (2004) Folate bioavailability. *Proc Nutr Soc* **63**: 529–536.
10. Fox TE, Van den Heuvel EGHM, Atherton CA, Dainty JR, Lewis DJ, Langford NJ, Crews HM, Luten JB, Lorentzen M, Sieling FW, van Aken-Schneyder P, Hoek M, Kotterman MJJ, van Dael P, Fairweather-Tait SJ (2004) Bioavailability of selenium from fish, yeast and selenate: a comparative study in humans using stable isotopes. *Europ J Clin Nutr* **58**: 343–349.
11. Kris-Etherton PM, Hecker KD, Bonanome A, Coval SM, Binkoski AE, Hilpert KF, Griel AE, Etherton TD (2002) Bioactive compounds in foods: their role in the prevention of cardiovascular disease and cancer. *Am J Med* **113** (9B): 71S–88S.
12. Lau FC, Shukitt-Hale B, Joseph JA (2005) The beneficial effects of fruit polyphenols on brain aging. *Neurobiology of Aging* **26S**: S128–S132.
13. Sun AY, Simonyi A, Sun GY (2002) The "French Paradox" and beyond: neuroprotective effects of polyphenols. *Free Radical Biology & Medicine* **32(4)**: 314–318.
14. Cheynier V (2005) Polyphenols in foods are more complex than often thought. *Am J Clin Nutr* **81**: 223S–229S.
15. Manach C, Scalbert A, Morand C, Remesy C, Jimenez L (2004) Polyphenols: food sources and bioavailability. *Am J Clin Nutr* **79**: 727–747.
16. Renaud S, de Lorgeril M (1992) Wine, alcohol, platelets, and the French paradox for coronary heart disease. *Lancet* **339**: 1523–1526.

17. Dell'Agli M, Busciala A, Bosisio E (2004) Vascular effects of wine polyphenols. *Cardiovascular Research* **63**: 593–602.
18. Manach C, Williamson G, Morand C, Scalbert A, Remesy C (2005) Bioavailability and bioefficacy of polyphenols in humans. I. Review of 97 bioavailability studies. *Am J Clin Nutr* **81**: 230S–242S.
19. Williamson G, Manach C (2005) Bioavailability and bioefficacy of polyphenols in humans. II. Review of 93 intervention studies. *Am J Clin Nutr* **81**: 243S–255S.
20. Brenes M, Garcia A, Dobarganes MC, Velasco J, Romero C (2002) Influence of thermal treatments simulating cooking processes on the polyphenol content in virgin olive oils. *J Agric Food Chem* **50**: 5962–5967.
21. Fleschhut J, Kratzer F, Rechkemmer G, Kulling SE (2006) Stability and biotransformation of various dietary anthocyanins in vitro. *Eur J Nutr* **45(1)**: 7–18.
22. Kurilich AC, Clevidence BA, Britz SJ, Simon PW, Novotny JA (2005) Plasma and urine responses are lower for acylated vs nonacylated anthocyanins from raw and cooked purple carrots. *J Agric Food Chem* **53**: 6537–6542.
23. Talcott ST, Brenes CH, Pires DM, Del Pozo-Insfran D (2003) Phytochemical stability and color retention of copigmented and processed muscadine grape juice. *J Agric Food Chem* **51**: 957–963.
24. Klopotek Y, Otto K, Böhm V (2005) Processing of strawberries to different products alters contents of vitamin C, total phenolics, total anthocyanins, and antioxidant capacity. *J Agric Food Chem* **53**: 5640–5646.
25. Shi J, Le Maguer M (2000) Lycopene in tomatoes: chemical and physical properties affected by food processing. *Crit Rev Food Sci Nutr* **40(1)**: 1–42.
26. Dewanto V, Wu X, Adom KK, Liu RH (2002) Thermal processing enhances the nutritional value of tomatoes by increasing total antioxidant activity. *J Agric Food Chem* **50**: 3010–3014.
27. Gahler S, Otto K, Böhm V (2003) Alterations of vitamin C, total phenolics, and antioxidant capacity as affected by processing tomatoes to different products. *J Agric Food Chem* **51**: 7962–7968.
28. Gärtner C, Stahl W, Sies H (1997) Lycopene is more bioavailable from tomato paste than from fresh tomatoes. *Am J Clin Nutr* **66(1)**: 116–122.
29. Clinton SK (2005) Tomatoes or lycopene: a role in prostate carcinogenesis. *J Nutr* **135**: 2057S–2059S.
30. Giovannucci E (2005) Tomato products, lycopene, and prostate cancer: a review of the epidemiological literature. *J Nutr* **135: 2030S–2031S.**
31. Castenmiller JJM, West CE, Linssen JPH, van het Hof KH, Voragen AGJ (1999) The food matrix of spinach is a limiting factor in determining the bioavailability of β-carotene and to a lesser extent of lutein in humans. *J Nutr* **129**: 349–355.
32. Ungar Y, Osundahunsi OF, Shimoni E (2003) Thermal stability of genistein and daidzein and its effect on their antioxidant activity. *J Agric Food Chem* **51**: 4394–4399.
33. Sancho AI, Rigby NM, Zuidmer L, Asero R, Mistrello G, Amato S, Gonzalez-Mancebo E, Fernandez-Rivas M, van Ree R, Mills ENC (2005) The effect of thermal processing on the IgE reactivity of the non-specific lipid transfer protein from apple, Mal d 3. *Allergy* **60**: 1262–1268.
34. Lajolo FM, Genovese MI (2002) Nutritional significance of lectins and enzyme inhibitors from legumes. *J Agric Food Chem* **50**: 6592–6598.
35. Fairweather-Tait SJ (2003) Human nutrition and food research: opportunities and challenges in the post-genomic era. *Phil Trans R Soc Lond B* **358**: 1709–1727.

4 Biological Activities of Maillard Reaction Products

Doris Marko *

Abstract

The typical brown colour, originating from Maillard reactions between reducing carbohydrates and amino acids, is, in addition to the unique aroma, a key factor in the consumer acceptance of thermally processed food. Despite intensive studies on this nonenzymatic browning reaction, little is known so far about the physiological relevance of the chromophores formed by Maillard reactions. We investigated cellular effects of a series of selected Maillard reaction products formed from carbohydrates under household heating conditions with special emphasis on growth modulating and/or potentially genotoxic mechanisms. One of the compounds, 4-hydroxy-5-methyl-2*H*-furan-3-one (**1**), commonly known as norfuraneol, was found to enhance the growth of human tumour cells *in vitro*. In contrast, several Maillard-type chromophores were identified as potent inhibitors of tumour cell growth. Two of these compounds, 2-(2-furyl)methylidene-4-hydroxy-5-methyl-2*H*-furan-3-one (**2**) and 3-hydroxy-4-[(*E*)-(2-furyl)- methylidene]methyl-3-cyclopentene-1,2-dione (**13**) were found to suppress the activity of the mitogen-activated protein kinase cascade, one of the major signalling pathways in the regulation of cell growth. In accordance with these results, the cells accumulated in the G_1-phase of the cell cycle. After 24 h of incubation the activation of caspase 3 activity and DNA fragmentation were observed as characteristic features for the onset of apoptosis. However, at higher substance concentrations the most potent compound with respect to growth inhibition, **13**, was found to interfere with microtubule assembly, raising the question whether the compound possesses genotoxic and mutagenic properties e.g. by interference with nuclear spindle formation during mitosis. In a respective micronucleus assay (V79 cells) no induction of micronucleus formation was observed up to a concentration of 1 μM for 24 h. At higher substance concentrations a substantial decrease in the cell division rate and a loss of cell viability prohibited further testing. In the hprt-assay (V79 cells) **13** did not affect the mutation rate in the *hprt* locus up to 30 μM for 24 h. Thus, it is likely to assume that due to the effective suppression of cell proliferation and the onset of apoptosis the potential interference of **13** with microtubule assembly is not of relevance for the maintenance of DNA integrity at least in V79 cells, lacking substantial expression of xenobiotic metabolising systems.

* Institute of Applied Biosciences, Section of Food Toxicology, University of Karlsruhe (TH), Fritz-Haber-Weg 2, 76744 Karlsruhe, Germany.

Thermal Processing of Food: Potential Health Benefits and Risks
DFG, Deutsche Forschungsgemeinschaft

ISBN: 978-3-527-31909-1

In summary, several Maillard reaction products were identified as potent modulators of tumour cell growth *in vitro*, raising the question whether normal cells are targeted as well. So far, potentially genotoxic and mutagenic properties, such as the interference with microtubule assembly appear not to be of toxicological relevance in V79 cells, but warrant further investigations with respect to cell systems competent in xenobiotic metabolism, especially non-transformed human cells. Furthermore, considering the potency of the single compounds, the question of potential combinatory effects of such Maillard-type chromophores in complex mixtures needs to be addressed.

4.1 Introduction

During thermal food processing a complex network of reactions between reducing carbohydrates and amino compounds, collectively referred to as the Maillard reaction, generates a spectrum of compounds important for the characteristic flavour and colour of many products of the daily diet such as coffee, maple syrup, bread crust or roasted meat. With respect to consumer's acceptance and expectations, these browning reactions are highly desirable and are directly associated with the image of a delicious high-grade product. In the last decades enormous efforts were made to unravel the structures and formation pathways of Maillard reaction products. However, the knowledge about the physiological relevance of these compounds is still quite limited. Potentially beneficial as well as adverse effects of Maillard products have been reported [1]. The formation of Maillard products possessing antioxidative properties has been observed in model systems but also in a spectrum of heat treated food such as bread crust, coffee or kilned and roasted malts [2–7]. Furthermore, Maillard products have been identified modulating the activity of phase I and II enzymes of the metabolism [2, 8, 9]. Especially the induction of detoxifying phase II enzymes raises expectations toward potential chemopreventive properties. However, depending on the precursors and heating conditions, the Maillard reaction is also known to yield genotoxic and mutagenic products [10–16]. To play a role in carcinogenesis, compounds do not necessarily have to act as genotoxic agents, but might also e.g. affect the regulation of cell growth by interference with cellular signalling pathways. Malignant transformation is associated with changes in the expression and/or activity of signalling elements involved in the regulation of cell growth and differentiation. The extracellular-signal-regulated/mitogen-activated protein kinase (ERK/MAPK) pathway represents one of the major intracellular signaling cascades in the control of cell proliferation [17, 18]. In our group we addressed the question whether Maillard-type chromophores affect the growth of human tumour cells and, if so, whether the ERK/MAPK pathway might represent a potential cellular target.

4.2 Modulation of Tumour Cell Growth *In Vitro*

We investigated the effect of a series of Maillard reaction products formed from carbohydrates under household heating conditions on the growth of human tumour cell lines, originating from the gastrointestinal tract (GXF251L, gastric carcinoma; CXF94L, colorectal carcinoma). Furthermore, the large cell lung carcinoma cell line LXFL529L, known to be sensitive to a panel of standard chemotherapeutic agents, was included in the testing. The growth modulating potential of the compounds **1–14** (Fig. 4.1) was determined using the sulforhodamine B assay [19]. In LXFL529L cells 4-hydroxy-5-methyl-3(2*H*)-furanone (**1**), commonly known as norfuraneol, strongly stimulated cell proliferation (Fig. 4.2). The growth promoting effect of **1** was less prominent, but still significant, in the gastric carcinoma cell line GXF251L (Fig. 4.3). In contrast, no stimulation of cell growth was mediated by the structurally closely related Maillard

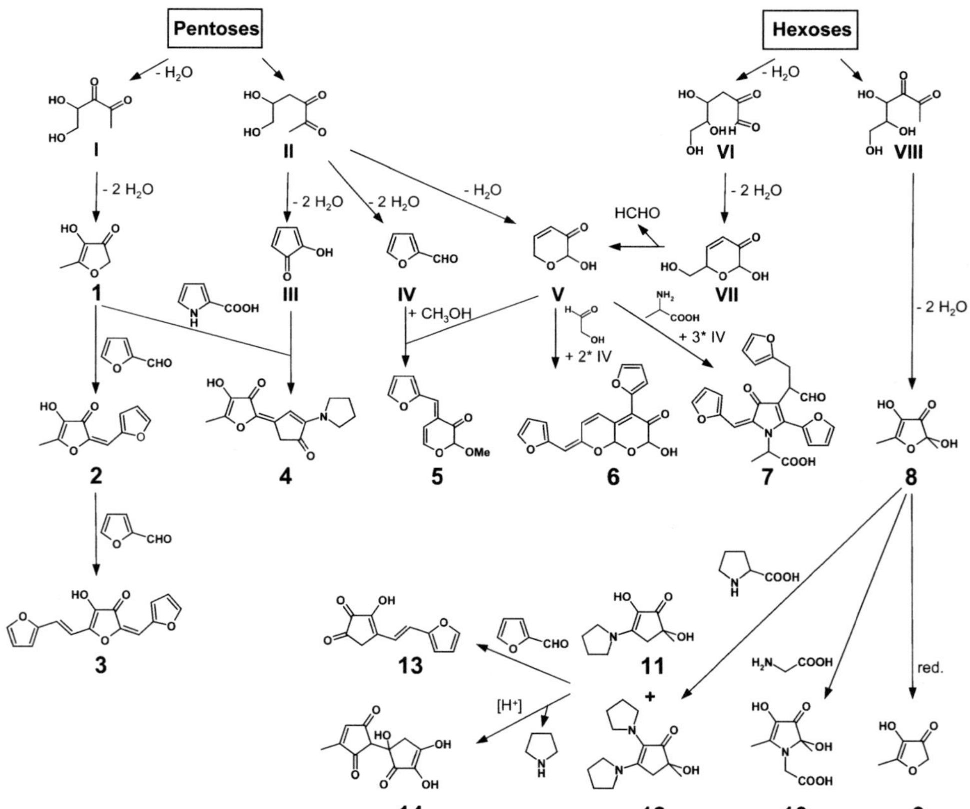

Figure 4.1: Route map of chromogenic reactions involved in non-enzymatic browning according to [20].

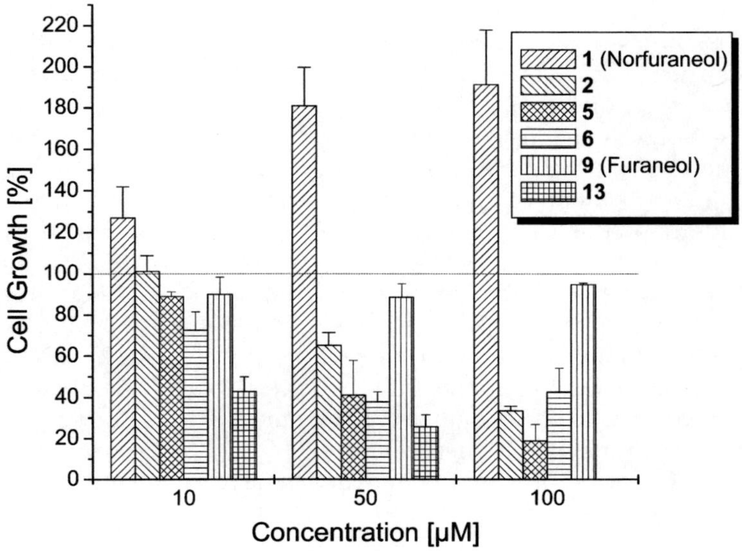

Figure 4.2: Modulation of tumour cell growth by selected Maillard reaction products in the sulforhodamine B assay adopted from [20]. LXFL529L cells were treated for 72 h with the respective compound in serum-containing medium. The data were calculated as protein content of treated cells over control cells ×100 (T/C). Values are presented as the mean ± SD of three independent experiments, each performed in quadruplicate.

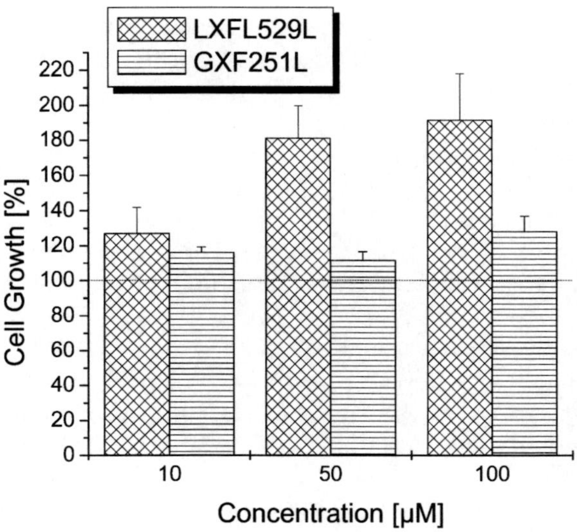

Figure 4.3: Comparison of the growth stimulating effect of norfuraneol on two different human xenograft tumour cell lines in the sulforhodamine B assay adopted from [20]. The cells were incubated for 72 h with norfuraneol in serum-containing medium. The data were calculated as protein content of treated cells over control cells ×100 (T/C). Values are presented as the mean ± SD of three independent experiments, each performed in quadruplicate.

products **8** (2,4-dihydroxy-2,5-dimethyl-3(2*H*)-furanone) and **9** (4-hydroxy-2,5-dimethyl-3(2*H*)-furanone; furaneol) [20]. The condensation of **1** with furan-2-aldehyde resulted in the formation of 2-(2-furyl)methylidene-4-hydroxy-5-methyl-2*H*-furan-3-one (**2**) (Fig. 4.1), exhibiting substantial growth inhibitory properties (Fig. 4.2). A further condensation with furan-2-alaehyde, leading to 2-[(2-furyl)-methylidene]-4-hydroxy-5-[(*E*)-(2-furyl)methylidene]-methyl-2*H*-furan-3-one (**3**) [21], was found to diminish the growth inhibitory effect. The intensively coloured Maillard products (*E*)-2-methoxy-4-[(2-furyl)methylene]-2*H*-pyran-3-one (**5**) [22] and (1S, 8aR)- and (1R,8aR)-4-(2-furyl)-7-[(2-furyl)methylidene]-2-hydroxy-2*H*,7*H*,8a*H*-pyrano[2,3-*b*]pyran-3-one (**6**) [21] inhibited tumour cell growth effectively. From the Maillard-type chromophores tested, 3-hydroxy-4-[(*E*)-(2-furyl)methylidene]methyl-3-cyclopentene-1,2-dione (**13**) exhibited by far the highest growth inhibitory potential (Fig. 4.2) [20]. In summary, from the Maillard products **1–14**, the compounds **2–6** and **13** were found to inhibit the growth of human tumour cell lines *in vitro* [20].

Further studies on the underlying mechanism of action focussed on the most potent compounds with respect to growth inhibition, **2** and **13**.

4.3 Modulation of Cell Cycle Distribution and Induction of Apoptosis

Analysis of characteristic modulations in cell cycle distribution provides a useful tool for first hints towards the potential mechanism of action. In gastric carcinoma cells (GXF251L), synchronized by serum deprivation, treatment with **2** or **13** led to a clear increase of cells in the G_1-phase of the cell cycle. These results indicate that **2** and **13** might interfere with cellular targets crucial for G_1 passage and/or G_1/S-transit. Indeed, both compounds, **2** and **13**, were found to suppress the phosphorylation of ERK1 and ERK2, as crucial elements of the mitogen-activated protein kinase cascade in HT29 cells (manuscript in preparation). Subsequently, **13** was found to diminish effectively the phosphorylation of the transcription factor Elk-1 [23].

Flow cytometry showed a marked increase of cells in the sub-G_1 area as a result of the incubation of GXF251L cells for 24 h with 100 μM of **2** or 50 μM of **13**, indicating the onset of apoptosis. In accordance, treatment of GXF251L cells with growth-inhibitory concentrations of **2** or **13** led to characteristic DNA fragmentation as well as the activation of caspase 3 [20]. Thus, it is likely to assume that **2** and **13** induce apoptosis in the gastric carcinoma cell line GXF251L.

4.4 Genotoxicity of Maillard-type Chromophores

In a cell-free system compound **13** was found to interfere with microtubule assembly (Fig. 4.4) [23]. However, the interference with microtubule assembly was not limited to the purified protein preparation. In MCF-7 mammary carcinoma cells treatment with 20 μM of **13** for 24 h led to the breakdown of the cytoskeleton [23]. Concomitantly, the cells exhibited a strong reduction of the cell volume together with a condensation of the chromatin, indicating the induction of apoptosis. The impairment of microtubule organisation can result in the onset of apoptosis [24, 25], a desirable effect in the treatment of tumour cells. However, from the toxicological point of view, the inhibition of microtubule assembly can also lead to adverse undesired results, namely genotoxic and mutagenic effects as a consequence of the disturbance of the mitotic spindle apparatus [26–28]. Thus, we investigated whether **13** may act as an aneugenic agent in cell culture. As an indication for an aneugenic mechanism of action the formation of CREST positive micronuclei was determined. V79 cells were incubated with **13** for 24 h under serum-free conditions. After a postincubation period of 16 h in serum-containing medium, to allow cell division, the cells were fixed, stained with DAPI and analyzed by fluorescence microscopy for micronucleus formation. Up to a concentration of 1 μM no effect on the formation of micronuclei was observed. However, an increase in substance concentration re-

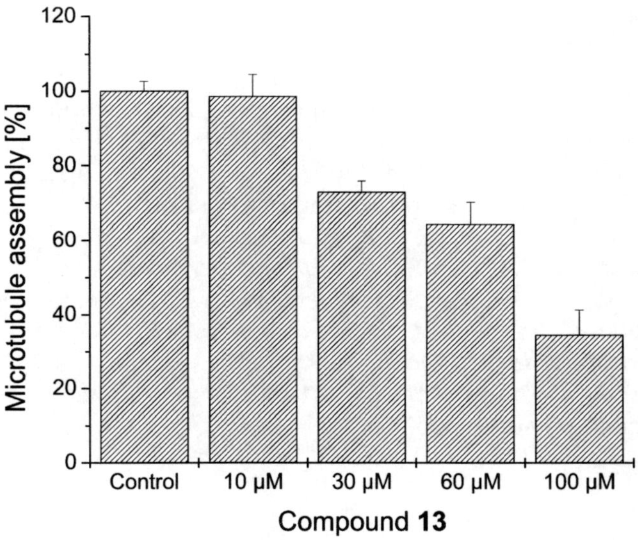

Figure 4.4: Inhibition of *in vitro* assembly of microtubule proteins purified from bovine brain (according to [23]). Microtubule assembly was initiated at 35 °C by addition of 0.5 mM GTP. The absorbance was determined at 350 nm. The data are presented as percentage of polymerisation after 20 min and represent the mean of at least two independent experiments performed in duplicate.

sulted in a substantial decrease in the cell division rate and a loss in viability, impeding further testing.

The mutagenic potency of **13** was determined in the hprt-assay. The incubation of V79 cells with 30 μM of **13** for 24 h did not affect the mutation rate in the *hprt* locus (publication in preparation). Also in this test system, a substantial loss in viability impeded the testing of higher concentrations. It is likely to assume that due to the effective suppression of cell proliferation and the onset of apoptosis the potential interference of **13** with microtubule assembly is not of relevance for the maintenance of DNA integrity, at least in V79 cells, lacking substantial expression of xenobiotic metabolising systems. Potential genotoxic effects on human cells, especially non-transformed human cells, require further testing.

4.5 Risk-benefit Evaluation

Several Maillard reaction products were identified as potent modulators of tumour cell growth *in vitro*. The inhibition of tumour cell growth *in vitro* raises the question whether these compounds might be of interest with respect to chemoprevention. However, so far no information is available whether normal human cells are targeted as well, which would represent a potentially undesired effect. So far, possible genotoxic and mutagenic properties, such as the interference with microtubule assembly, appear not be of toxicological relevance in V79 cells, but warrant further investigations with respect to cell systems competent in xenobiotic metabolism, especially non-transformed human cells.

Furthermore, it has to be considered that growth modulating effects were observed in the micromolar range (Figs. 4.2 and 4.3). Knowledge about the occurrence of these compounds in the diet is quite limited. As a source for Maillard products not only thermal processing of food under household conditions has to be considered, but also the use of reaction flavours in the industrial production. Data on the occurrence of bioactive Maillard products and respective intake estimations are needed to enable a reliable risk-benefit evaluation. Moreover, considering the potency of the single compounds, the question of potential combinatory effects of such Maillard-type chromophores in complex mixtures needs to be addressed.

References

1. Friedman M. Biological effects of Maillard browning products that may affect acrylamide safety in food: biological effects of Maillard products. *Adv Exp Med Biol.* 2005; **561**:135–156.
2. Somoza V, Wenzel E, Lindenmeier M, Grothe D, Erbersdobler HF, Hofmann T. Influence of feeding malt, bread crust, and a pronylated protein on the activity of chemopreventive enzymes and antioxidative defense parameters in vivo. *J Agric Food Chem.* 2005; **53(21)**: 8176–8182.
3. Sun Y, Hayakawa S, Chuamanochan M, Fujimoto M, Innun A, Izumori K. Antioxidant effects of Maillard reaction products obtained from ovalbumin and different D-aldohexoses. *Biosci Biotechnol Biochem.* 2006; **70(3)**: 598–605.
4. Antony SM, Han IY, Rieck JR, Dawson PL. Antioxidative effect of maillard reaction products formed from honey at different reaction times (dagger). *J Agric Food Chem.* 2000; **48(9)**: 3985–3989.
5. Samaras TS, Camburn PA, Chandra SX, Gordon MH, Ames JM. Antioxidant properties of kilned and roasted malts. *J Agric Food Chem.* 2005; **53(20)**: 8068–8074.
6. Dittrich R, El-Massry F, Kunz K, Rinaldi F, Peich CC, Beckmann MW, Pischetsrieder M. Maillard reaction products inhibit oxidation of human low-density lipoproteins in vitro. *J Agric Food Chem.* 2003; **51(13)**: 3900–3904.
7. Yanagimoto K, Lee KG, Ochi H, Shibamoto T. Antioxidative activity of heterocyclic compounds found in coffee volatiles produced by Maillard reaction. *J Agric Food Chem.* 2002; **50(19)**: 5480–5484.
8. Lindenmeier M, Faist V, Hofmann T. Structural and functional characterization of pronyl-lysine, a novel protein modification in bread crust melanoidins showing in vitro antioxidative and phase I/II enzyme modulating activity. *J Agric Food Chem.* 2002; **50(24)**: 6997–7006.
9. Faist V, Lindenmeier M, Geisler C, Erbersdobler HF, Hofmann T. Influence of molecular weight fractions isolated from roasted malt on the enzyme activities of NADPH-cytochrome c-reductase and glutathione-S-transferase in Caco-2 cells. *J Agric Food Chem.* 2002; **50(3)**: 602–606.
10. Glosl S, Wagner KH, Draxler A, Kaniak M, Lichtenecker S, Sonnleitner A, Somoza V, Erbersdobler H, Elmadfa I. Genotoxicity and mutagenicity of melanoidins isolated from a roasted glucose-glycine model in human lymphocyte cultures, intestinal Caco-2 cells and in the Salmonella typhimurium strains TA98 and TA102 applying the AMES test. *Food Chem Toxicol.* 2004; **42(9)**: 1487–1495.
11. Taylor JL, Demyttenaere JC, Abbaspour Tehrani K, Olave CA, Regniers L, Verschaeve L, Maes A, Elgorashi EE, van Staden J, de Kimpe N. Genotoxicity of melanoidin fractions derived from a standard glucose/glycine model. *J Agric Food Chem.* 2004; **52(2)**: 318–323.
12. Janzowski C, Glaab V, Samimi E, Schlatter J, Eisenbrand G. 5-Hydroxymethylfurfural: assessment of mutagenicity, DNA-damaging potential and reactivity towards cellular glutathione. *Food Chem Toxicol.* 2000; **38(9)**: 801–809.
13. Brands CM, Alink GM, van Boekel MA, Jongen WM. Mutagenicity of heated sugar-casein systems: effect of the Maillard reaction. *J Agric Food Chem.* 2000; **48(6)**: 2271–2275.
14. Hiramoto K, Nasuhara A, Michikoshi K, Kato T, Kikugawa K. DNA strand-breaking activity and mutagenicity of 2,3-dihydro-3,5-dihydroxy-6-methyl-4H-pyran-4-one (DDMP), a Maillard reaction product of glucose and glycine. *Mutat Res.* 1997; **395(1)**: 47–56.
15. Hiramoto K, Sekiguchi K, Ayuha K, Aso-o R, Moriya N, Kato T, Kikugawa K. DNA breaking activity and mutagenicity of soy sauce: characterization of the active components and identification of 4-hydroxy-5-methyl-3(2*H*)-furanone. *Mutat Res.* 1996; **359(2)**: 119–132.

16. Chen CC, Tseng TH, Hsu JD, Wang CJ. Tumor-promoting effect of GGN-MRP extract from the Maillard reaction products of glucose and glycine in the presence of sodium nitrite in C3H10T1/2 cells. *J Agric Food Chem.* 2001; **49(12)**: 6063–6067.
17. Lewis TS, Shapiro PS, Ahn NG. Signal transduction through MAP kinase cascades. *Adv Cancer Res.* 1998; **74**: 49–139.
18. Marais R, Marshall CJ. Control of the ERK MAP kinase cascade by Ras and Raf. *Cancer Surv.* 1996; **27**: 101–125.
19. Skehan P, Storeng R, Scudiero D, Monks A, McMahon J, Vistica D, Warren JT, Bokesch H, Kenney S, Boyd MR. New colorimetric cytotoxicity assay for anticancer-drug screening. *J Natl Cancer Inst.* 1990; **82(13)**: 1107–1112.
20. Marko D, Habermeyer M, Kemeny M, Weyand U, Niederberger E, Frank O, Hofmann T. Maillard reaction products modulating the growth of human tumor cells in vitro. *Chem Res Toxicol.* 2003; **16(1)**: 48–55.
21. Hofmann T. Characterization of the most intense coloured compounds from Maillard reactions of pentoses by application of colour dilution analysis (CDA). *Carbohydr Res.* 1998; **313**: 203–213.
22. Hofmann T. Studies on the influence of the solvent on the contribution of single Maillard reaction products to the total color of browned pentose/alanine-solutions – A quantitative correlation by using the color activity concept. *J Agric Food Chem.* 1998; **46**: 3912–3917.
23. Marko D, Kemeny M, Bernady E, Habermeyer M, Weyand U, Meiers S, Frank O, Hofmann T. Studies on the inhibition of tumor cell growth and microtubule assembly by 3-hydroxy-4-[(E)-(2-furyl)methylidene]methyl-3-cydopentene-1,2-dione, an intensively coloured Maillard reaction product. *Food Chem Toxicol.* 2002; **40(1)**: 9–18.
24. Yamazaki Y, Tsuruga M, Zhou D, Fujita Y, Shang X, Dang Y, Kawasaki K, Oka S. Cytoskeletal disruption accelerates caspase-3 activation and alters the intracellular membrane reorganization in DNA damage-induced apoptosis. *Exp.* 2000; **259(1)**: 64–78.
25. Haldar S, Basu A, Croce CM. Bcl2 is the guardian of microtubule integrity. *Cancer Res.* 1997; **57(2)**: 229–233.
26. Tsutsui T, Tamura Y, Hagiwara M, Miyachi T, Hikiba H, Kubo C, Barrett JC Induction of mammalian cell transformation and genotoxicity by 2-methoxyestradiol, an endogenous metabolite of estrogen. *Carcinogenesis* 2000; **21(4)**: 735–740.
27. Wallin M, Friden B, Billger M. Studies of the interaction of chemicals with microtubule assembly in vitro can be used as an assay for detection of cytotoxic chemicals and possible inducers of aneuploidy. *Mutat Res.* 1988; **201(2)**: 303–311.
28. Wallin M, Hartley-Asp B. Effects of potential aneuploidy inducing agents on microtubule assembly in vitro. *Mutat Res.* 1993; **287(1)**: 17–22.

5 Risk Assessment of Acrylamide

Erik Dybing *

Abstract

Acrylamide has been detected in certain fried, baked and deep-fried foods, as well as in coffee. The major determinants of acrylamide formation during heat treatment of food are the presence of asparagine and reducing sugars or reactive carbonyls. Reported mean acrylamide intakes in Europe and USA are between 0.28 and 0.71 μg/kg bodyweight/day in adults and about 1.5-fold higher or more among children and adolescents. Maximal intakes in Norwegian 13 year olds have been estimated to 8.0 μg/kg bodyweight/day. Acrylamide is oxidised in the body to the reactive glycidamide, which is mutagenic. Acrylamide is carcinogenic to rats and mice, causing tumours at multiple organ sites in both species when given in drinking water or by skin application. Acrylamide administered in drinking water to rats, consistently induced peritesticular mesotheliomas, thyroid follicular cell tumours and mammary gland tumours. An increase in primary brain tumours was also evident when all such tumours were included in data analysis. Acrylamide is also a neurotoxicant with a NOAEL for induction of morphological changes in nerves of rats as detected by electron microscopy of 0.2 mg/kg bodyweight/day. It also induces reproductive and developmental effects with a NOAEL of 2 mg/kg bodyweight/day. Acrylamide presumably causes cancer in laboratory animals through a genotoxic mechanism via glycidamide. Glycidamide-derived DNA adducts have been found in all tissues of mice and rats examined after i.p. administration of acrylamide, including in the testis and mammary gland. There has been no evidence of increases in cancer in studies of occupational populations exposed to acrylamide. There is conflicting epidemiological evidence with most studies showing no association between acrylamide food exposure and increased cancer. One study has reported an increased risk in breast cancer related to earlier consumption of potato chips (French fries) at preschool age. Most of the existing epidemiological information is limited by insufficient power to detect modest increases in tumour types such as those observed in the rodent studies. Given the results of the carcinogenicity and mutagenicity studies in animals, it is prudent to regard acrylamide as a probable human carcinogen acting through a genotoxic mechanism. Using the default linear extrapolation methods LED10 and T25, the lifetime cancer hazard after lifelong exposure to 1 μg acrylamide/kg body

* Division of Environmental Medicine, Norwegian Institute of Public Health, PO Box 4404 Nydalen, NO-0403 Oslo, Norway.

Thermal Processing of Food: Potential Health Benefits and Risks
DFG, Deutsche Forschungsgemeinschaft

ISBN: 978-3-527-31909-1

weight/day, has been calculated to be, on average, 1.3×10^{-3}. JECFA has derived a lowest BMDL of 0.30 mg/kg bodyweight/day for total mammary tumours in rats. By comparing this value to their estimate of an average acrylamide intake of 1 μg/kg bodyweight/day, a margin of exposure (MOE) of 300 has been calculated. For intakes in high consumers (4 μg/kg bodyweight/day), JECFA reported a MOE of 75.

5.1 Introduction

Studies at the University of Stockholm had in the late 1990s revealed a conspicuous background level of the haemoglobin adduct *N*-(2-carbamoylethyl)valine in persons apparently without known exposure to acrylamide [1, 2]. The researchers speculated that this adduct originated from acrylamide formed during cooking and were able to show that rats fed a standard animal diet that was fried, showed a strong increase in the level of the haemoglobin adduct [2]. In order to verify that the human adduct was related to acrylamide formation from intake of heated food, the research group was able to identify the formation of acrylamide from heated carbohydrate-rich foods such as potato chips (French fries), other fried potato products and potato crisps (chips) [3]. Later, other heat-treated foods including coffee were also found to contain significant levels of acrylamide [4]. Since acrylamide has been classified as a Group 2A carcinogen by the International Agency for Research on Cancer [5] and a Category 2 carcinogen and Category 2 mutagen by the European Union (http://ecb.jr.it/classification-labelling/), these findings caused worldwide concern [4, 6, 7]. Thereafter, a number of scientific studies and regulatory activities related to acrylamide in food have been undertaken. An important question to answer for the scientific community and food regulatory agencies is, whether or not the intake of acrylamide-containing food represents a health risk, and if it does, what the probability and consequences of such a risk may be.

5.2 Formation of Acrylamide in Food

It is now generally agreed that acrylamide is formed in food by the Maillard reaction from the condensation of the amino acid asparagine with reducing sugars such as fructose or glucose upon heating at temperatures above 120 °C [8, 9]. The reaction between the two major determinants of acrylamide formation is not very efficient. Reaction yields on a molar basis of around 0.1–0.3% of the initial asparagine content were repeatedly found under optimal model reaction conditions [8–11]. Similar conversion rates have been reported from food model systems such as in bread crust after spiking the dough with different amounts of asparagine [12]. Under certain model conditions, significant amounts of acrylamide were reported to be formed from ammonia and acrolein in the absence of asparagine, a pathway that was suggested to play a role in lipid-rich foods [11]. Acrolein can be formed by heat decomposition of triglycerides present in frying oil. This route of acrylamide formation via acrolein does, however, not seem to be involved in the formation of acrylamide in fried potatoes [10].

5.3 Exposure Assessment of Acrylamide in Food

Extensive amounts of data on levels of acrylamide have been collected by European authorities, the Confederation of the Food and Drink Industries in the EU (CIAA) and the US Food and Drug Administration (FDA). A website has been created to allow the rapid compilation of data, organised by WHO/JIFSAN (http://www.acrylamide-food.org/). More than 4500 entries on acrylamide levels in the most important food matrices can be found in the European Monitoring Database (http://www.irmm.jrc.be), a summary of the database content as of June 2005 is presented in Table 5.1.

After the initial finding of considerable acrylamide contents in various foods, several independent studies have estimated overall exposure levels of various consumer groups. The foods that contribute most to acrylamide exposure vary depending upon the population's eating habits and the way the foods are processed and prepared. Generally, the most important categories of food appear to be: fried potato products such as potato chips (French fries) and

Table 5.1: Acrylamide levels [μg/kg] in food as collected by the EU Joint Research Center (updated June 2005) (from http://www.irmm.jrc.be/).

Food commodity	*N*	Min	25%	Median	75%	Max
Potato chips (French fries)	1004	5	80	173	339	4653
Potato crisps (chips)	684	5	351	570	960	3370
Potato fritter	98	10	173	362	709	2916
Fine bakery ware	814	5	3324	50	134	333
Gingerbread	546	5	130	269	594	7834
Crispbread	380	5	79	248	514	2838
Infant food	171	5	910	41	90	214
Diabetics cakes + biscuits	349	5	106	260	664	3044
Breakfast cereals	242	5	15	53	126	1540
Coffee roasted	134	70	214	285	387	1112
Coffee substitutes	85	116	537	788	1250	2955

Table 5.2: Exposure estimates of acrylamide in food (from [14, 15]).

Exposure assessment	Daily intake [µg/kg bw/day]	
	Mean (age group)	$95^{th}/90^{th}$ percentile
FAO/WHO, 2000	0.3–0.8	
SNFA, Sweden, 2002	0.45 (18–74)	1.03
SNT, Norway, 2002	0.49 (males)	1.01 [a]
	0.46 (females)	0.86 [a]
	0.36 (9, boys)	0.72 [a]
	0.32 (9, girls)	0.61 [a]
	0.52 (13, boys)	1.35 [a]
	0.49 (13, girls)	1.2 [a]
	0.53 (16–30, males)	
	0.50 (16–30, females)	
EU, SCF, 2002	0.2–0.4	
BfR, Germany, 2002	1.1 (15–18)	3.4
BAG, Switzerland, 2002	0.28 (16–57)	
AFSSA, France, 2002	0.5 (>15)	1.1
	1.4 (2–14)	2.9
FDA, 2002	0.7	
FDA, 2003	0.37 (>2)	0.81 [a]
	1.00 (2–5)	2.15 [a]
NFCS, The Netherlands, 2003	0.48 (1–97)	0.60
	1.04 (1–6)	1.1
	0.71 (7–18)	0.9
FDA, 2004	0.43 (>2)	0.92 [a]
	1.06 (2–5)	2.31 [a]
JECFA, 2004	1 [b]	
	4 [c]	

[a] 90th percentile. [b] Children included. [c] High consumers, children included.

crisps (chips), ready-to-eat breakfast cereals, baked goods such as cookies, pies and cakes, brewed coffee and breads. In Norway, potato products contributed approximately 30% to the daily acrylamide intake, whereas coffee and bread contributed 28% and 20%, respectively [13]. A summary of exposure estimates of acrylamide in food is presented in Table 5.2.

5.4 Hazard Identification of Acrylamide

5.4.1 Toxicokinetics

The bioavailability of acrylamide after oral administration is high with 68–90% of the dose being absorbed in rats [16, 17], 73% in dogs [18] and 99% in miniature pigs [18], respectively. In a human volunteer study, 34% of an oral dose appeared in the urine during the first 24 hours [19].

Acrylamide due to its hydrophilicity, is widely distributed in tissues, but is also metabolised by the action of P450 2E1 to the epoxide glycidamide [20–22]. In addition, glycidamide is subject to nucleophilic attack by ring opening of the epoxide by sulfhydryl or amino groups, or by water. Interaction of glycidamide also occurs with the nucleophilic nitrogens in DNA, producing DNA adducts [23], which suggests that glycidamide may be more important for carcinogenic and genotoxic properties than the parent compound acrylamide, whose reaction with DNA constituents is much slower. However, acrylamide is believed to be the more responsible component for neurotoxicity than glycidamide because of its potent ability to react with proteins [24]. Oral administration of acrylamide of mice leads to higher relative internal exposure to glycidamide compared to intravenous dosing. Further, the conversion of acrylamide to glycidamide appears to be more efficient at lower doses [25].

An outline of the major metabolic routes for acrylamide is presented by Dybing and co-workers [14] (see Fig. 5.1). Direct conjugation of acrylamide

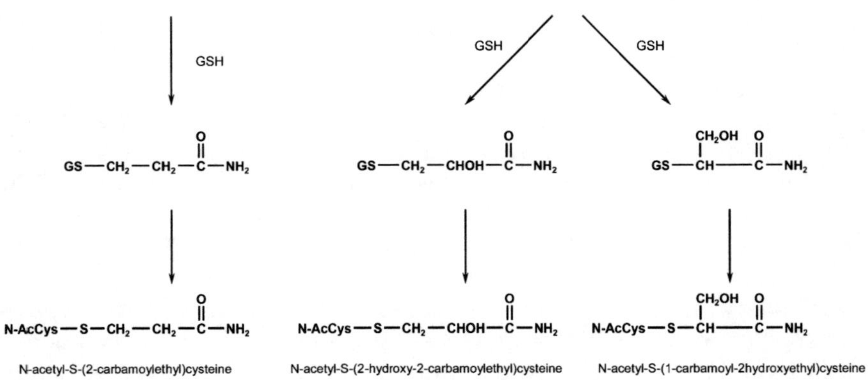

Figure 5.1: Major metabolic routes of acrylamide.

with glutathione results in the subsequent formation of a mercapturic acid propionamide (*N*-acetyl-*S*-(2-carbamoylethyl)cysteine). Glycidamide, which is produced by oxygenation by P450, undergoes further metabolism either by hydrolysis to glyceramide, or glutathione conjugation, resulting in the subsequent formation of two mercapturic acid products (*N*-acetyl-*S*-(2-hydroxy-2-carbamoylethyl)cysteine and *N*-acetyl-*S*-1-carbamoyl-2-hydroxyethyl)cysteine).

5.4.2 Genotoxicity

Whereas acrylamide at best is a weak mutagen, the metabolite glycidamide is mutagenic in a number of *in vitro* and *in vivo* systems [26–31]. The main DNA-adduct formed when mice are exposed to acrylamide is N-7(2-carbamoyl-2-hydroxyethyl)guanine [32]. Glycidamide-derived DNA adducts of adenine and guanine were shown to be formed in all tissues examined of rats and mice dosed with acrylamide, including in the target tissues identified in rodent carcinogenicity studies [23]. These results provide strong support for a genotoxic mechanism of acrylamide carcinogenicity in rodents.

5.4.3 Carcinogenicity

Two long-term drinking water studies of possible carcinogenic effects of acrylamide have been conducted in F344 rats [33, 34]. Acrylamide induced statistically significant increases in peritesticular mesotheliomas, thyroid follicular cell tumours and mammary gland tumours (Tabs. 5.3 and 5.4). There was also a statistically significant increase in brain tumours when all such tumours were included in the data analysis [35]. In addition, increases were seen in one of the studies in adrenal pheochromocytomas, adenomas of pituitary and clitoral glands, papillomas of the oral cavity and adenocarcinomas of the uterus (Tab. 5.3) [33].

In a series of skin painting experiments with mice, it was found that acrylamide initiates the formation of skin tumours and lung adenomas in mice [36–38]. Acrylamide also induces skin tumours in mice after drinking water administration [37].

The US National Center for Toxicological Research (NCTR) is currently undertaking three types of carcinogenicity bioassays for acrylamide and glycidamide. First, a shorter-term assay will be conducted in a neonatal mouse model that is a sensitive indicator of direct-acting genotoxic carcinogens. Second, NCTR will also conduct a two-year chronic rodent carcinogenicity bioassay in rats and mice under the auspices of the US National Toxicology Program. This assay will include study of dose-response relationships, histopathology, and correlation of adduct levels in target tissues with tumour incidence. The chronic carcinogenicity bioassay is expected to be completed in 2008. Third, NCTR will conduct a neonatal mouse carcinogenicity study, in which pregnant mice are exposed perinatally to acrylamide in drinking water (http://www.cfsan.fda.gov/~dms/acrypla3.html).

Table 5.3: Numbers of Fischer 344 rats with tumours after receiving acrylamide in the drinking water for two years (data from [33] as compiled by [35]).

Type of tumour	Sex	Dose [mg/kg bw/day]				
		0	0.01	0.1	0.5	2.0
Thyroid gland, follicular adenomas	M	1/60	0/58	2/59	1/59	7/59 *
Peritesticular mesotheliomas	M	3/60	0/60	7/60	11/60 *	10/60 *
Adrenal gland, pheochromocytomas	M	3/60	7/59	7/60	5/60	10/60 *
Mammary tumours	F	10/60	11/60	9/60	19/58	23/61 *
Central nervous system, glial tumours	F	1/60	2/59	1/60	1/60	9/61 *
Thyroid gland, follicular adenomas or adenocarcinomas	F	1/58	0/59	1/59	1/58	5/60 *
Oral cavity, squamous papillomas	F	0/60	3/60	2/60	1/60	7/61 *
Uterus, adenocarcinomas	F	1/60	2/60	1/60	0/59	5/60 *
Clitoral gland, adenomas[a]	F	0/2	1/3	3/4	2/4	5/5 *
Pituitary adenomas	F	25/59	30/60	32/60	27/60	32/60

[a] Only clitoral glands with gross lesions were examined histologically.

* $p<0.05$; pair-wise Mantel-Haenszel comparison with the control group adjusted for mortality.

Table 5.4: Numbers of Fischer 344 rats with tumours after receiving acrylamide in the drinking water for two years (data from [34]).

Type of tumour	Sex	Dose [mg/kg bw/day]			
		0	0.1	0.5	2.0
Thyroid gland, follicular adenomas	M	3/204	9/203	5/101	12/75 *
Thyroid gland, follicular adenomas	F	0/100	–	7/100 *	16/100 *
Thyroid gland, follicular adenocarcinomas	F	2/100	–	3/100	7/100
Peritesticular mesotheliomas	M	8/204	9/204	8/102	13/75 *
Mammary gland, fibroadenomas	F	9/96	–	20/94 *	26/95 *
Mammary gland, adenocarcinomas	F	2/96	–	2/94	4/95

* $p<0.05$; Fischer's exact test.

5.4.4 Neurotoxicity

Acrylamide is neurotoxic both in humans and experimental animals. Repeated administration of acrylamide in rats has been shown to induce nerve terminal degeneration in brain areas critical for learning, memory and other cognitive functions. A growing body of evidence indicates that the nerve terminal is a primary site of acrylamide action and that inhibition of corresponding membrane-fusion processes impairs neurotransmitter release and promotes eventual degeneration (reviewed in [24]).

5.4.5 Reproduction Toxicity

Acrylamide at low doses decreases litter size in reproductive studies, with rats more sensitive than mice. At higher doses, sperm morphology and motility are affected and neurotoxicity induced, which decreases mating frequency. Acrylamide does not affect female reproduction directly, whereas females exhibit neurotoxicity and dominant lethal mutations cause decreased newborn litter size. The mechanism of action appears to be acrylamide/glycidamide binding to spermatid protamines and acrylamide binding to motor proteins (reviewed in [39]).

5.4.6 Epidemiological Studies of Cancer in Humans

Collins and co-workers [40] investigated causes of death in four factories, three in the United States and one in The Netherlands, in order to assess possible cancer risks after exposure to acrylamide. The investigation included 8854 males employed in the factories between 1925 and 1976. Marsh and co-workers [41] have presented an update for the period 1984–1994 for this cohort. In the exposed workers, no tissue-specific cancer risk was significantly elevated and cancer overall was not increased [40–42].

Mucci and co-workers [43] have reanalysed a population-based Swedish case-control study encompassing cases with cancer of the large bowel ($n = 591$), bladder ($n = 263$) and kidney ($n = 133$) and 538 healthy controls. They assessed dietary acrylamide by linking extensive food frequency data with acrylamide levels reported by the Swedish National Food Agency [4]. There was a consistent lack of an excess risk, or any convincing trend of cancer of the large intestine and rectum, urinary bladder or kidney in high consumers of 14 different food items with a high or moderate acrylamide content.

A reanalysis of data from the Swedish component of an international collaborative population-based case-control study of renal cell cancer has been performed by Mucci and co-workers [44]. Food frequency data were linked with national food databases on acrylamide content and risk of renal cell cancer was evaluated for intake of food items with elevated acrylamide levels and for total daily acrylamide dose. There was no evidence that food items with elevated acrylamide were associated with a higher risk of renal cancer. Furthermore, there was no association between estimated daily acrylamide intake and cancer.

The relation between intake of fried/baked potatoes and cancer risk has been analysed from a series of hospital-based case-control studies conducted in

Italy and Switzerland [45]. The odds ratios for the highest vs. the lowest tertile of intake ranked between 0.8 and 1.1 for cancer of the oral cavity and pharynx, oesophagus, larynx, large bowel, breast or ovary.

Mucci and co-workers [46] have prospectively studied whether acrylamide through exposure via the diet was associated with higher risk of breast cancer in a cohort of 43404 Swedish women. Participants completed a semi-quantitative food frequency questionnaire and the Swedish national acrylamide database was used to determine median levels in various food items. The study participants were categorised into quintiles based on the distribution of acrylamide in the cohort. Compared with the lowest quintile of acrylamide intake, there was no significantly increased risk of breast cancer in the higher quintiles. For the highest quintile compared with the lowest quintile, the relative risk was 1.19 (95% confidence interval, 0.91–1.55).

A recent nested case-control study from two US prospective cohort studies of nurses has reported an increased risk of breast cancer associated with previous frequent consumption of potato chips (French fries) during preschool ages 3–5 years. For one additional serving of potato chips per week, the odds ratio for breast cancer adjusted for adult life breast cancer risk factors was 1.27 (95% confidence interval 1.12–1.44) [47].

5.5 Hazard Characterisation of Acrylamide

The most sensitive no-observed-adverse-effect level (NOAEL) of acrylamide neurotoxicity in rats was morphological nerve changes established at 0.2 mg/kg bw/day by electron microscopy [48] and 0.5 mg/kg bw/day by light microscopy [33, 34]. With respect to reproductive toxicity, a NOAEL of 2 mg/kg bw/day has been determined for male-mediated implantation losses [49].

The tumorigenic dose indicators LED10 [50] and T25 [51, 52] have been used to characterise the dose-response relationships of acrylamide carcinogenic-

Table 5.5: Comparison of LED10 and T25 values estimated from acrylamide drinking water studies in rats (from [13]).

Study	LED10 Males (testicular mesotheliomas) [mg/kg bw/day]	LED10 Females (mammary fibroadenomas) [mg/kg bw/day]	T25 Males (testicular mesotheliomas) [mg/kg bw/day]	T25 Females (mammary fibroadenomas) [mg/kg bw/day]
Johnson et al., 1986 [33]	0.66	0.40	0.89	0.64
Friedman et al., 1995 [34]	0.84	0.86	3.7	1.9

ity in rats [13]. The differences in the estimates were small (Tab. 5.5), with values somewhat lower for mammary gland fibroadenomas compared to peritesticular mesotheliomas. Therefore, LED10- and T25-values for mammary gland fibroadenomas in the most sensitive study [33] have been used to calculate lifetime cancer hazards after exposure to acrylamide in humans. Given the assumption that there is a non-threshold and linear dose–response relationship for these tumour types, the corresponding human lifetime hazard after exposure to 1 µg acrylamide/kg bw/day was estimated to $1.0 \cdot 10^{-3}$ and $1.6 \cdot 10^{-3}$ using LED10 and T25, respectively, which were averaged to $1.3 \cdot 10^{-3}$ [13].

5.6 Risk Assessment of Acrylamide in Food

Using the mean daily intake doses of acrylamide in foods determined for the Norwegian adult population of 0.46 to 0.49 µg/kg bodyweight and the lifetime hazard after lifelong exposure to 1 µg/kg bw/day of $1.3 \cdot 10^{-3}$, a lifetime cancer risk of $0.6 \cdot 10^{-3}$ (corresponding to six cancer cases per 10000 individuals) was calculated by Dybing and Sanner [13]. When the intake estimates presented by JECFA [15] of 1 µg/kg bw/day (mean) and 4 µg/kg bw/day (high) are used, the life time cancer risk becomes $1.3 \cdot 10^{-3}$ and $5.2 \cdot 10^{-3}$, respectively. Risk assessments based on the same methodology have been published from The Netherlands [53] and Belgium [54]. However, JECFA [15] did not present point risk estimates in their 2004 evaluation of acrylamide in food. Rather, they determined benchmark doses (BMD) and their 95% lower confidence values (BMDL) for the various tumour sites and used the lowest BMDL value for total mammary tumours, 0.30 mg/kg bw/day, to calculate a margin of exposure (MOE). With the average acrylamide intake of 1 µg/kg bw/day, a MOE of 300 was estimated, whereas a MOE of 75 was calculated for high consumers (4 µg/kg bw/day) [15].

It is assumed that the potential risks for neurotoxic and reproductive toxic effects have a dose threshold, where the NOAEL is used as a surrogate for the dose threshold. Using the average exposure estimate from JECFA [15] of 1 µg/kg bw/day and NOAELs for neurotoxicity of 0.2 mg/kg bw/day and 2 mg/kg bw/day for reproductive toxicity, the margins of safety (MOS) become 200 and 2000, respectively. For high consumers with an estimated intake of 4 µg/kg bw/day, the corresponding MOS values become 50 and 500, respectively [15].

5.7 Discussion

The present database on acrylamide toxicity in experimental animals points to cancer as the most sensitive endpoint with mammary fibroadenomas and peritesticular mesotheliomas as consistent tumour types in rats [33, 34]. Since the acrylamide metabolite glycidamide is clearly mutagenic [26–31] and because glycidamide-derived DNA adducts are present in the target tissues [23], it seems probable that the tumour responses are a reflection of a genotoxic mechanism of action. However, since acrylamide produced tumours in endocrine-sensitive organs (mammary gland, testis and thyroid), an endocrine, non-genotoxic mechanism may also play a role.

The point estimates of lifetime cancer risk presented by Dybing and Sanner [13] assume linearity in the dose-response relationships from high to low doses. There are several examples in the literature of genotoxic carcinogens that show a linear dose-response in the low dose range [55–57]. On the other hand, tumour site data from the NCI-NTP bioassay programmes were more often consistent with a quadratic response than with a linear response, suggesting that the use of linear extrapolations will often overestimate the risks [58]. Thus, the point estimates may be seen as an upper bound of the cancer risk derived from the results of animal experiments.

An interesting alternative to the linear, low-dose extrapolation for cancer risk assessment of acrylamide in food is the application of a margin of exposure (MOE) concept as has been done by JECFA [15]. The margin was calculated to be 300 for the average consumer and 75 for the high consumer. JECFA [15] considered these MOEs to be low for a compound that is genotoxic and carcinogenic, and that they may indicate a human health concern. This can be contrasted with MOEs of 25 000 and 10 000 for average and high exposures to polycyclic aromatic hydrocarbons (PAHs) in food, using surrogate marker levels of 4 ng benzo[a]pyrene/kg bw/day and 10 ng benzo[a]pyrene/kg bw/day, respectively [15].

A ratio of 50 between acrylamide exposure in high consumers and the NOAEL for morphological changes in nerves detected by electron microscopy [48] indicates an erosion of the necessary safety margin with respect to neurotoxicity in such consumers. The safety margin with respect to reproductive toxicity appears to be large enough (500 in high consumers) to avoid such effects.

The epidemiological evidence from occupationally exposed populations does not indicate that acrylamide is a potent carcinogen in humans [41]. However, the value of the epidemiological information from the occupational cohorts is limited by insufficient power to detect modest increases in tumour types such as those observed in the rodent studies. A point to further consider is that there may be differences in glycidamide target doses between occupational and dietary exposures, as indicated from the observed first pass effect after oral administration in mice [25]. Case control studies of dietary acrylamide exposure and cancers of the large bowel, kidney and bladder have also not shown any evidence of associations, but these studies have been small and could suffer from considerable non-differential misclassification of exposure to acrylamide [59]. Based on animal evidence, it has been assumed that the relative risk for cancer could be 1.05 for those 2% of the Swedish study population with the highest intake compared to the mean intake. From a statistical point of view, a study would have to include about 470000 cases and half as many controls in order to significantly show a relative risk of 1.05 [59].

The cohort study of Mucci and co-workers [46] examining any association between dietary acrylamide exposure and breast cancer, also did not find evidence of an increased risk. The strengths of this study were the use of a validated food frequency questionnaire, the prospective study design and complete cohort follow-up. However, also this study was not able to ascertain an increased relative risk of 1.05 or smaller [46]. In contrast with this, Michels and co-workers have recently reported an increased risk in breast cancer related to earlier consumption of potato chips (French fries) at preschool age [47]. Since the information on preschool dietary habits was based on questionnaires completed by the mothers of the cases and controls, this study could suffer from differential recall.

5.8 Conclusions

Consumption of heat-treated foods results in exposure to acrylamide in the order of 0.5–1 µg per kg body weight per day for the average consumer. High consumers may be exposed to 4–8 µg per kg per day. Acrylamide is mutagenic through its metabolite glycidamide and glycidamide-derived DNA adducts are found in all rodent tissues examined, including the target organs in carcinogenicity studies. In addition, acrylamide is a neurotoxicant and reproductive toxicant. Using linear extrapolation methods, lifetime cancer risks have been calculated from the animal data to be in the order of $1 \cdot 10^{-3}$ and $5 \cdot 10^{-3}$ for the average and high consumer, respectively. Using a margin of exposure approach resulted in values of 300 and 75 for the two consumer groups, respectively. Both of these assessments point to a health concern from dietary exposure to acrylamide. There is conflicting epidemiological evidence with most studies showing no association between acrylamide exposure and increased cancer risks. However, most of the existing epidemiological information is limited by insufficient power to reveal moderate increases in risk. Overall, it seems prudent to regard acrylamide as a probable human carcinogen acting through a genotoxic mechanism. Judged from the experimental data in animals, there does not appear to be any appreciable concern for neurotoxic or reproductive toxic effects from exposure to acrylamide.

References

1. Bergmark, E. (1997) Hemoglobin adducts of acrylamide and acrylonitrile in laboratory workers, smokers, and nonsmokers. *Chem. Res. Toxicol.* **10**, 78–84.
2. Tareke, E., Rydberg, P., Karlsson, P., Eriksson, S., Törnqvist, M. (2002) Acrylamide: a cooking carcinogen? *Chem. Res. Toxicol.* **13**, 517–522.
3. Tareke, E., Rydberg, P., Karlsson, P., Eriksson, S., Tornqvist, M. (2002) Analysis of acrylamide, a carcinogen formed in heated foodstuffs. *J. Agric. Fd. Chem.* **50**, 4998–5006.
4. Swedish National Food Agency (2002). http://www.slv.se
5. International Agency for Research on Cancer (IARC) (1994) Acrylamide. *IARC Monogr. Eval. Carcinog. Risks Hum.* **60**, 389–433.
6. World Health Organization (WHO) (2002) FAO/WHO Consultation on the health implications of acrylamide in food. Geneva, 25–27 June 2002. www.who.int
7. European Commission (2002) Opinion of the Scientific Committee on Food on new findings regarding the presence of acrylamide in food. http://europa.eu.int/comm/food/fs/sc/scf/out131_enpdf
8. Stadler, R.H., Blank, I., Varga, N., Robert, F., Hau, J., Guy, P.A., Robert, M.C., Riediker, S. (2002) Acrylamide from Maillard reaction products. *Nature* **419**, 449.
9. Mottram, D.S., Wedzicha, B.L., Dodson, A.T. (2002) Acrylamide is formed in the Maillard reaction. *Nature* **419**, 448–449.
10. Becalski, A., Lau, B.P., Lewis, D., Seaman, S.W., Hayward, S., Sahagian, M., Ramesh, M., Leclerc, Y. (2004) Acrylamide in French fries: influence of free amino acids and sugars. *J. Agric. Fd. Chem.* **52**, 3801–3806.
11. Yashura, A., Tanaka, Y., Hengel, M., Shibamoto, T. (2003) Gas chromatographic investigation of acrylamide formation in browning model systems. *J. Agric. Fd. Chem.* **51**, 3999–4003.
12. Surdyk, N., Rosen, J., Andersson, R., Aman, P. (2004) Effects of asparagine, fructose, and baking conditions on acrylamide content in yeast-leavened wheat bread. *J. Agric. Fd. Chem.* **52**, 2047–2051.
13. Dybing, E., Sanner, T. (2003) Risk assessment of acrylamide in foods. *Toxicol. Sci.* **75**, 7–15.
14. Dybing, E., Farmer, P.B., Andersen, M., Fennell, T.R., Lalljie, S.P.D., Müller, D.J.G., Olin, S., Petersen, B.J., Schlatter, J., Scholz, G., Scimeca, J.A., Slimani, N., Törnqvist, M., Tuijtelaars, S., Verger, P. (2005) Human exposure and internal dose assessments of acrylamide in food. *Fd. Chem. Toxicol.* **43**, 365–410.
15. Joint FAO/WHO Expert Committee on Food Additives (JECFA) (2005) Summary and conclusions of sixty-fourth meeting, Rome, 8–17 February 2005. http://www.fao.org/es./esn/jecfa/acrylamide_en.stm
16. Miller, M.J., Carter, D.E., Sipes, I.G. (1982) Pharmacokinetics of acrylamide in Fischer-344 rats. *Toxicol. Appl. Pharmacol.* **25**, 36–44.
17. Kadry, A.M., Friedman, M.A., Abdel-Rahman, M.S. (1999) Pharmacokinetics of acrylamide after oral administration in male rats. *Env. Toxicol. Pharmacol.* **7**, 127–133.
18. Ikeda, G.J., Miller, E., Sapienza, P.P., Michel, T.C., Inskeep, P.B. (1987) Comparative tissue distribution and excretion of [1-^{14}C]acrylamide in beagle dogs and miniature pigs. *Fd. Chem. Toxicol.* **25**, 871–875.
19. Fennell, T.R., Snyder, R., Burgess, J.P., Friedman, M.A. (2004) Metabolism and haemoglobin adducts of [1,2,3-$^{13}C_3$]acrylamide in humans. *Toxicologist*, 173.
20. Calleman, C.J., Bergmark, E., Costa, L.G. (1990) Acrylamide is metabolized to glycidamide in the rat: Evidence from haemoglobin adduct formation. *Chem. Res. Toxicol.* **3**, 406–412.
21. Sumner, S.C., MacNeela, J.P., Fennell, T.R. (1992) Characterization and quantitation of urinary metabolites of [1,2,3-^{13}C]acrylamide in rats and mice using ^{13}C nuclear magnetic resonance spectroscopy. *Chem. Res. Toxicol.* **5**, 81–89.

22. Sumner, S. C., Fennell, T. R., Moore, T. A., Chanas, B., Gonzalez, F., Ghanayem, B. I. (1999) Role of cytochrome P450 2E1 in the metabolism of acrylamide and acrylonitrile in mice. *Chem. Res. Toxicol.* **12**, 1110–1116.
23. Doerge, D. R., Costa, G. G., McDaniel, L. P., Churchwell, M. I., Twaddle, N. C., Beland, F. A. (2005) DNA adducts derived from administration of acrylamide and glycidamide to mice and rats. *Mutat. Res.* **580**, 131–141.
24. LoPachin, R. M. (2004) The changing view of acrylamide neurotoxicity. *Neurotoxicology* **25**, 617–630.
25. Doerge, D. R., Young, J. F., McDaniel, L. P., Twaddle, N. C., Churchwell, M. I. (2005) Toxicokinetics of acrylamide and glycidamide in B6C3F1 mice. *Toxicol. Appl. Pharmacol.* **202**, 258–267.
26. Hashimoto, K., Tanii, H. (1985) Mutagenicity of acrylamide and its analogues in *Salmonella typhimurium*. *Mutat. Res.* **158**, 129–133.
27. Barfknecht, T. R., Mecca, D. J., Naismith, R. W. (1988) The genotoxic activity of acrylamide. *Environ. Mol. Mutagen.* **11** (Suppl.), 11.
28. Butterworth, B. E., Eldridge, S. R., Sprankle, C. S., Working, P. K., Bentley, K. S., Hurtt, M. E. (1992) Tissue-specific genotoxic effects of acrylamide and acrylonitrile. *Environ. Mol. Mutagen.* **20**, 148–155.
29. Generoso, W. M., Sega, G. A., Lockhart, A. M., Hughes, L. A., Cain, K. T., Cacheiro, N. L., Shelby, M. D. (1996) Dominant lethal mutations, heritable translocations, and unscheduled DNA synthesis induced in male mouse germ cells by glycidamide, a metabolite of acrylamide. *Mutat. Res.* **371**, 175–183.
30. Adler, I. D., Baumgartner, A., Gonda, H., Friedman, M. A., Skerhut, M. (2000) 1-Aminobenzotriazole inhibits acrylamide-induced dominant lethal effects in spermatids of male mice. *Mutagenesis* **15**, 133–136.
31. Abramsson-Zetterberg, L. (2003) The dose-response relationship at very low doses of acrylamide is linear in the flow cytometer-based mouse micronucleus assay. *Mutat. Res.* **535**, 215–222.
32. Segerbäck, D., Calleman, C. J., Schroeder, J. L., Costa, L. G., Faustman, E. M. (1995) Formation of N-7-(2-carbamoyl-2-hydroxyethyl)guanine in DNA of the mouse and the rat following intraperitoneal administration of [^{14}C]acrylamide. *Carcinogenesis* **16**, 1161–1165.
33. Johnson, K. A., Gorzinski, S. J., Bodner, K. M., Bampbell, R. A., Wolf, C. H., Friedman, M. A., Mast, R. W. (1986) Chronic toxicity and oncogenicity study on acrylamide incorporated in the drinking water of Fischer 344 rats. *Toxicol. Appl. Pharmacol.* **85**, 154–168.
34. Friedman, M. A., Dulak, L. H., Stedham, M. A. (1995) A lifetime oncogenicity study in rats with acrylamide. *Fundam. Appl. Toxicol.* **27**, 95–105.
35. Rice, J. M. (2005) The carcinogenicity of acrylamide. *Mutat. Res.* **580**, 3–20.
36. Bull, R. J., Robinson, M., Laurie, R. D., Stoner, G. D., Greisiger, E., Meier, J. R., Stober, J. (1984) Carcinogenic effects of acrylamide in Sencar and A/J mice. *Cancer Res.* **44**, 107–111.
37. Bull, R. J., Robinson, M., Stober, J. A. (1984) Carcinogenic activity of acrylamide in the skin and lung of Swiss-ICR mice. *Cancer Lett.* **24**, 209–212.
38. Robinson, M., Bull, R. J., Knutsen, G. L., Shields, R. P., Strober, J. A. (1986) A combined bioassay utilizing both the lung adenoma and skin papilloma protocols. *Environ. Hlth. Perspect.* **68**, 141–145.
39. Tyl, R. W., Friedman, M. A. (2003) Effects of acrylamide on rodent reproductive performance. *Reprod. Toxicol.* **17**, 1–13.
40. Collins, J. J., Swaen, G. M. H., Marsh, G. M., Utidjian, H. M. D., Caporossi, J. C., Lucas, L. J. (1989) Mortality patterns among workers exposed to acrylamide. *J. Occup. Med.* **31**, 614–617.
41. Marsh, G. M., Lucas, L. J., Youk, A. O., Schall, L. C. (1999) Mortality patterns among workers exposed to acrylamide: 1994 follow up. *Occup. Environ. Med.* **56**, 181–190.
42. Erdreich, L. S., Friedman, M. A. (2004) Epidemiologic evidence for assessing the carcinogenicity of acrylamide. *Reg. Toxicol. Pharmacol.* **39**, 150–157.

43. Mucci, L. A., Dickman, P. W., Steineck, G., Adami, H. O, Augustsson, K. (2003) Dietary acrylamide and cancer of the large bowel, kidney, and bladder: Absence of an association in a population-based study in Sweden. *Br. J. Cancer* **88**, 84–89.
44. Mucci, L. A., Lindblad, P., Steineck, G., Adami, H. O. (2004) Dietary acrylamide and risk of renal cell cancer. *Int. J. Cancer* **109**, 774–776.
45. Pelucchi, C., Franceschi, S., Levi, F., Trichopoulos, D., Bosetti, C., Negri, E., LaVecchia, C. (2003) Fried potatoes and human cancer. *Int. J. Cancer* **105**, 558–560.
46. Mucci, L. A., Sandin, S., Bälter, K., Adami, H. O., Magnusson, C., Weiderpass, E. (2005) Acrylamide intake and breast cancer risk in Swedish women. *J. Amer. Med. Assoc.* **293**, 1326–1327.
47. Michels, K. B., Rosner, B. A., Chumlea, W. C., Colditz, G. A., Willett, W. C. (2005) Preschool diet and adult risk of breast cancer. *Int. J. Cancer*, online 10 August.
48. Burek, J. D., Albee, R. R., Beyer, J. E., Bell, T. J., Carreon, R. M., Morden, D. C., Wade, C. E., Hermann, E. A., Gorzinski, S. J. (1980) Subchronic toxicity of acrylamide administered to rats in the drinking water followed by up to 144 days of recovery. *J. Environ. Pathol. Toxicol.* **4**, 157–182.
49. Tyl, R. W., Friedman, M. A., Losco, P. E., Fisher, L. C., Johnson, K. A., Strother, D. E., Wolf, C. H. (2000) Rat two-generation reproduction and dominant lethal study of acrylamide in drinking water. *Reprod. Toxicol.* **14**, 385–401.
50. U. S. Environmental Protection Agency (1996) Proposed guidelines for carcinogen risk assessment. *Fed. Reg.* **61**, 17960–18011.
51. Dybing, E., Sanner, T., Roelfzema, H., Kroese, D., Tennant, R. W. (1997) T25: A simplified carcinogenic potency index: Description of the system and study of correlations between carcinogenic potency and species/site specificity and mutagenicity. *Pharmacol. Toxicol.* **80**, 272–279.
52. Sanner, T., Dybing, E., Willems, M. I., Kroese, E. D. (2001) A simple method for quantitative risk assessment of non-threshold carcinogens based on the dose descriptor T25. *Pharmacol. Toxicol.* **88**, 331–341.
53. Konings, E. J. M., Baars, A. J., van Klaveren, J. D., Spanjer, M. C., Rensen, P. M., Hiemstra, M., van Kooij, J. A., Peters, P. W. J. (2003) Acrylamide exposure from foods of the Dutch population and an assessment of the consequent risks. *Fd. Chem. Toxicol.* **41**, 1569–1579.
54. Matthys, C., Bilau, M., Govaert, Y., Moons, E., De Henauw, S., Willems, J. L. (2005) Risk assessment of dietary acrylamide intake in Flemish adolescents. *Fd. Chem. Toxicol.* **43**, 271–278.
55. Wogan, G. N., Paglialunga, S., Newberne, P. M. N. (1974) Carcinogenic effects of low dietary levels of aflatoxin B_1 in rats. *Fd. Cosmet. Toxicol.* **12**, 681–685.
56. Littlefield, N. A., Farmer, J. H., Gaylor, P. W., Sheldon, W. G. (1979) Effects of dose and time in a long-term, low-dose carcinogenesis study. *J. Environ. Pathol. Toxicol.* **3**, 17–34.
57. Peto, R., Gray, R., Brantom, P., Grasso, P. (1991) Effects on 4080 rats of chronic ingestion of N-nitrosodiethylamine or N-nitrosodimethylamine: A detailed dose-response study. *Cancer Res.* **51**, 6415–6451.
58. Hoel, D. G., Portier, C. J. (1994) Nonlinearity of dose-response functions for carcinogenicity. *Environ. Hlth. Perspect.* **102** (Suppl. 1), 109–113.
59. Hagmar, L., Törnqvist, M. (2003) Inconclusive results from an epidemiological study on dietary acrylamide and cancer. *Br. J. Cancer* **89**, 774–775.

6 An Ex-vivo Approach to Assess Low Dose Effects of Acrylamide

Matthias Baum *, Daniel Bertow *, Evelyne Fauth *, Silke Thielen *, and Gerhard Eisenbrand *

Abstract

Acrylamide (AA) is a carcinogen as demonstrated in animal experiments with rats after long term oral application of 1–2 mg AA/kg bodyweight. AA has been found inactive in mutagenicity tests with bacteria in the presence or absence of activating systems. AA induces sister chromatid exchanges and chromosomal aberrations *in vitro*. In most of these studies, millimolar to molar AA concentrations were used. Some studies reported AA to induce chromosomal mutations in rats and mice, again at doses that in most cases were relatively high. In total, there is limited evidence for a direct mutagenic and genotoxic potential of AA. Glycidamide (2,3-epoxypropanamide, GA) is generated metabolically from AA by cytochrome P450 IIE1. GA induces mutations in bacterial mutagenicity tests. In animal experiments, application of AA resulted in GA-DNA adducts primarily at N^7 of guanine. GA is considered therefore as the ultimate carcinogen generated from AA, inducing mutations by covalently binding to DNA. AA and GA are inactivated in part by glutathione conjugation and by binding to proteins like hemoglobin (Hb) or albumin.

The biological activity of AA thus appears to strongly depend on the balance between toxifying and detoxifying metabolic pathways in the organism.

In this study we used human blood as model to study concentration dependent interactions in lymphocytes with DNA, hemoglobin and other blood components. Blood is an easily accessible substrate to measure genotoxicity and protein binding. Concentration–response relationships of genotoxic effects were established to explore the blood concentration range at which DNA damage in lymphocytes becomes detectable. To approach a biomarker-guided dosimetry of genotoxic effects in blood, concentration dependent formation of Hb adducts was compared to DNA damage in lymphocytes.

Significant binding of AA and GA to the *N*-terminal valine in hemoglobin was observed for AA already at 0.3 μM (4 h) and for GA at 3 μM (4 h). AA did not show genotoxic activity in blood cells under our test conditions. GA was genotoxic in human blood (comet assay after additional treatment with the

* University of Kaiserslautern, Erwin-Schroedinger-Str. 52, 67663 Kaiserslautern, Germany.

Thermal Processing of Food: Potential Health Benefits and Risks
DFG, Deutsche Forschungsgemeinschaft

ISBN: 978-3-527-31909-1

DNA repair enzyme FPG), exerting significant effects already at 10 μM. For comparison, the known carcinogens benzo[a]pyrene-7,8-dihydrodiol-epoxide (BPDE) and *α*-acetoxy-*N*-nitroso-diethanolamine (*α*-A-NDELA) were active in the same concentration range.

The results of the study show that sensitive biomarkers for exposure to AA are available, becoming significantly positive under our experimental conditions at blood concentrations as low as 300 nM. The study also shows clearly that AA itself is not genotoxic in human blood even at very high concentrations. In the case of GA, reaction with hemoglobin became detectable at 3 μM, thus reflecting its lower reactivity to hemoglobin compared to AA. Genotoxicity was detectable at 10 μM as lowest concentration, as evidenced by the comet assay. With this detection system, genotoxic effects of GA thus become detectable at concentrations about tenfold higher than those inducing significant Hb adduct formation.

6.1 Introduction

Acrylamide (AA) is a carcinogen as demonstrated in animal experiments [1, 2]. The mechanism of carcinogenicity and its relevance for the human situation is still under debate. Several studies demonstrated the ability of AA to induce chromosomal mutations in animals as well as sister chromatid exchanges and chromosomal aberrations *in vitro*, but in most of the studies relatively high acrylamide concentrations were used [3–6]. Moreover, acrylamide is inactive in mutagenicity tests with bacteria in the presence or absence of activating systems [7–9].

There is substantial evidence that glycidamide (2,3-epoxypropanamide, GA) acts as the ultimate carcinogen. GA is generated from AA *in vivo* primarily by CYPIIE1 mediated epoxidation. As evidenced by animal experiments, AA does not show substantial reactivity towards DNA [10]. In contrast, GA interacts effectively with DNA bases, predominantly forming N^7-adducts with guanine and to a minor extent N^3-adducts with adenine [10–13]. AA and GA can be inactivated in phase II reactions, especially by conjugation with glutathione and degradation to mercapturic acids to be excreted via urine [14]. Accordingly, *N*-acetyl-*S*-(2-carbamoylethyl)cysteine (AAMA) and *N*-acetyl-*S*-(2-aminocarbonyl-2-hydroxyethyl)-cysteine (GAMA) have been found in human urine [15, 16]. In addition, AA and GA might be detoxified effectively by interaction with nucleophilic centers in proteins like amino-groups or sulfhydryl-groups [17–19] (Fig. 6.1).

In the present study we used human blood as easily accessible body compartment reflecting absorption of AA through the gut to investigate concentration dependent interactions of AA and its ultimate metabolite GA with DNA in lymphocytes, with hemoglobin in erythrocytes and with other blood components. Genotoxicity was determined by the modified alkaline single cell gel

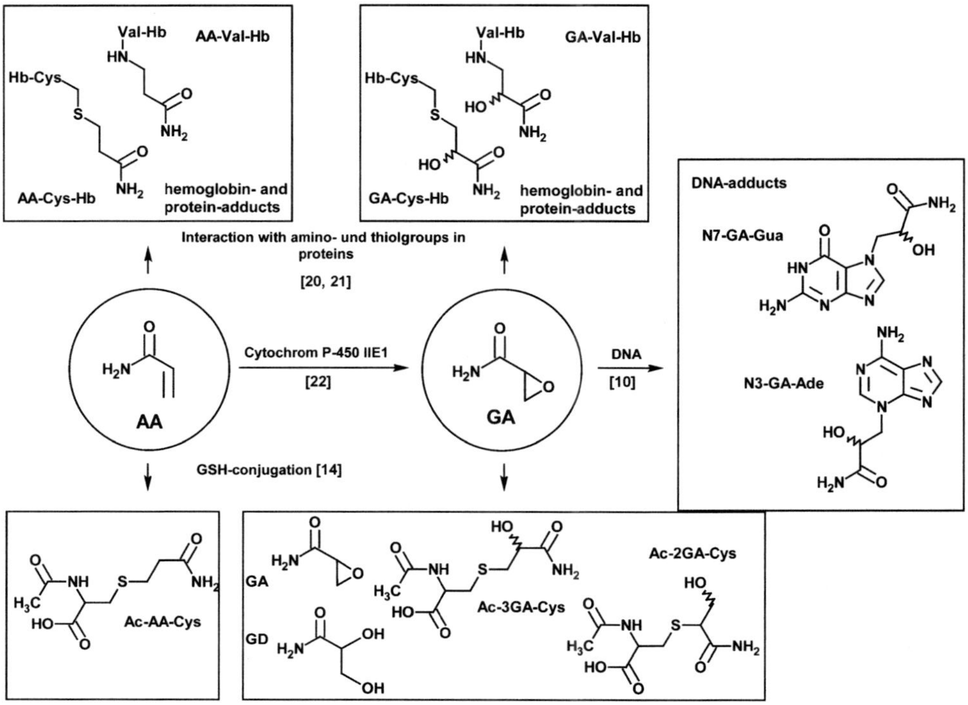

Figure 6.1: Metabolism of acrylamide (abbreviations: AA-/GA-Cys-/Val-Hb: hemoglobin-adducts of AA and GA at *N*-terminal valin or cysteine; Ac-AA-Cys: *N*-acetyl-*S*-(2-aminocarbonylethyl)-cysteine; Ac-3GA-Cys: *N*-acetyl-*S*-(2-aminocarbonyl-2-hydroxyethyl)-cysteine; Ac-2GA-Cys: *N*-acetyl-*S*-(1-aminocarbonyl-2-hydroxyethyl)-cysteine; GD: glyceramide; N7-GA-Gua, N3-GA-Ade: DNA-adducts of GA at N^7 of guanine and N^3 of adenine).

electrophoresis (SCGE, comet assay). Lymphocytic DNA was additionally treated with the DNA-repair-enzyme formamido-pyrimidine-DNA-glycosylase (FPG). FPG recognizes apurinic and apyrimidinic sites as well as ring-opened pyrimidines (formamido-pyrimidines) [23] and oxidized purines, rendering these lesions detectable by inducing additional DNA-strand breaks. Mutagenic activity was investigated using the *hPRT* gene mutation assay in V79 mammalian cells. In comparison to AA and GA we investigated the genotoxic activity of the known carcinogens (±)-anti-benzo[a]pyrene-7,8-dihydrodiol-9,10-epoxide (BPDE) and *α*-acetoxy-*N*-nitroso-diethanolamine (*α*-A-NDELA).

Additionally, we investigated the potential of AA and GA to induce chromosomal mutations in human blood by detecting sister chromatid exchange, being indicative for enhanced DNA repair after genotoxic events.

A concentration dependent formation of Hb adducts was investigated to approach a biomarker guided dosimetry of genotoxic effects in blood. The goal is to establish potential correlations between biomarker of exposure (Hb adducts) and biomarker of effect (DNA damage) under conditions simulating as closely as possible AA exposure from the diet.

6.2 Material and Methods

6.2.1 Sister Chromatid Exchange

Blood samples were taken from six male healthy donors by venipuncture using heparinized blood tubes. 800 μL blood were mixed with 22 μg/mL phytohemagglutinine and 8 mL RPMI 1640 medium supplemented with 15% FCS, 100 U/mL penicillin and 100 μg/mL streptomycin and incubated for 71 h at 37 °C, 5% CO_2 and 95% saturated atmospheric humidity. After 42 h 3 μg/mL bromodesoxyuridine were added and after 48 h blood was incubated with AA and GA for 23 h. 2 h prior to lymphocyte-harvest 0.2 μg/mL colcemide was supplemented. 50 cells in metaphase of the second cell-cycle were determined.

6.2.2 Alkaline Single Cell Gel Electrophoresis (SCGE, Comet Assay)

Blood samples were taken from three male healthy donors by venipuncture using heparinized blood tubes. Blood (990 μL) was incubated with substance solution (10 μL, DMSO).

The alkaline comet assay (Fig. 6.2) was performed according to Singh et al. [38] with slight modifications. Additional treatment of lymphocyte DNA with FPG was conducted as described elsewhere by Collins et al. [24].

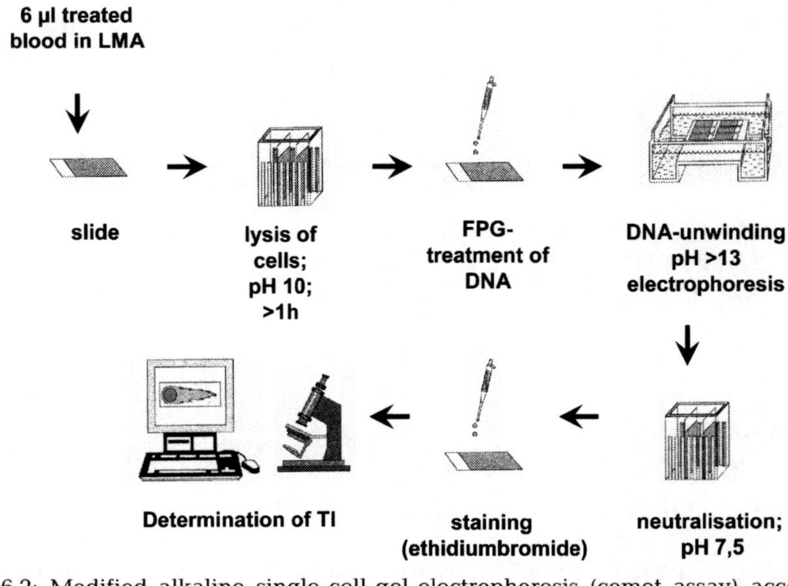

Figure 6.2: Modified alkaline single-cell-gel-electrophoresis (comet assay) according to Singh et al. [38]. Collins et al. [24] LMA: low melting agarose; FPG: formamido-pyrimidine-DNA-glycosylase.

Microscopic analysis was done with a Zeiss Axioskop 20, equipped with a filter set 15 (excitation BP 546/12; emission, LP 590). Tail intensity (TI) of cells was quantified by computer-assisted microscopy using Comet II®-software (Perceptive Instruments, Haverhill, UK).

6.2.3 hPRT-Test with V79 Cells

Cells were maintained in Dulbecco's modified Eagle's medium (DMEM), supplemented with 10% fetal calf serum, 1 mM sodium pyruvate, 100 U/mL penicillin and 100 μg/mL streptomycin at 37 °C, 5% CO_2 and 95% saturated atmospheric humidity.

The *hPRT*-Test was performed according to Bradley et al., with slight modifications [25]. Mutant frequencies (mutants/10^6 cells) were determined in triplicate and cloning efficiencies were scored in duplicate as described [26]. Mutagenic potential of substances was compared by calculated D_{3C}-values, reflecting the concentration that induced a threefold increase in mutagenicity compared to the solvent control.

Cytotoxicity was monitored in cell suspension by determination of trypan blue exclusion.

6.2.4 Adduct Formation at the N-terminal Valine of Hemoglobin

Blood samples were incubated with AA or GA (0.3–10.0 μM) for 1–6 h. Adducts formed were isolated using the *N*-alkyl Edman method as pentafluorophenyl thiohydantoine (PFPTH) derivatives and quantified by GC/MS (see Poster IV-13).

6.2.5 Distribution of ^{14}C-AA in Human Blood

Blood samples were incubated with ^{14}C-labelled acrylamide (^{14}C-AA) with a finalconcentration of 30 μM. The blood was fractionated and the distribution of radiolabel was measured using a scintillation counter (see Poster IV-13).

6.3 Results

6.3.1 Induction of Sister Chromatid Exchange (SCE)

AA was inactive up to 2.5 mM while GA induced SCE dose-dependently already at the lowest concentration tested (50 μM) (Fig. 6.3). Mitomycin (positive control) significantly elevated SCE-frequency.

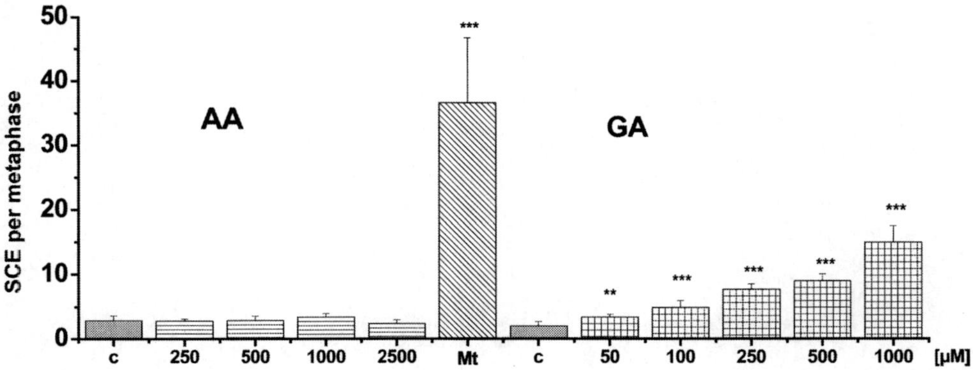

Figure 6.3: Sister chromatid exchanges in lymphocytes after incubation of human blood with AA and GA (1–4 h) (mitomycin (positive control) (1.5 μM); c: solvent control; (*** $p<0.005$; values represent means and SD of 3 experiments).

6.3.2 Induction of DNA Strand Breaks (Single Cell Gel Electrophoresis; Comet Assay)

AA did not induce DNA-damage under our conditions tested up to 6 mM [27]. In contrast, all other compounds tested were time- and concentration-dependently active. GA significantly induced DNA-damage at ≥ 300 μM (4 h). BPDE and *α*-A-NDELA were active in a tenfold lower concentration-range at ≥10–30 μM (1 h).

With additional FPG-treatment of lymphocyte-DNA, increased formation of DNA-strandbreaks was only observed for GA (10 μM; 4 h) while the amount of strand-breaks induced by BPDE and *α*-A-NDELA was not enhanced by FPG (Fig. 6.4).

6.3.3 Induction of hPRT Mutations in V79 Cells

The induction of *hPRT* gene mutations at the *hPRT* locus in V79 Chinese hamster lung fibroblasts was investigated of AA and GA in comparison to BPDE and *α*-A-NDELA.

BPDE significantly induced *hPRT* mutations beginning at 3 μM (MF: 20±4; D_{3C}: 7 μM). *α*-A-NDELA significantly induced *hPRT* mutations at ≥10 μM (MF: 19±2; D_{3C}: 9 μM). GA was significantly active in a 100-fold higher concentration range at 800 μM and beyond. AA was inactive up to 5 mM [27] (Fig. 6.5).

Under our test conditions AA, BPDE and *α*-A-NDELA had no influence on cell viability. In contrast, GA was cytotoxic at 800 μM and higher with an IC_{50} of 2000 μM (data not shown). As a further control, cloning efficiency (CE) was monitored after day 5 and was not found to be influenced.

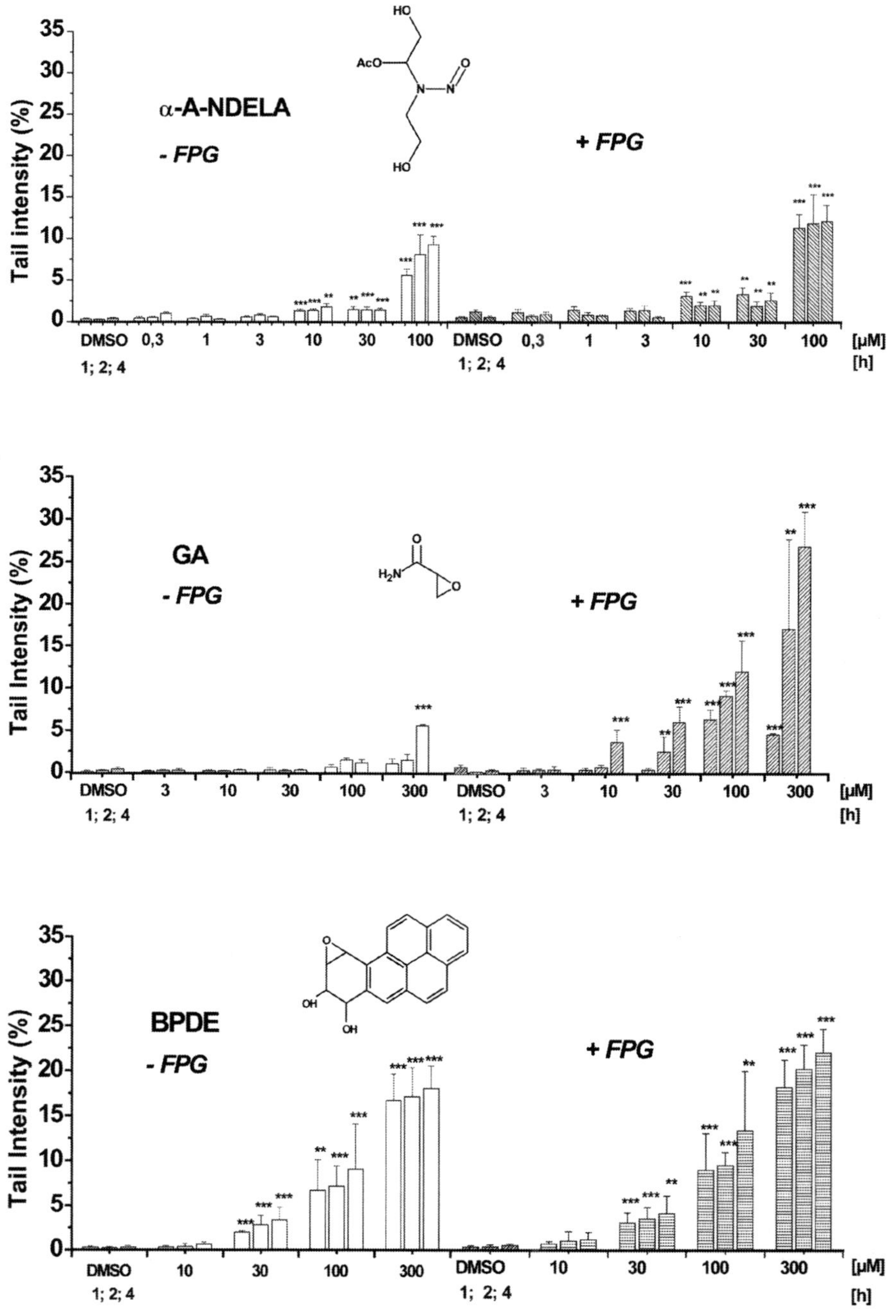

Figure 6.4: DNA damage in lymphocytes after incubation of human blood with GA, α-A-NDELA and BPDE (1–4 h) with and without FPG-treatment ($^{*}p<0.5$; $^{**}p<0.05$; $^{***}p<0.005$ (one-sided paired student t-test); values represent means and SD of three experiments).

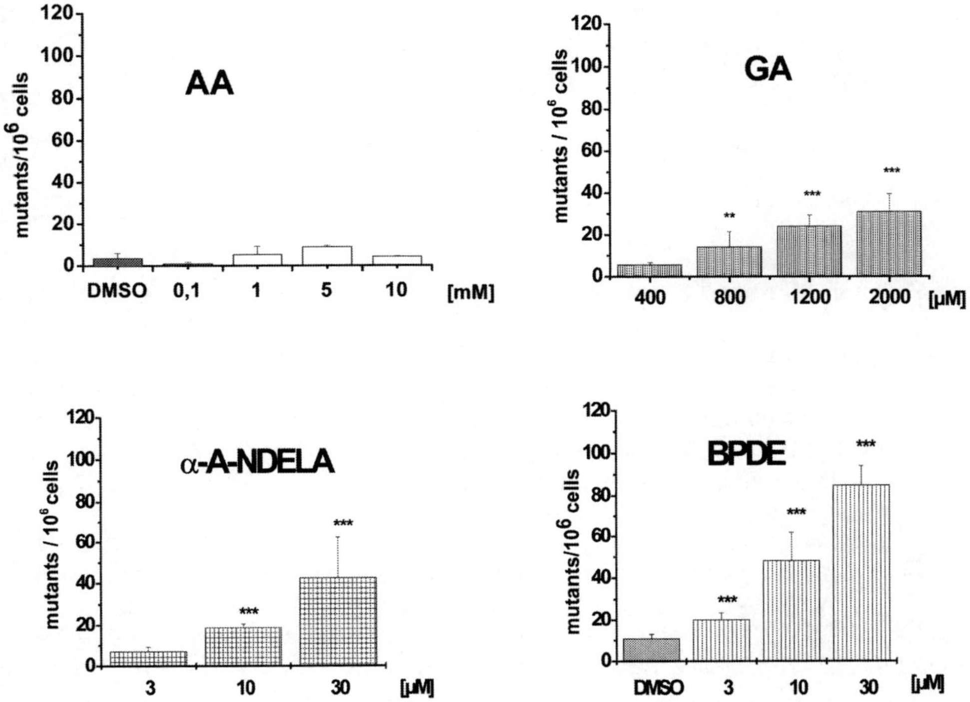

Figure 6.5: Induction of *hPRT* mutations in V79 cells by AA, GA, BPDE and *α*-A-NDELA (incubation time: 24 h) (mutations are selected by treatment of cells with 6-thioguanine; ** $p<0.05$, *** $p<0.005$ (one-sided paired student *t*-test); values represent means and SD of three experiments).

6.3.4 Adduct Formation at the *N*-terminal Valine of Hemoglobin

An average background valine adduct level in the range of 20–65 pmol/g Hb with AA and in the range of 6–46 pmol/g Hb for GA was found in blood of non-smoking probands. Incubation experiments showed that adduct levels of AA with *N*-terminal valine in hemoglobin significantly increased after incubation with 0.3 µM AA (4 h), respectively 3 µM (1 h) while GA-valine adduct formation became significant after 4 h incubation with 3 µM GA. It was found that GA reacts between two and three times more slowly with hemoglobin in whole blood compared to AA (see Poster IV-13).

6.3.5 Distribution of ^{14}C-AA in Human Blood

Incubation of human whole blood with 30 µM ^{14}C-AA resulted in about 35–45% of the label found in the erythrocyte fraction and about 60% found in plasma. Binding to hemoglobin counted for 10% of the radio label, binding to plasma proteins 10–20% (acetone precipitation). In total 25–35% radioactivity are bound to proteins or cell fragments (see Poster IV-13).

6.4 Discussion

The aim of this study was to investigate the concentration dependent induction of genotoxic effects by AA and its metabolite GA in human blood as a model system monitoring at the same time the formation of Hb-adducts. Significant binding of AA and GA to the *N*-terminal valine in hemoglobin was observed at 3 µM for each compound after 1–4 h incubation. Concerning the distribution and total binding of ^{14}C radiolabelled AA in human blood after 4 h we found 5–10% radioactivity bound to hemoglobin. In total 25–35% of the radiolabel were bound to plasma proteins, cell fragments or hemoglobin.

In lymphocytes from AA or GA treated blood, the genotoxic potency of AA and GA was investigated by detecting sister chromatid exchange and measuring DNA damage by single cell gel electrophoresis (comet assay). Comparison was made to the potent carcinogens BPDE and *α*-A-NDELA. In V79 mammalian cells we investigated the ability of AA and GA to induce mutations at the *hPRT* locus in comparison to the mutagenic potential of BPDE and *α*-A-NDELA.

AA did not show genotoxic activity. While GA became mutagenic in V79 cells at 800 µM, BPDE and *α*-A-NDELA showed mutagenic activity in a 100-fold lower concentration range (3–10 µM). In human blood, DNA damage induced by GA became detectable at ≥300 µM. By additional treatment with the DNA repair enzyme FPG, genotoxicity of GA became significant already at 10 µM. BPDE and *α*-A-NDELA were directly active in the same concentration range. In these cases additional treatment with FPG did not increase rate of DNA strand-breaks.

While AA does not show reactivity towards DNA-bases, GA effectively interacts with DNA by primarily forming N^7-adducts of guanine. These adducts yield 5-*N*-alkyl-2,6-diamino-4-hydroxyformamidopyrimidine (alkyl-FAPy-G lesions) after ring-opening. Alternatively apurinic sites (AP sites) might be generated via depurination. Due to the alkaline comet assay conditions, AP sites are converted rapidly into strand breaks. In contrast, alkyl-FAPy-G lesions are stable in DNA under physiological conditions and need to be excised by special repair enzymes, e.g., FPG [28]. However, a certain proportion of AP sites remains undetected under the alkaline comet conditions [29], as they are not converted into DNA strand breaks, yet recognized by FPG. It can therefore be concluded that the enhancement of DNA strand breaks observed after FPG treatment is in line with the premise that GA induces both kinds of lesions.

BPDE interacts covalently with DNA, preferentially binding to exocyclic amino-groups of deoxyguanosine (dG) or deoxyadenosine (dA) thereby forming

N^2-dG bulky adducts (80–90% of the total adducts induced) predominantly [30]. These adducts are stable and are not recognized by FPG. DNA strand breaks induced by (±)-BPDE in the comet assay without FPG treatment might be due to formation of alkali-labile sites [31] and repair processes, especially as a consequence of excision repair [32].

The hydroxyethylating intermediate which is formed after hydrolysis of *α*-A-NDELA reacts with DNA bases like guanine at the positions N^7 [33] or O^6 [34]. It might, however, also interact with the sugar phosphate backbone to generate instable 2-hydroxyethyl-phosphotriesters, which are converted into DNA strand breaks by rapid spontaneous decomposition [35–37].

The *hPRT* gene mutation assay in V79 cells responds at similar sensitivity towards (±)-BPDE and *α*-A-NDELA, whereas GA is 100-fold less potent [27]. Since the *hPRT* assay encompasses about 10 cell division rounds after mutagen challenge, lesions induced by GA might be more effectively repaired or are less promutagenic.

6.5 Conclusion

In human blood, the lowest GA concentration inducing significant genotoxicity was 10 μM, suggesting GA to exert genotoxic effects in a concentration range comparable to potent carcinogens like BPDE or *α*-A-NDELA. Binding of AA and GA to hemoglobin was observed at 3 μM. Considering diet related exposure to AA, even assuming a worst case scenario, AA or GA blood concentrations exceeding 1 μM are highly improbable. While the most sensitive biomarker, the GA-Val adduct, became significant at 3 μM, genotoxicity was observed at 10 μM. We choose human blood as a model system to mimic the situation following the absorption of AA from the gastrointestinal tract. The results are to be considered taking into account that incubation times were limited to 4 h. This does not account for long term exposure as might be expected from dietary uptake. However, hemoglobin adduct formation might be utilized as a guiding biomarker to be exploited for comparison to long term exposures.

Further *in vivo* investigations are mandatory to approach the aim of evaluating potential effects of dietary long term exposure to acrylamide by monitoring DNA damage and Hb adduct formation concomitantly.

Acknowledgments

This work was supported by grants of AIF 108, Bundesministerium für Wirtschaft und Arbeit via Forschungskreis der Ernährungsindustrie and Bund für Lebensmittelrecht und Lebensmittelkunde e.V. (BLL).

Abbreviations

AA, acrylamide; AAMA, *N*-Acetyl-*S*-(2-aminocarbonylethyl)cysteine; *α*-A-NDELA, *α*-acetoxy-*N*-nitroso-diethanolamine; BER, base excision repair; (±)-BPDE, (±)-anti-benzo[a]pyrene-7,8-dihydrodiol-9,10-epoxide; CE, cloning efficiency; DMEM, Dulbecco's modified Eagle's medium; FPG, formamido-pyrimidine-DNA-glycosylase; GA, glycidamide; GAMA, *N*-Acetyl-*S*-(2-carbamoyl)-2-hydroxyethyl)cysteine; *hPRT*, hypoxanthine-phosphoribosyl-transferase; MF, mutant frequency; SCE, sister chromatid exchange; SCGE, single cell gel electrophoresis; SSB, single-strand breaks; TI, Tail Intensity.

References

1. K. A. Johnson, S. J. Gorzinski, K. M. Bodner, R. A. Campbell, C. H. Wolf, M. A. Friedman, R. W. Mast, Chronic toxicity and oncogenicity study on acrylamide incorporated in the drinking water of Fischer 344 rats, *Toxicol. Appl. Pharmacol.* **85**(2) (1986) 154–168.
2. M. A. Friedman, L. H. Dulak, M. A. Stedham, A lifetime oncogenicity study in rats with acrylamide, *Fundam. Appl. Toxicol.* **27**(1) (1995) 95–105.
3. K. L. Dearfield, G. R. Douglas, U. H. Ehling, M. M. Moore, G. A. Sega, D. J. Brusick, Acrylamide: a review of its genotoxicity and an assessment of heritable genetic risk, *Mutat. Res.* **330**(1–2) (1995) 71–99.
4. D. Adler, T. E. Schmid, A. Baumgartner, Induction of aneuploidy in male mouse germ cells detected by the sperm-FISH assay: a review of the present data base, *Mutat. Res.* **504**(1–2) (2002) 173–182.
5. R. W. Tyl, M. A. Friedman, Effects of acrylamide on rodent reproductive performance, *Reprod. Toxicol.* **17**(1) (2003) 1–13.
6. S. Madle, L. Broschinski, O. Mosbach-Schulz, G. Schönig, A. Schulte, Zur aktuellen Risikobewertung von Acrylamid in Lebensmitteln, *Bundesgesundheitsblatt-Gesundheitsforschung-Gesundheitsschutz* **46** (2003) 405–415.
7. E. Zeiger, B. Anderson, S. Haworth, T. Lawlor, K. Mortelmans, W. Speck, Salmonella mutagenicity tests. III. Results from the testing of 255 chemicals, *Environ. Mutagen.* **9** (Suppl. 9) (1987) 1–110.
8. G. Knaap, P. G. Kramers, C. E. Voogd, W. G. Bergkamp, M. G. Groot, P. G. Langebroek, H. C. Mout, J. J. van der Stel, H. W. Verharen, Mutagenic activity of acrylamide in eukaryotic systems but not in bacteria, *Mutagenesis* **3**(3) (1988) 263–268.

9. H. Tsuda, C.S. Shimizu, M.K. Taketomi, M.M. Hasegawa, A. Hamada, K.M. Kawata, N. Inui, Acrylamide; induction of DNA damage, chromosomal aberrations and cell transformation without gene mutations, *Mutagenesis* **8**(1) (1993) 23–29.
10. G. Da Costa, M.I. Churchwell, P. Hamilton, L.S. von Tungeln, F.A. Beland, M.M. Marques, D.R. Doerge, DNA adduct formation from acrylamide via conversion to glycidamide in adult and neonatal mice, *Chem. Res. Toxicol.* **16** (2003) 1328–1337.
11. D.R. Doerge, G. Gamboa da Costa, L.P. McDaniel, M.I. Churchwell, N.C. Twaddle, F.A. Beland, DNA adducts derived from administration of acrylamide and glycidamide to mice and rats. *Mutat. Res.* **580**(1–2) (2005) 131–141.
12. B.I. Ghanayem, L.P. McDaniel, M.I. Churchwell, N.C. Twaddle, R. Snyder, T.R. Fennell, D.R. Doerge, Role of CYP2E1 in the epoxidation of acrylamide to glycidamide and formation of DNA and hemoglobin adducts. *Toxicol. Sci.* **88**(2) (2005) 311–318.
13. T. Maniere, T. Godard, D.R. Doerge, M.I. Churchwell, M. Guffroy, M. Laurentie, J.M. Poul, DNA damage and DNA adduct formation in rat tissues following oral administration of acrylamide. *Mutat. Res.* **580**(1–2) (2005) 119–129.
14. S.C.J. Summer, J.P. MacNeela, T.R. Fennell, Characterization and quantitation of urinary metabolites of [1,2,3-^{13}C]acrylamide in rats and mice using ^{13}C nuclear magnetic resonance spectroscopy, *Chem. Res. Toxicol.* **5** (1992) 81–89.
15. M.I. Boettcher, H.M. Bolt, H. Drexler, J. Angerer, Excretion of mercapturic acids of acrylamide and glycidamide in human urine after single oral administration of deuterium-labelled acrylamide. *Arch. Toxicol.* (2005) 1–7.
16. M.I. Boettcher, T. Schettgen, B. Kutting, M. Pischetsrieder, J. Angerer, Mercapturic acids of acrylamide and glycidamide as biomarkers of the internal exposure to acrylamide in the general population. *Mutat. Res.* **580**(1–2) (2005) 167–176.
17. E. Bergmark, Hemoglobin adducts of acrylamide and acrylonitrile in laboratory workers, smokers and nonsmokers, *Chem. Res. Toxicol.* **10**(1) (1997) 78–84.
18. B. Paulsson, J. Grawe, M. Tornqvist, Hemoglobin adducts and micronucleus frequencies in mouse and rat after acrylamide or N-methylolacrylamide treatment. *Mutat. Res.* **516**(1–2) (2002) 101–111.
19. Q. Xie, H. Sun, Y. Liu, X. Ding, D. Fu, K. Liu, Adduction of biomacromolecules with acrylamide (AA) in mice at environmental dose levels studied by accelerator mass spectrometry. *Toxicol. Lett.* **163**(2) (2006) 101–108.
20. C.J. Calleman, E. Bergmark, L.G. Costa, Acrylamide is metabolized to glycidamide in the rat: evidence from hemoglobin adduct formation, *Chem. Res. Toxicol.* **3**(5) (1990) 406–412.
21. E. Bergmark, C.J. Calleman, F. He, L.G. Costa, Determination of hemoglobin adducts in humans occupationally exposed to acrylamide, *Toxicol. Appl. Pharmacol.* **120**(1) (1993) 45–54.
22. S.C.J. Sumner, T.R. Fennell, T.A. Moore, B. Chanas, F. Gonzalez, B.I. Ghanayem, Role of cytochrome P450 2E1 in the metabolism of acrylamide and acrylonitrile in mice, *Chem. Res. Toxicol.* **12**(11) (1999) 1110–1116.
23. S. Boiteux, R.T. O'Connor, F. Lederer, A. Gouyette, J. Laval, Homogeneous Escherichia coli FPG protein. A DNA glycosylase which excises imidazole ring-opened purines and nicks DNA at apurinic/apyrimidinic sites. *J. Biol. Chem.* **265**(7) (1990) 3916–3922.
24. R. Collins, Measurement of Oxidative DNA Damage using the Comet Assay, in: Lunec J., Griffiths H.R. (eds.) Measuring in vivo Oxidative Damage: A Practical Approach, John Wiley and Sons (2000) 83–94.
25. M.O. Bradley, B. Bhuyan, M.C. Francis, R. Langenbach, A. Peterson, E. Huberman, Mutagenesis by chemical agents in V79 chinese hamster cells: a review and analysis of the literature. A report of the Gene-Tox Program, *Mutat. Res.* **87** (1981) 81–142.
26. H.-R. Glatt, in: Fahrig R. (ed.) Mutationsforschung und genetische Toxikologie, Wissenschaftliche Buchgesellschaft, Darmstadt (1993) 243–262.
27. M. Baum, E. Fauth, S. Fritzen, A. Herrmann, P. Mertes, K. Merz, M. Rudolphi, H. Zankl, G. Eisenbrand, Acrylamide and glycidamide: genotoxic effects in V79 cells and human blood, *Mutat. Res.* **580** (2005) 61–69.
28. K.S. Gates, T. Nooner, S. Dutta, Biologically Relevant Chemical Reactions of N7-Alkylguanine Residues in DNA. *Chem. Res. Toxicol.* **17**(7) (2004) 840–56.

29. R. Collins, The comet assay for DNA damage and repair-principles, applications, and limitations. *Molec. Biotech.* **26** (2004) 249–261.
30. S.C. Cheng, B.D. Hilton, J.M. Roman, A. Dipple, DNA adducts from carcinogenic and noncarcinogenic enantiomers of benzo[a]pyrene dihydrodiol epoxide. *Chem. Res. Toxicol.* **2**(5) (1989) 334–340.
31. E. Sage, W.A. Haseltine, High ratio of alkali-sensitive lesions to total DNA modification induced by benzo(a)pyrene diol epoxide. *J. Biol. Chem.* **259**(17) (1984) 11098–11102.
32. G. Speit, S. Hanelt, R. Helbig, A. Seidel, A. Hartmann, Detection of DNA effects in human cells with the comet assay and their relevance for mutagenesis. *Toxicol. Lett.* **88**(1–3) (1996) 91–98.
33. R.N. Loeppky, F. Gu, P. Goelzer, J. Shi, The preparation, decomposition, and DNA adduct chemistry of α-acetoxy-N-nitroso-diethanolamine. Book of Abstracts, 214th ACS National Meeting, Las Vegas, NV, September 7–11, 1997.
34. M.K. Dennehy, R.N. Loeppky, Mass spectrometric methodology for the determination of glyoxaldeoxyguanosine and O^6-hydroxyethyldeoxyguanosine DNA adducts produced by nitrosamine bident carcinogens. *Chem. Res. Toxicol.* **18**(3) (2005) 556–565.
35. J. Conrad, N. Mueller, G. Eisenbrand, Studies on the stability of trialkyl phosphate and di-(2′-deoxythymidine) phosphotriesters in alkaline and neutral solutions. A model study for hydrolysis of phosphotriesters in DNA and on the influence of a β-hydroxyethyl ester group. *Chem. Biol. Interact.* **60** (1986) 57–65.
36. G. Eisenbrand, N. Mueller, E. Denkel, W. Sterzel, DNA adducts and DNA damage by antineoplastic and carcinogenic N-Nitrosocompounds. *J. Canc. Res. Clin. Oncol.* **112** (1986) 196–204.
37. N. Müller, J. Conrad, G. Eisenbrand, Alkylation of 2′-deoxythymidylyl(3′–5′)-2′-deoxythymidine with 2-hydroxyethylnitrosourea and stability of the resulting phosphotriester. *IARC Sci. Publ.* **84** (1987) 194–196.
38. N.P. Singh, M.T. McCoy, R.R. Tice, E.L. Schneider, A simple technique for quantitation of low levels of DNA damage in individual cells. *Exp. Cell Res.* **175** (1988) 184–191.

7 Risk Assessment of Furan

Josef Schlatter *

7.1 Introduction

Furan (C_4H_4O) (CAS-Nr. 110-00-9) is a clear, colourless liquid with an ethereal odour, is highly volatile (melting point –85 °C, boiling point 31 °C) and has a low polarity. It is an intermediate in the chemical synthesis and preparation of numerous polymers [1]. Because of its large production volume and use and the potential for widespread human exposure to a variety of furan-containing compounds, furan was evaluated for its carcinogenic potential by the U.S. National Toxicology Program [1]. Based on these data, furan was classified by the International Agency for Research on Cancer as possibly carcinogenic to humans (group 2B [2]). Because furan and some derivatives of furan were found occasionally in space craft and space station atmosphere, the National Academy of Sciences (NAS) evaluated furan to derive Spacecraft Maximum Allowable Concentrations [3].

In May 2004, the U. S. Food and Drug Administration (FDA) published the results of a survey of furan in a range of canned and jarred foods that undergo heat treatment. Furan levels up to 125 µg/kg were reported and the FDA issued a request for data and information [4]. That furan may occur in heat treated food was not new at that time as levels up to 4000 µg/kg were previously detected under experimental conditions in canned meat and other heated proteins such as soy protein or fish protein [5]. All the available data on occurrence of furan in food together with the data on the toxic potential of furan triggered the panel on contaminants in the food chain of the European Food Safety Authority (EFSA) to compile an initial report on the state-of-the-art on exposure and adverse effects of furan [6].

The present paper "risk assessment of furan", presented at the symposium "Thermal Processing of Food: Potential Health Benefits and Risks" of the Senate Commission on Food Safety (SKLM) of the Deutsche Forschungsgemeinschaft (DFG) is primarily based on the reports of the EFSA and the NAS.

* Swiss Federal Office of Public Health, Food Toxicology Section, Stauffacherstr. 101, CH-8004 Zürich, Switzerland.

Thermal Processing of Food: Potential Health Benefits and Risks
DFG, Deutsche Forschungsgemeinschaft

ISBN: 978-3-527-31909-1

7.2 Occurrence Data in Food and Assessment of Dietary Exposure

As mentioned above, furan is a volatile molecule that occurs in a variety of heated foods such as coffee, canned and jarred foods including baby food containing meat, and various vegetables, which suggests that there are probably multiple routes of formation. Fresh vegetables do not contain furan. Only a limited set of data on occurrence of furan in various food categories is available. The levels of furan in food ranged from non detectable to about 200 µg/kg [6]. The limit of quantitation is matrix dependent and was usually in the range 0.5–2 µg/kg [7]. Details of the occurrence data can be found in the Annex of the EFSA report in which individual data collected by the US FDA, the Swiss Federal Office of Public Health, the Chemisches und Veterinäruntersuchungsamt Stuttgart, Germany, and the Kantonales Laboratorium Basel-Stadt, Switzerland are listed [8]. The US FDA data is also available on the internet (http://www.cfsan.fda.gov/~dms/furandat.html). Based on the limited initial data on occurrence of furan in food, DiNovi and Mihalov [9] estimated a mean human exposure of 0.3 µg/kg body weight (bw) with a 90th percentile of 0.6 µg/kg bw. For infants (0–1 year old), the estimated values were 0.4 and 1 µg/kg bw for the mean and the 90th percentile, respectively. In the EFSA opinion only ranges of intake from individual food categories were listed and no dietary exposure assessment was performed as the data available was considered to be too incomplete for drawing any conclusion on dietary exposure [6]. Furthermore, the furan levels reported in food were not considered to be an indicator of the actual furan intake via food because most data reported do not reflect the levels of furan in food as consumed. As an example, initial data have shown that 10 minutes after brewing coffee a strong reduction of the furan levels is observed in this beverage as a result of its high volatility [10, 11]. Interestingly, there was no correlation of the furan levels in food with the respective levels of acrylamide in the same food [11].

7.3 Toxicological Data on Furan

Furan can easily pass through biological membranes and is readily and extensively absorbed from the lung or intestine. It is rapidly metabolised by P-450 enzymes, primarily by CYP 2E1. While the major metabolite was carbon dioxide, *cis*-2-butene-1,4-dial has been identified as a reactive and cytotoxic key metabolite of furan. Furan is widely distributed in the body but the rapid hepatic metabolism seems to limit its systemic delivery [3, 6]. Within the first 24 hours after a single oral application of furan (8 mg/kg bw) to male rats, 80% of the total radioactivity were eliminated via the lung, urine and faeces [12]. About 14% of the applied furan dose were eliminated in the exhaled air as the parent compound and

26% as carbon dioxide. In urine, 20% were found as the sum of more than ten different compounds. About 22% were found in the faeces.

The toxicity database of furan is incomplete: only qualitative data are available on the acute toxicity and there are no data on its reproductive toxicity and developmental toxicity. However as already mentioned, toxicology and carcinogenesis studies were conducted with furan by the US National Toxicology Program because of its large production volume and the potential for widespread human exposure to a variety of furan-containing compounds. In these studies, furan (purity >99%) was administered in corn oil by gavage to groups of F344/N rats and B6C3F1 mice of each sex for 16 days, 13 weeks, and two years [1]. The main findings of the two year studies are summarised below (for more details see [1]):

Rats of each sex ($n=70$) were administered furan at 0 (controls), 2, 4, or 8 mg/kg bw five days per week for two years with interim kills of ten rats per group after 9 and 15 months. Furan was cytotoxic and the liver was found to be the primary target organ. Survival of male and female rats in the highest dose groups was lower than in the control groups. Non-neoplastic liver lesions in treated rats included biliary tract fibrosis, hyperplasia, chronic inflammation, proliferation, hepatocyte cytomegaly, cytoplasmic vacuolization, degeneration, nodular hyperplasia, and necrosis. Furthermore, the severity of nephropathy increased with dose and the incidence of forestomach hyperplasia was increased.

Cholangiocarcinoma of the liver occurred in all groups of dosed rats (males: 0/50, 43/50, 48/50, 49/50; females: 0/50, 49/50, 50/50, 48/50) and was already seen in rats of each sex at the 9- and 15-month interim evaluations (9-month: males – 0/10, 5/10, 7/10, 10/10; females – 0/10, 4/10, 9/10, 10/10; 15-month: males – 0/10, 7/10, 9/10, 6/10; females – 0/10, 9/10, 9/10, 7/10). Hepatocellular adenomas or carcinomas (combined) were significantly increased in male and female rats after 2 years (males: 1/50, 5/50, 22/50, 35/50; females: 0/50, 2/50, 4/50, 7/50). Hepatocellular neoplasms were not observed at the 9- or 15-month interim evaluations. The incidence of mononuclear cell leukemia was increased in male and female rats that received 4 or 8 mg/kg furan (males: 8/50, 11/50, 17/50, 25/50; females: 8/50, 9/50, 17/50, 21/50).

Mice of each sex ($n=50$) received doses of furan of 0 (controls), 8 or 15 mg/kg bw five days per week for two years. The liver was again the primary target organ. Survival of low- and high-dose male and high-dose female mice was lower than controls. The non-neoplastic hepatocellular lesions included hepatocyte cytomegaly, degeneration, necrosis, multifocal hyperplasia, and cytoplasmic vacuolization and biliary tract dilatation, fibrosis, hyperplasia, and inflammation. The incidence of focal inflammation and papillary hyperplasia of the forestomach were increased in male mice and focal hyperplasia of the adrenal medulla were increased in low- and high-dose male and in high-dose female mice.

The incidences of hepatocellular adenomas and carcinomas (combined) were significantly increased (males: 26/50, 44/50, 50/50; females: 7/50, 34/50, 50/50).

A 2-year stop exposure study was conducted by NTP in which 50 male rats were administered 30 mg/kg furan in corn oil by gavage 5 days per week for 13 weeks and then maintained for the remainder of the 2 years without further treatment [1]. Groups of ten animals were evaluated for the presence of treatment-related lesions at the end of the 13-week period of furan administration and at 9 and 15 months. Cholangiocarcinoma of the liver occurred with an overall incidence of 100% and hepatocellular carcinoma occurred with an overall incidence of 15%. Cholangiocarcinoma was observed in all ten males at both the 9-month and 15-month interim evaluations. Hepatocellular carcinoma was first observed in two males at the 15-month interim evaluation.

According to the U.S. FDA [4], a preliminary report from a second 2-year bioassay in female mice found an increased incidence and multiplicity of hepatic tumours and a decreased tumour latency in mice dosed with 4 or 8 mg/kg bw furan, but not in mice dosed with 0.5, 1.0, or 2.0 mg/kg bw furan [13].

Preweaning male B6C3F1 mice were treated by intraperitoneal administration of either a single dose of 400 mg/kg bw or six doses of 200 mg/kg bw. The mice were scarified between 28 and 95 weeks after dosing [14]. In the single-dose group, the furan treated mice showed a statistically non-significant increase of the overall frequency of hepatocytic neoplasia and of the overall liver tumour multiplicity, compared with the vehicle control group. No treatment-related histopathological lesions other than development of liver tumours were found. In mice given six doses of 200 mg/kg furan, there was a statistically significant increased incidence of hepatocellular adenomas and carcinomas and tumour multiplicity compared with the vehicle control group. The relative frequency of *Hras1* activation was 82% in the 28 tumours analysed from the single-dose group and 32% in the 28 tumours analysed from the multiple-dose group.

Reynolds et al. [15] also found activated oncogenes in mouse liver tumours obtained from the NTP study. While the frequencies of activated *H-ras* and *K-ras* oncogenes were similar in liver tumours from furan treated mice and vehicle controls, the spectrum of mutations in the *H-ras* gene differed significantly and novel mutations in eight liver tumours were seen. The unique *ras* oncogene mutational profile was interpreted by the authors as a suggestion that the novel mutations could have been due to a genotoxic effect of furan.

Furan was mainly negative in the following test systems for genotoxicity: *Salmonella typhimurium* (strains TA100, TA1535, TA1537 and TA98 with and without S9 metabolic activation), sex-linked recessive lethal mutation test in *Drosophila melanogaster*, sister chromatid exchange (SCE) *in vivo* in mouse bone marrow cells (up to 350 mg/kg bw i.p.) and unscheduled DNA synthesis in mouse or rat hepatocytes *in vivo* after a single oral dose of 200 or 100 mg/kg bw, respectively [1, 6]. On the other hand, furan was mutagenic in the thymidine kinase locus of L5178Y mouse lymphoma cells (with and without S9 metabolic activation), induced chromosomal aberrations and SCE in Chinese hamster ovary cells (with and without S9 activation), produced DNA double-strand breaks in isolated rat hepatocytes and induced structural chromosome aberrations in mice bone marrow cells *in vivo* (250 mg/kg bw by i.p.).

The furan metabolite *cis*-2-butene-1,4-dial was directly mutagenic at non-toxic concentrations in the *S. typhimurium* strain TA104, a strain sensitive to aldehydes, caused DNA single-strand breaks and DNA cross-links in CHO cells [16] and produced DNA adducts *in vitro* [17]. The adducts resulted from the addition of *cis*-2-butene-1,4-dial to the exo- and endo-cyclic nitrogens of 2′-deoxyguanosine and 2′-deoxyadenosine.

7.4 Conclusions

Furan administered for two years by gavage in corn oil is clearly carcinogenic to rats and mice. The presently available data on the mode of action of furan, including the demonstrated mutagenic potential of *cis*-2-butene-1,4-dial, indicate that furan-induced carcinogenicity is probably mediated by a genotoxic mechanism but that chronic toxicity with secondary cell proliferation may indirectly amplify the tumour response [6]. Based on the very limited data available, it appears that there might be relative small difference between possible human exposures and the doses that produce carcinogenic effects in rodents. However, for a reliable risk assessment, further data on both toxicity and exposure would be needed.

The significance of furan in the context of substituted furans should also be clarified as a number of furan-substituted aliphatic hydrocarbons, alcohols, aldehydes, ketones, carboxylic acids and related esters, sulfides, disulfides and ethers may occur in food as flavours. Several of these compounds were positive in genotoxicity tests [18], which raises the question as to whether these could show similar toxicological properties to the parent molecule – furan.

References

1. NTP (National Toxicology Program): Toxicology and carcinogenesis studies of furan (CAS No. 110-00-9) in F344/N rats and B6C3Fl mice (gavage studies), NTP Technical Report No. 402, U.S. Department of Health and Human Services, Public Health Service, National Institutes of Health, Research Triangle Park, NC (1993). Available at http://ntp.niehs.nih.gov/ntp/htdocs/LT_rpts/tr402.pdf
2. IARC (International Agency for Research on Cancer): Monographs on the Evaluation of Carcinogenic Risks to Humans, Volume 63. Dry Cleaning, Some Chlorinated Solvents and Other Industrial Chemicals, IARC Lyon, France, 3194–3407 (1995).
3. NAS (National Academy of Sciences): Spacecraft Maximum Allowable Concentrations for Selected Airborne Contaminants: Volume 4, B14, Furan, page 307–329 (2000).
4. FDA (U.S. Food and Drug Administration): Furan in Food. Thermal Treatment; Request for Data and Information (2004). Available at http://www.fda.gov/OHRMS/DOCKETS/98fr/04n-0205-nrd0001.pdf

5. Maga J.A.: Furans in Food. *Critical Reviews in Food Science and Nutrition* **4**, 355–399 (1979).
6. EFSA (European Food Safety Authority): Report of the Scientific Panel on Contaminants in the Food Chain on furan in food (Question N° EFSA-Q-2004-109), adopted on 7 December 2004. The EFSA Journal 137, 1-20 (2004). Available at http://www.efsa.eu.int/science/contam/contam_documents/760/furan10feb051.pdf
7. Reinhard H., Sager F., Zimmermann H. and Zoller O.: Furan in Foods on the Swiss Market – Method and Results. *Mitt. Lebensm. Hyg.* **95**, 532–535 (2004).
8. EFSA (European Food Safety Authority): Annex to the Report of the Scientific Panel on Contaminants in the Food Chain on furan in food (2004). Available at http://www.efsa.eu.int/science/contam/contam_documents/760/furan_annex1.pdf
9. DiNovi M. and Mihalov J.: An Exposure Assessment for Furan from the Consumption of Adult and Baby Foods (2004). Available at http://www.fda.gov/ohrms/dockets/ac/04/slides/2004-4045s2-03-Mihalov.pdf
10. Zoller O. Personal communication (2004).
11. Fritz H., Gatermann R., Hoenicke K. and Weidemann S.: Analysis of Furan in Different Foods using Gas Chromatography Mass Spectrometry. Poster presented at the lebensmittelchemische Regionaltagung, Hamburg, Germany (March 8, 2005).
12. Burka L.T., Washburn K.D., and Irwin R.D.: Disposition of [14C]-furan in the male F344 rat. *J. Toxicol. and Environ. Health* **34**, 245–257 (1991).
13. Goldsworthy T.L., Goodwin R., Burnett R.M., King P., El-Sourady H., Moser G., Foley J. and Maronpot R.: Dose response relationships between furan induced cytotoxicity and liver cancer. Society of Toxicologic Pathology Annual Conference, Orlando, USA (2001). Cited in U.S. FDA: Furan in Food, Thermal Treatment; Request for Data and Information (2004).
14. Johansson E., Reynolds S., Anderson M. and Maronpot R.: Frequency of Ha-ras gene mutations inversely correlated with furan dose in mouse liver tumours. *Mol. Carcinog.* **18**, 199–205 (1997).
15. Reynolds S., et al. (1987) bitte ergänzen.
16. Marinari U.M., Ferro M., Sciaba F., Finollo R., Bassi A.M. and Brambilla G.: DNA-damaging activity of biotic and xenobiotic aldehydes in Chinese hamster ovary cells. *Cell Biochem. Funct.* **2**, 243–248 (1984).
17. Byrns M.C., Predecki D.P. and Peterson L.A.: Characterization of nucleoside adducts of cis-2-butene-1,4-dial, a reactive metabolite of furan. *Chem. Res. Toxicol.* **15**, 373–379 (2002).
18. WHO (World Health Organization): Summary and conclusions of the 65th meeting of the Joint FAO/WHO Expert Committee on Food Additives (JECFA), p. 3 (2005). Available at ftp://ftp.fao.org/es/esn/jecfa/jecfa65_summary.pdf.

8 Heterocyclic Aromatic Amines: Potent Genotoxicants Formed in Cooked Meats

Robert J. Turesky * and Ricky D. Holland **

Abstract

Heterocyclic aromatic amines (HAAs) are formed in grilled meats and tobacco smoke condensate, and they also occur in diesel exhaust. The concentrations of HAAs in cooked meats can vary greatly and are dependent upon the type of animal protein, and the temperature and duration of cooking. Many HAAs are potent genotoxicants in bacterial and mammalian cell assays and induce cancer in experimental laboratory animals. HAAs undergo oxidation at the exocyclic amine group by cytochrome P450 (P450) enzymes, to form the genotoxic *N*-hydroxylated-HAA metabolites; these metabolites may react with DNA or undergo further metabolism by phase II enzymes, to produce unstable esters that adduct to DNA. Recent investigations using tandem mass spectrometry have led to the discovery of previously unreported HAAs in grilled meats: these novel HAAs are also detected in human urine following consumption of cooked meat. One novel HAA also has been identified in urine of vegetarians. The contribution of these newly discovered HAAs to the genotoxic burden and health risk of dietary HAAs remains to be determined. Continued improvements in the sensitivity of mass spectrometric instrumentation should allow the measurement of biomarkers in humans, such as HAA-protein and HAA-DNA adducts, that may be used to refine exposure and risk assessments of these genotoxicants.

8.1 Introduction

Humans are continually exposed to various chemicals in food, air, and water that are mutagenic and/or carcinogenic in experimental systems [1–3]. Heterocyclic aromatic amines (HAAs) are one class of hazardous chemicals that have received widening attention as a risk factor for human cancer. These chemicals

* Correspondance to: Robert J. Turesky, Division of Environmental Disease Prevention, Wadsworth Center, New York State, Department of Health, Albany, NY 12201-0509, USA.

** National Center for Toxicological Research, Jefferson, USA.

Thermal Processing of Food: Potential Health Benefits and Risks
DFG, Deutsche Forschungsgemeinschaft

ISBN: 978-3-527-31909-1

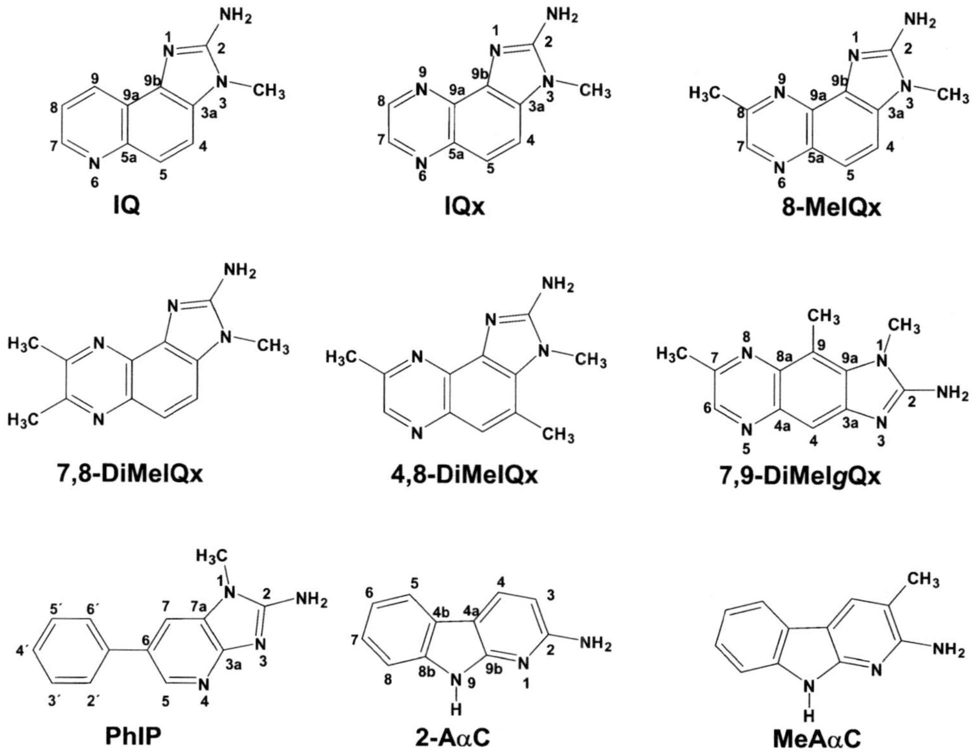

Figure 8.1: Chemical structures of representative HAAs formed in cooked meats. At least 16 HAAs have been identified in cooked meats [3].

were first identified more than 25 years ago by Professor Sugimura and colleagues in Japan [4]. HAAs are formed during the cooking of muscle foods, and more than 16 HAAs have been identified in cooked meats, fish, and poultry (Fig. 8.1) [3, 5–7]. 2-Amino-1-methyl-6-phenylimidazo[4,5-*b*]pyridine (PhIP) [8], 2-amino-9*H*-pyridole[2,3-*b*]indole (2-A*α*C), and 2-amino-3-methyl-9*H*-pyridole [2,3-*b*]indole (MeA*α*C) also occur in tobacco smoke condensate and form in diesel exhaust [9, 10].

HAAs are strong bacterial mutagens [3], and they provoke mutations in mammalian cells, where the potency of the HAAs is markedly influenced by the expression of phase I and II enzymes involved in bioactivation of these compounds [11–13]. Many HAAs are carcinogenic in rodents and induce tumors at multiple sites, including: the liver, stomach, lung, colorectum, and prostate and breast glands [3, 14]. 2-Amino-3-methylimidazo[4,5-*f*]quinoline (IQ) is also a powerful liver carcinogen in non-human primates [15]. Since human tissues are able to bioactivate HAAs into genotoxicants [16–18], there is concern that frequent consumption of grilled meats containing HAAs can contribute to an increased risk of cancer.

8.1.1 Epidemiological Studies and Cancer Risk of HAAs

Epidemiological studies have linked frequent consumption of red meat with increased risk of colorectal cancer [19, 20]. A higher risk for colon cancer has been observed in individuals who frequently eat meat grilled well-done; this meat presumably contains elevated levels of HAAs [21–26]. Moreover, the highest risk for individuals to develop colon cancer is conferred in those subjects who are both rapid cytochrome P450 1A2 *N*-oxidizers and rapid *N*-acetylators; both of these phenotypes are associated with enzymes that bioactivate HAAs. Increased risk for cancer development of the breast [27–30] and esophagus [31] has also been reported to be associated with frequent consumption of meats grilled well-done. These epidemiological observations have led to the hypothesis that HAAs are the causative agents in grilled meats for the induction of such cancers, because HAAs undergo bioactivation by these human tissues [16, 32–35] and induce tumors in analogous sites in experimental animals [3, 36]. However, the molecular epidemiological data on the associations among drug metabolism activity, consumption of cooked meat, and cancer development are inconsistent, and the findings of some other studies have questioned the role of HAAs in cancer risk [37–39]. The elevated cancer risk may be discerned only in individuals who are both rapid metabolizers of HAAs (bioactivation) and who are frequent consumers of foods that contain high levels of HAAs [21, 26, 40]. Cooked meats additionally contain a variety of carcinogens at low concentrations; these include polycyclic aromatic hydrocarbons, *N*-nitroso compounds, lipid peroxides, other pro-oxidative agents, and fungal products. The carcinogenic potency of grilled meats and the associated health risks may be related not only to HAAs, but also to this complex mixture of genotoxic compounds [41, 42]. Biomarkers of exposure and genetic damage such as urinary metabolites [20, 43–46], protein [47], and DNA adducts [48, 49] of these carcinogens may help to identify the critical genotoxicants in grilled meats that are involved in cancer risk.

8.2 HAA Formation in Grilled Meats

8.2.1 Mechanism of Formation of HAAs

The formation of HAAs in cooked meats is dependent upon the type of meat, and the temperature and duration of cooking; these factors result in concentrations of HAAs that vary by more than 100-fold [7, 50, 51]. Prolonged cooking time and high-temperature cooking surfaces produce the highest concentrations of HAAs [50, 52–55]. Two classes of HAAs are formed in cooked meats. HAAs that contain the *N*-methyl-2-aminoimidazole moiety are proposed to form through the reaction of pyridine or pyrazines, which are heat-catalyzed degradation products of amino acids, with sugars and creatine, a key precursor present in muscle-meats, to produce the IQ- and IQx-type compounds [6, 56]. These compounds can form at appreciable concentrations in meats heated to 150 °C or higher temperature, and their formation has been characterized in model systems [56, 57]. The second class of HAAs, which include 2-AαC, MeAαC, and the glutamic acid and tryptophan pyrolysate mutagens, are formed directly from the respective amino acids or proteins heated at high temperature (>250 °C) [4, 58, 59].

The concentrations of HAAs formed in meats prepared under common household cooking practices occur at the low parts per billion (ppb) range level. However, the concentration of HAAs in meats, fish or poultry that has been cooked well-done or charred [50, 51, 58, 60], or the concentration in grilled pan scrapings that are often used for gravy, can be as high as 500 ppb [54, 55, 61]. 8-MeIQx and PhIP are the most abundant of the HAAs formed in grilled beef, bacon, fish, and poultry [7, 50, 54, 55, 61–63]. 2-AαC also forms in appreciable quantities in some meats cooked well-done at elevated temperatures (>250 °C) [58, 64]. IQ and its methylated homologue, 2-amino-3,4-dimethylimidazo[4,5-*f*] quinoline (MeIQ), the most potent among all HAAs tested in bacterial mutagenicity assays and animal carcinogen bioassays [65], are present in broiled salmon in the low ppb range [66] and occur at several hundred ppb in beef extract [67–69]. However, neither compound is formed at an appreciable concentration in grilled ground beef (<0.1 ppb) [5, 70].

8.2.2 Kinetics of HAA Formation in Grilled Meats as a Function of Cooking Time and Temperature; Detection of Novel HAAs

The identification and quantification of HAAs such as 8-MeIQx, PhIP, 4,8-DiMeIQx in grilled meats have been performed by HPLC with UV diode array or fluorescence detection [6, 50, 53, 54, 71]. More recently, HPLC in combination with electrospray ionization tandem mass spectrometry (LC-ESI-MS/MS) has been employed to measure HAAs in cooked meats, and this method of detection provides a greater degree of confidence in the identities of the analytes [7, 55, 64, 72]. We quantified ten known HAAs by LC/ESI-MS/MS and also detected eight previously uncharacterized HAAs in cooked meats [55]. Based upon their product ion mass spectra, we found that these compounds contained imidazoquinoline or imidaquinoxaline skeletons [55]. Two of the novel HAAs are isomers of 2-amino-3-methylimidazo[4,5-*f*]quinoxaline (IQx) and 8-MeIQx, whose chemical structures remain to be determined. UV and mass spectroscopic data suggest that both

Figure 8.2: HAA formation in ground beef patties. (A) Kinetics of HAAs formation in ground beef cooked at 165 °C as a function of time, (B) HAA formation in ground beef cooked for 10 min/side at three different temperatures, (C) HAA formation in grilled scrapings of ground beef cooked at 165 °C for 10 min/side.

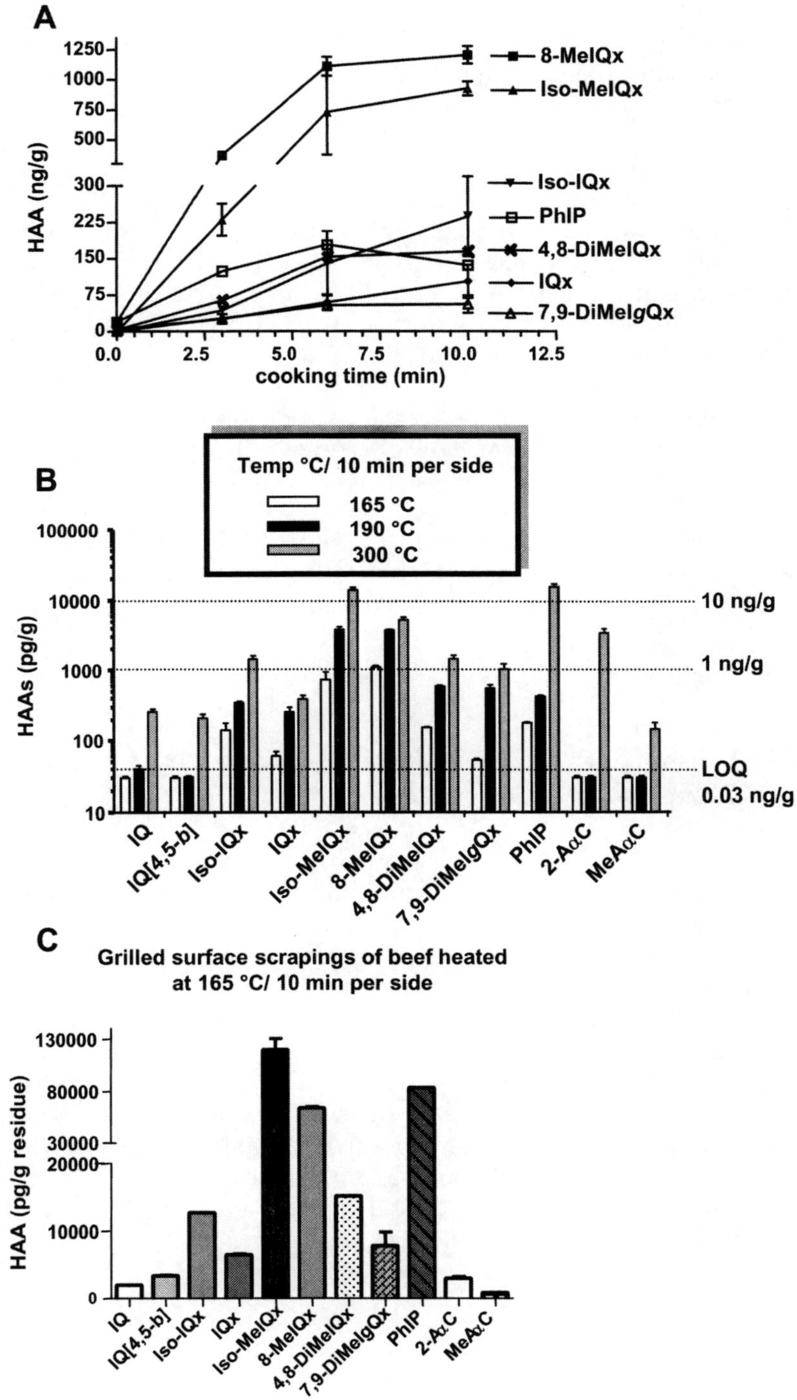
A
HAA (ng/g)
1250
1000
750
500
300
225
150
75
0
0.0
2.5
5.0
7.5
10.0
12.5
cooking time (min)
8-MeIQx
Iso-MeIQx
Iso-IQx
PhIP
4,8-DiMeIQx
IQx
7,9-DiMeIgQx
B
Temp °C/ 10 min per side
165 °C
190 °C
300 °C
HAAs (pg/g)
100000
10000
1000
100
10
10 ng/g
1 ng/g
LOQ
0.03 ng/g
IQ
IQ[4,5-b]
Iso-IQx
IQx
Iso-MeIQx
8-MeIQx
4,8-DiMeIQx
7,9-DiMeIgQx
PhIP
2-AαC
MeAαC
C
Grilled surface scrapings of beef heated
at 165 °C/ 10 min per side
HAA (pg/g residue)
130000
80000
30000
20000
10000
0
IQ
IQ[4,5-b]
Iso-IQx
IQx
Iso-MeIQx
8-MeIQx
4,8-DiMeIQx
7,9-DiMeIgQx
PhIP
2-AαC
MeAαC

compounds are linear tricyclic ring structures containing the 2-aminoimidazo[4,5-*g*]quinoxaline ring skeleton (unpublished data, R. Turesky) [55, 73]. The linear isomer of IQ, 2-amino-1-methylimidazo [4,5-*b*]quinoline (IQ[4,5-*b*]), was also discovered for the first time in grilled meats and pan scrapings [55].

We examined the kinetics of HAA formation as a function of time in ground beef cooked on a cafeteria grill at a temperature of 165 °C by LC/ESI-MS/MS, using stable, isotopically labeled internal standards for quantification [55]. After 3 min of cooking, the concentrations of HAAs were relatively modest, but the amounts increased over time. After 10 min of cooking, 8-MeIQx was the most abundant HAA, followed by the novel isomer of 8-MeIQx (Fig. 8.2), while the amount of PhIP formed was 5- to 10-fold lower. These beef patties were cooked well-done by 10 min, but still palatable; this is the preferred degree of doneness for some consumers. The formation of HAAs in grilled ground beef was also examined as a function of cooking temperature. As previously reported by Felton and co-workers [52, 54, 71, 74], formation of HAAs, PhIP in particular, increased greatly with increasing temperature. It is noteworthy that the newly discovered isomer of 8-MeIQx is also formed at much higher concentrations than many of the other HAAs at all temperatures examined (Fig. 8.2B). This compound is also the most abundant HAA found in grilled pan scrapings of ground beef cooked at 165 °C (Fig. 8.2C).

8.3 Metabolism of HAAs in Experimental Laboratory Animals and Humans

HAAs undergo extensive metabolism in experimental animal models and humans [75–78]; these pathways are depicted in Figure 8.3. Oxidation at the heterocyclic ring and at the methyl and exocyclic amine groups is carried out primarily by P450 1A2, which is mainly expressed in liver [79, 80]. P450 1A2 is the principal enzyme involved in HAA bioactivation, which occurs through N-oxidation. This pathway is also catalyzed by P450s 1A1 and 1B1 in extrahepatic tissues [16, 81–85]. Direct conjugation of the exocyclic amino groups by glucuronosyl transferases (UGTs) and sulfotransferases (SULTs) occurs, to form detoxicated products. *N*-Acetylation, which is an important mechanism of detoxication of primary arylamines [86, 87], is not a prominent pathway of detoxication of HAAs containing the *N*-methyl-aminoimidazole moiety, although NATs do catalyze the detoxication of several pyrolysate HAA mutagens [78].

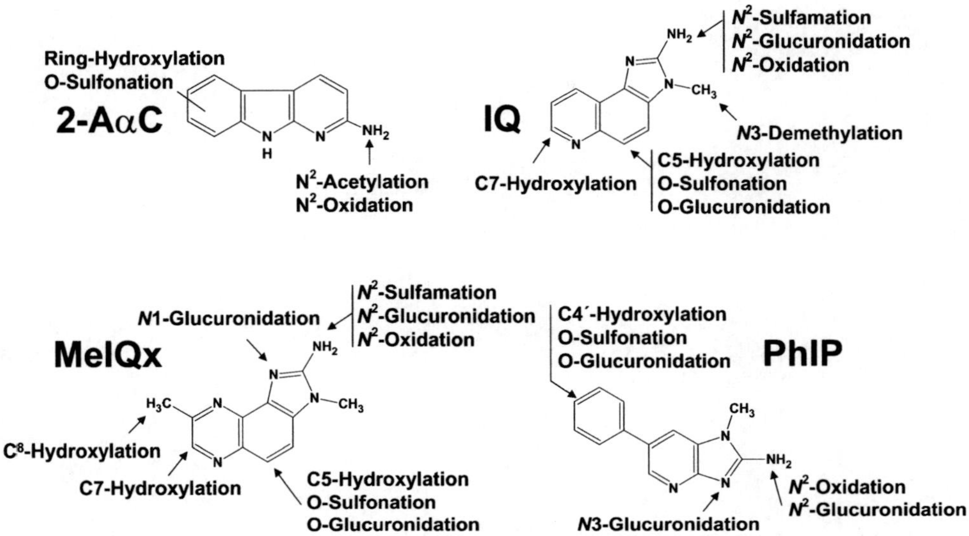

Figure 8.3: Major pathways of HAA metabolism in experimental animals and humans.

8.3.1 Metabolic Activation of HAAs and Pathways of DNA Adduct Formation

The *N*-hydroxy-HAAs metabolites that are produced by P450s may react with DNA or may undergo further transformation by SULTs or *N*-acetyltransferases (NATs) to produce unstable esters that adduct to DNA [88, 89]. The major HAA-DNA adducts occur through the reaction of the activated exocyclic amine groups of the HAAs and the C-8 atom of deoxyguanosine (dG), to produce dG-C8-HAA adducts [88–90]. In the cases of IQ and 8-MeIQx, DNA adducts also form at the N^2 group of dG and the C-5 atoms of the heterocyclic ring structures, indicating charge delocalization of the nitrenium ion over the heteronuclei of these respective HAAs (Fig. 8.4) [91]. Although the quantity of dG-N^2 adducts formed by biomimetic methods is low relative to the quantity of dG-C8 isomers, the dG-N^2 adducts persist in vivo to become the prominent lesions in slowly dividing tissues of rats and non-human primates during chronic exposure to IQ [91–95], and the dG-N^2 adduct of MeIQx also is a prominent lesion in liver of rats treated with this carcinogen [96].

8.3.2 Interspecies Differences in P450 Catalytic Activity and Regioselectivity of HAA Metabolism

There are important differences between human and rat P450s in terms of catalytic activity and regioselectivity of HAA oxidation; these characteristics affect the toxicological properties of the molecules [18, 97]. The catalytic efficiency of

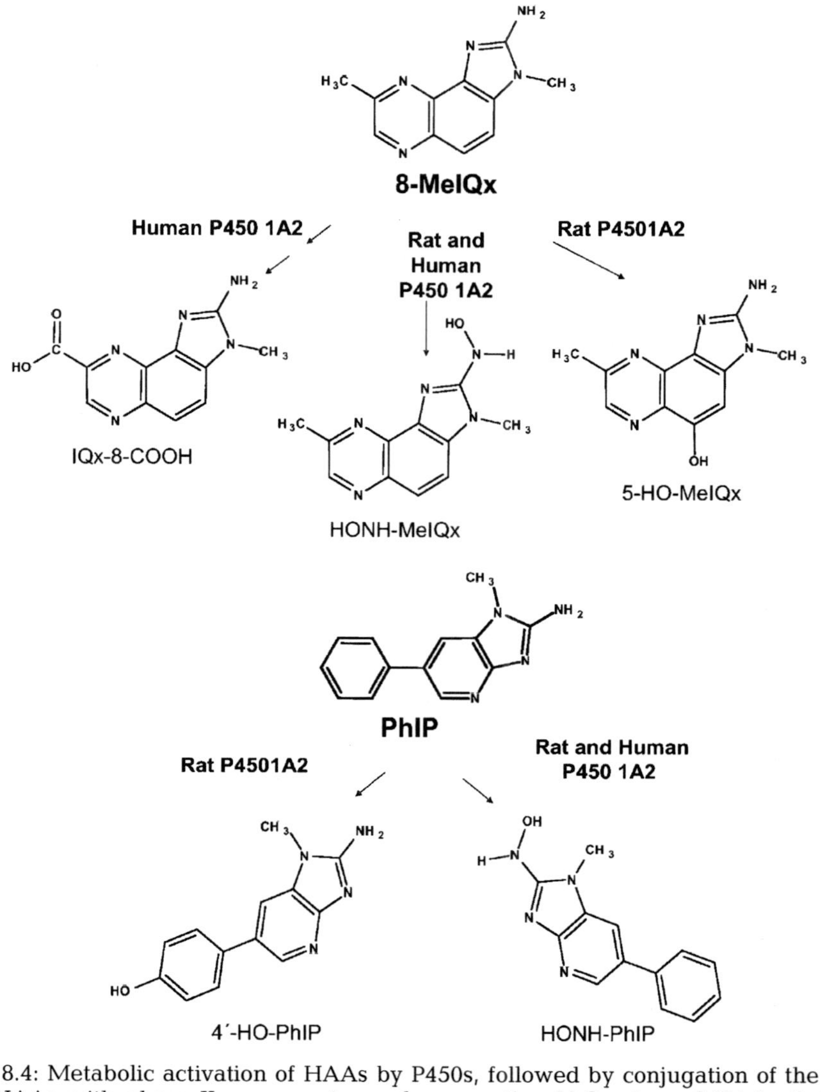

Figure 8.4: Metabolic activation of HAAs by P450s, followed by conjugation of the *N*-hydroxy-HAAs with phase II enzymes to produce reactive *N*-O-esters of HAAs that form isomeric dG-C8-HAA and dG-N^2-HAA adducts.

human P450 1A2 is superior to that of rat P450 1A2, in *N*-oxidation of MeIQx and PhIP. Recombinant human P450 1A2 shows a 13-fold lower K_m for PhIP *N*-oxidation, and a 15-fold higher velocity (nmol product/nmol P450/min) is observed in MeIQx *N*-oxidation than does rat P450 1A2. In contrast, the kinetic parameters for human and rat P450 1A2 are similar for *O*-demethylation of methoxyresorufin [18, 98]. Comparable kinetic parameters have been measured with human and rat liver microsomal samples. Thus, species differences in P450 1A2 activity are highly dependent upon the structure of the substrate.

Important species differences also exist in the regioselectivity of P450 1A2-mediated oxidation of HAAs. For example, human P450 1A2 is unable to efficiently catalyze the detoxication of PhIP through 4′-hydroxylation, while the rat orthologue readily carries out this reaction (Fig. 8.5) [18, 83, 97, 98]. However, human P450 1A2 does catalyze the oxidation of the C-8-methyl group of 8-MeIQx to form the carboxylic acid, 2-amino-3-methylimidazo[4,5-*f*]quinoxaline-8-carboxylic acid (IQx-8-COOH); this is the major pathway of metabolism and detoxication of 8-MeIQx in humans [44, 99]. Rat P450 1A2 does not catalyze IQx-8-COOH formation, but it does catalyze the detoxication of 8-MeIQx through C-5 hydroxylation [44, 99, 100].

Significant species and interindividual differences in the level of expression of P450 1A2 have been reported. The amount of P450 1A2 varies by more than 50-fold among human liver samples from different individuals, and the level of protein expression in human liver is generally at least several-fold and up to 10-fold higher than the expression level in rodent liver [18, 101]. The interspecies variations in catalytic activities, regioselectivities, and levels of expression of P450s must be considered, when human risk assessments of genotoxicants are conducted [102].

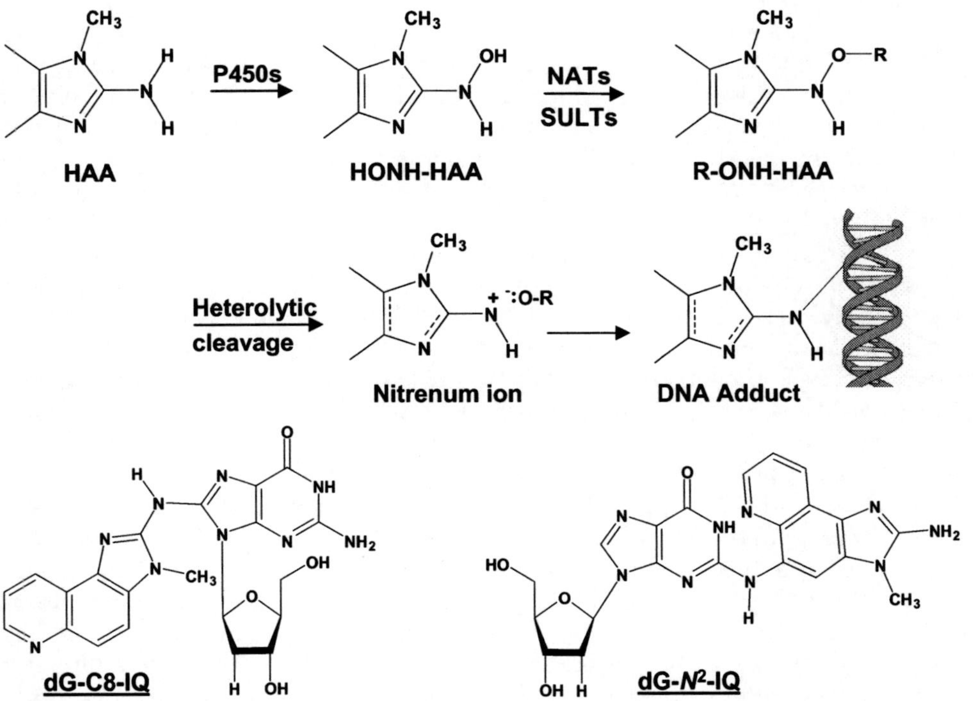

Figure 8.5: Species differences in regioselectivity of P450 1A2 oxidation (bioactivation and detoxication) of 8-MeIQx and PhIP.

8.4 Human Biomonitoring of HAAs

8.4.1 DNA and Protein Adducts of HAAs

Reliable data on the exposure and bioavailability of HAAs in epidemiological studies are required if we are to assess the role of these compounds in human cancers. For many large, population-based studies, the extent of HAA consumption is unknown, and biomarkers must be used to assess exposure and genetic damage [103]. Both DNA and protein adducts of HAAs have been characterized in experimental animal models. HAA-DNA adducts have been detected by ^{32}P-postlabeling, LC/ESI-MS, accelerator mass spectrometry (AMS), and LC/MS techniques in multiple tissues of experimental laboratory animals [89, 90, 96, 104–109]; the dG-C8-HAA adducts are shown to be the predominant lesions. The dG-*N*^{2}-IQ and dG-*N*^{2}-MeIQx adducts are also present, and appear to be more persistent lesions than are the respective dG-C8-HAA adducts [94, 96]. It is noteworthy that HAA-DNA adducts form in tissues of animals that do not develop tumors, indicating that the presence of DNA adducts alone is insufficient for the induction of cancer [89].

There are several reports of 8-MeIQx- and PhIP-DNA adduct formation in human colon, and breast tissue, and white blood cells [104, 110–114]. AMS studies have shown that PhIP and MeIQx form adducts in human breast and colorectal tissues when radiolabeled compounds are administered at dose levels comparable to those found in the daily human diet [110, 111, 114]. A ^{32}P-postlabeling study revealed the presence of the dG-C8-MeIQx adduct in some colon and kidney tissues, at levels of several adducts per 10^9 DNA bases [104]. Alkaline hydrolysis of DNA, followed by derivatization and gas chromatography/electron capture detection to measure PhIP, has been used as an indirect measure of the base-labile dG-C8-PhIP adduct in colorectal mucosa DNA of several individuals [115]. Another study, using alkaline-treated DNA, reported the presence of the putative dG-C8-PhIP adduct in white blood cells of subjects, at levels approaching ~3 adducts per 10^8 DNA bases, by LC/ESI-MS [112]; however, a second study failed to identify PhIP adducts in white blood cells using this approach [45]. DNA adducts of PhIP, presumably the dG-C8-PhIP lesion, were also detected in human breast at levels of ~1 adduct per 10^7 bases based upon immunohistochemical methods [116]. Thus, even at low concentrations, HAAs formed in grilled meats are capable of provoking damage to DNA in humans.

HAA metabolites appear to bind more avidly to serum albumin (SA) than to hemoglobin (Hb) in both rodents and humans [111, 117–121]. The cysteine [34] residue of SA of the rodents has been reported to bind to *N*-oxidized metabolites of IQ [117], 8-MeIQx [119], and PhIP [122], to form sulfinamide or sulfenamide derivatives; these may serve as a potential biomarker of HAAs in humans. One pilot study reported acid-labile SA adducts of PhIP in humans were 10-fold higher in meat-eaters than in vegetarians [123]. These data suggest that PhIP-SA adducts can be used in human biomonitoring studies; however, the structure of the adduct attributed to the acid-labile lesion remains to be elucidated.

8.4.2 Measurement of HAAs in Urine and Identification of Novel HAAs

Urine is a useful biological fluid for the measurement of exposure to various classes of carcinogens since large quantities may be obtained noninvasively [43]. Although urinary measurements of HAAs or their metabolites do not shed light on DNA damage, the measurement of HAAs and their metabolites in urine can assess the capacity of an individual to bioactivate and detoxicate these carcinogens and thereby assess the impact of xenobiotic enzyme polymorphisms on human health risk [124]. HAAs are rapidly absorbed from the gastrointestinal tract and eliminated in urine as multiple metabolites, with several percent of dose present as the unmetabolized parent compounds, within 24 h of consumption of grilled meats [44, 46, 125, 126]. In the case of 8-MeIQx, lower levels of the compound were found in urine of individuals with high P450 1A2 activity, indicating that P450 1A2 is an important enzyme in the metabolism of this HAA in vivo [126, 127]. The relationship among rapid P450 1A2 activity, HAA exposure, and cancer risk remains to be clarified.

HPLC with UV or fluorescence detection has been used to measure MeIQx, PhIP, and tryptophan pyrolysate mutagens in urine of healthy volunteers on a normal diet [128]. The extraction method was tedious and required multiple chromatographic purification steps before the HAAs could be identified. A simpler extraction scheme was developed for the isolation of HAAs from urine, followed by chemical derivatization and gas chromatography/negative ion chemical ionization mass spectrometry (GC/NICI/MS) for the measurement of MeIQx, 4,8-DiMeIQx, and PhIP [125, 129, 130]. Other investigators have used immunoaffinity chromatography, followed by analysis of HAAs with LC-ESI-MS/MS detection [131] or HPLC with fluorescence detection [132] or GC/NICI/MS [133]. Immunoaffinity purification provides clean extracts that facilitate HAA detection and quantification; however, extensive cleanup of the urine is required prior to use of the antibody. Moreover, the analysis is restricted to HAAs that are recognized by the antibody. Recently, we developed a rapid tandem solvent solid phase extraction procedure to isolate a variety of HAAs from urine; we employed LC-ESI-MS/MS for quantification of HAAs in the low ppt range, using less than 1 mL of urine for analysis [73]. LC-ESI-MS/MS chromatograms of urine from an individual who abstained from consuming grilled meat, and of urine of the same individual collected 10 hours following consumption of beef grilled well-done [45, 73], are shown in Figure 8.6.

HAAs were below the limit of detection (<2 ppt) in the control urine sample, with only residual quantities of HAAs detected, due to the impurity of the isotopically labeled internal standards (99% isotopic purity for HAAs, except for [^2H_3C]-8-MeIQx at 96% purity). Following consumption of grilled meat, peaks were readily identified in urine that corresponded to 8-MeIQx, 4,8-DiMeIQx, PhIP, and 2-AαC. Full product ion spectra were acquired in order to corroborate the identities of these analytes, and these spectra were in excellent agreement with those of the synthetic chemicals [73].

A major peak corresponding to the same nominal mass as 8-MeIQx (MH^+ at m/z 214) and with the same transition employed for screening [$MH^+ \rightarrow MH^+$

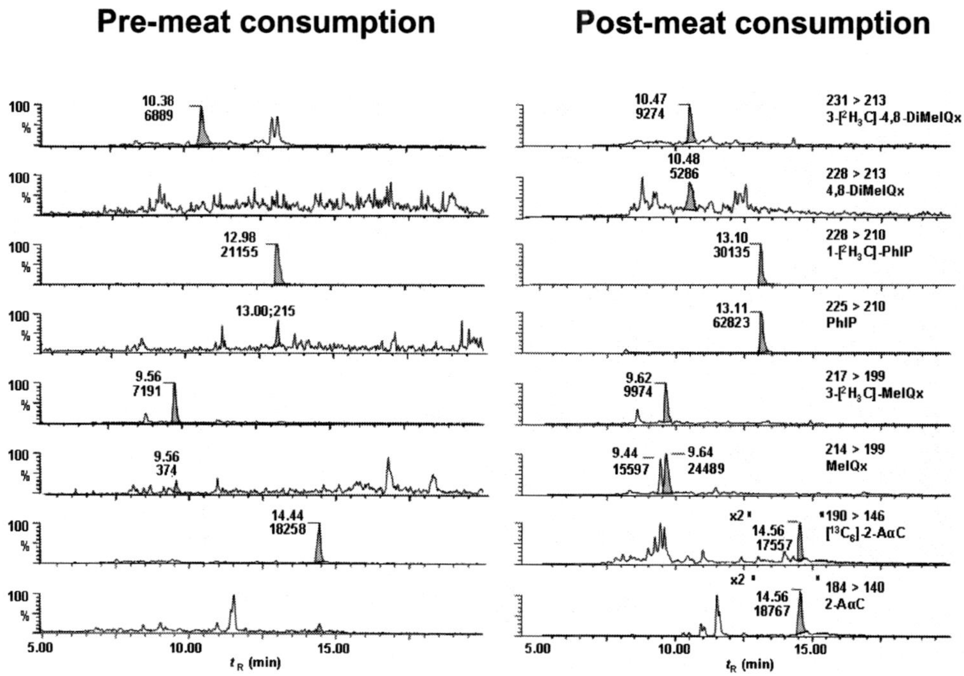

Figure 8.6: LC-ESI-MS/MS analysis of HAAs in urine of a subject before and after consumption of grilled beef. Peaks of internal isotopically labeled internal standards and HAAs are shaded in gray. Experimental details of the study are described in [45, 73].

$-CH_3^{\bullet}]$ was identified in urine of the subject following consumption of grilled meat at t_R 9.44 min. The product ion spectrum of the unknown analyte, along with accurate mass measurements, confirmed that this analyte is indeed an isomer of 8-MeIQx [73]; it was subsequently detected as one of the most abundant HAAs formed in grilled meats and pan scrapings (Fig. 8.2) [55].

Using the constant neutral loss scan mode (CNL), we monitored urine for other putative HAAs that display the transition in MS/MS $[MH^+ \rightarrow MH^+ - CH_3^{\bullet}]$ that is common to HAAs containing the *N*-methyl-aminoimidazole moiety (Fig. 8.7 A). In addition to PhIP, 8-MeIQx, the novel isomer of 8-MeIQx, we detected IQ[4,5-*b*], a linear isomer of the powerful rodent carcinogen IQ, based upon the strong similarities in the full scan product ion spectrum and online UV spectrum of the analyte and synthetic compound (Figs. 8.7 B and C). The amounts of IQ[4,5-*b*] measured in urine of human volunteers who consumed grilled beef ranged from 15 to 135% of the ingested dose, while the amounts of 8-MeIQx and PhIP excreted in urine were on average <2% of the ingested dose [70]. Base treatment of urine at 70 °C increased the concentrations of 8-MeIQx and PhIP by as much as 6-fold, indicating the presence of phase II conjugates; however, the amount of IQ[4,5-*b*] increased by more than 100-fold. Furthermore, IQ[4,5-*b*] was also detected in the urine of vegetarians following

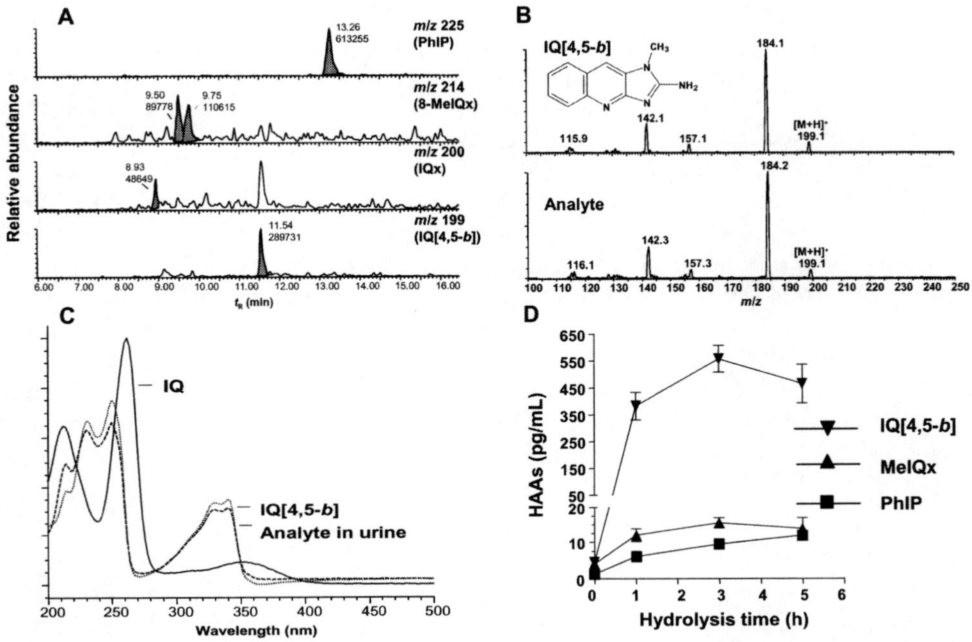

Figure 8.7: Identification of known and novel HAAs in human urine using the constant neutral loss scan mode $[M+H \rightarrow M+H\text{–}CH_3^{\bullet}]^+$. (A) Identification of 8-MeIQx, iso-MeIQx, PhIP, and the linear isomer of IQ, IQ[4,5-*b*], in urine of meat-eaters, (B) Product ion spectra of the analyte and synthetic IQ[4,5-*b*] standard, (C) HPLC-UV spectra of urinary analyte and synthetic standards of IQ[4,5-*b*] and IQ, (D) Kinetics of HAA formation in urine of a meat-eater as a function of base hydrolysis (1 N NaOH) over time at 70 °C. Figures adapted from [70].

base hydrolysis [70]. The formation of IQ[4,5-*b*], but not IQ, 8-MeIQx, or PhIP, also occurred in urine incubated at 37 °C. We proposed that creatinine and 2-aminobenzaldehyde are likely precursors of IQ[4,5-*b*] [70]. The detection of IQ[4,5-*b*] in urine of both meat eaters and vegetarians suggests either that this HAA is present in nonmeat food, or that IQ[4,5-*b*] formation occurs endogenously in the urinary bladder or other biological fluids.

8.5 Conclusions and Future Research

Recent advances in the sensitivity of MS of instrumentation have enabled the detection of HAAs and their metabolites at trace levels in urine of humans [44, 46, 70, 73, 125, 130]. The findings have shown both similarities, as well as important differences between humans and animal species in the enzymology of HAA metabolism, which influence the toxicological properties of these genotoxicants.

The identification of novel HAAs, including linear isomers of IQ, putative linear isomers of IQx, and 8-MeIQx, and the detection of additional previously uncharacterized HAAs in grilled meats [55, 70, 73], reveal that humans are exposed to a number of HAAs whose toxicological properties are unknown. The chemical characterization of these newly discovered HAAs, the quantification of their concentrations in cooked meats, and the biochemical toxicology of these novel HAAs merit investigation. A recent study reported that elevated urinary mutagenicity ascribed to HAAs is positively associated with an increase in colorectal adenoma risk [24], thereby further reinforcing the hypothesis that HAAs contribute to human cancer. Preliminary analytical data of a pilot study of human subjects suggest that SA protein adducts and white blood cell DNA adducts of PhIP are useful biomarkers [111, 112, 123] and may be used to assess the interactive effects of HAA exposure, DNA damage, genetic polymorphisms in xenobiotic metabolism enzymes, and cancer risk. Further studies on biomarkers of PhIP, along with the development of biomarkers to monitor other HAAs may refine the human risk assessment of these genotoxicants in the etiology of dietary cancers.

References

1. Sugimura, T. (1992) Multistep carcinogenesis: A 1992 perspective. *Science*, **258**, 603–607.
2. Wogan, G.N., Hecht, S.S., Felton, J.S., Conney, A.H., and Loeb, L.A. (2004) Environmental and chemical carcinogenesis. *Semin. Cancer Biol.*, **14**, 473–486.
3. Sugimura, T., Wakabayashi, K., Nakagama, H., and Nagao, M. (2004) Heterocyclic amines: Mutagens/carcinogens produced during cooking of meat and fish. *Cancer Sci.*, **95**, 290–299.
4. Sugimura, T., Nagao, N., Kawachi, T., Honda, M., Yahagi, T., Seino, Y., Stao, S., Matsukura, N., Matsushima, T., Shirai, A., Sawamura, M., and Matsumoto, H. (1977) Mutagen-carcinogens in food, with special reference to highly mutagenic pyrolytic products in broiled foods. In Hiatt, H.H., Watson, J.D., and Winstein, J.A. (eds.) *Origins of Human Cancer, Book C.* Cold Spring Harbor Laboratory, Cold Spring Harbor, NY, pp 1561–1577.
5. Felton, J.S., Jagerstad, M., Knize, M.G., Skog, K., and Wakabayashi, K. (2000) Contents in Foods, Beverages and Tobacco. In Nagao, M. and Sugimura, T. (eds.) *Food Borne Carcinogens Heterocyclic Amines.* John Wiley & Sons Ltd., Chichester, England, pp 31–71.
6. Skog, K.I., Johansson, M.A., and Jagerstad, M.I. (1998) Carcinogenic heterocyclic amines in model systems and cooked foods: a review on formation, occurrence and intake. *Food Chem. Toxicol.*, **36**, 879–896.
7. Guy, P.A., Gremaud, E., Richoz, J., and Turesky, R.J. (2000) Quantitative analysis of mutagenic heterocyclic aromatic amines in cooked meat using liquid chromatography-atmospheric pressure chemical ionisation tandem mass spectrometry. *J. Chromatogr. A*, **883**, 89–102.
8. Manabe, S., Tohyama, K., Wada, O., and Aramaki, T. (1991) Detection of a carcinogen, 2-amino-1-methyl-6-phenylimidazo[4,5-*b*]pyridine, in cigarette smoke condensate. *Carcinogenesis*, **12**, 1945–1947.
9. Manabe, S., Wada, O., and Kanai, Y. (1990) Simultaneous determination of amino-alpha-carbolines and amino-gamma-carbolines in cigarette smoke condensate by high-performance liquid chromatography. *J. Chromatogr.*, **529**, 125–133.
10. Manabe, S., Izumikawa, S., Asakuno, K., Wada, O., and Kanai, Y. (1991) Detection of carcinogenic amino-alpha-carbolines and amino-gamma-carbolines in diesel-exhaust particles. *Environ. Pollut.*, **70**, 255–265.

11. Wu, R.W., Panteleakos, F.N., and Felton, J.S. (2003) Development and characterization of CHO repair-proficient cell lines for comparative mutagenicity and metabolism of heterocyclic amines from cooked food. *Environ. Mol. Mutagen.*, **41**, 7–13.
12. Muckel, E., Frandsen, H., and Glatt, H.R. (2002) Heterologous expression of human N-acetyltransferases 1 and 2 and sulfotransferase 1A1 in Salmonella typhimurium for mutagenicity testing of heterocyclic amines. *Food Chem. Toxicol.*, **40**, 1063–1068.
13. Glatt, H., Pabel, U., Meinl, W., Frederiksen, H., Frandsen, H., and Muckel, E. (2004) Bioactivation of the heterocyclic aromatic amine 2-amino-3-methyl-9H-pyrido[2,3-b] indole (MeAalphaC) in recombinant test systems expressing human xenobiotic-metabolizing enzymes. *Carcinogenesis*, **25**, 801–807.
14. Shirai, T., Sano, M., Tamano, S., Takahashi, S., Hirose, M., Futakuchi, M., Hasegawa, R., Imaida, K., Matsumoto, K., Wakabayashi, K., Sugimura, T., and Ito, N. (1997) The prostate: A target for carcinogenicity of 2-amino-1-methyl-6-phenylimidazo[4,5-*b*]pyridine (PhIP) derived from cooked foods. *Cancer Res.*, **57**, 195–198.
15. Adamson, R.H., Thorgeirsson, U.P., Snyderwine, E.G., Reeves, J., Dalgard, D.W., Takayama, S., and Sugimura, T. (1990) Carcinogenicity of 2-amino-3-methylimidazo [4,5-*f*]quinoline in nonhuman primates: induction of tumors in three macaques. *Jpn. J. Cancer Res.*, **81**, 10–14.
16. Turesky, R.J., Lang, N.P., Butler, M.A., Teitel, C.H., and Kadlubar, F.F. (1991) Metabolic activation of carcinogenic heterocyclic aromatic amines by human liver and colon. *Carcinogenesis*, **12**, 1839–1845.
17. Rich, K.J., Murray, B.P., Lewis, I., Rendell, N.B., Davies, D.S., Gooderham, N.J., and Boobis, A.R. (1992) *N*-Hydroxy-MeIQx is the major microsomal oxidation product of the dietary carcinogen MeIQx with human liver. *Carcinogenesis*, **13**, 2221–2226.
18. Turesky, R.J., Constable, A., Richoz, J., Varga, N., Markovic, J., Martin, M.V., and Guengerich, F.P. (1998) Activation of heterocyclic aromatic amines by rat and human liver microsomes and by purified rat and human cytochrome P450 1A2. *Chem. Res. Toxicol.*, **11**, 925–936.
19. Willett, W.C., Stampfer, M.J., Colditz, G.A., Rosner, B.A., and Speizer, F.E. (1990) Relation of meat, fat, and fiber intake to the risk of colon cancer in a prospective study among women. *N. Engl. J. Med.*, **323**, 1664–1672.
20. Giovannucci, E., Rimm, E.B., Stampfer, M.J., Colditz, G.A., Ascherio, A., and Willett, W.C. (1994) Intake of fat, meat, and fiber in relation to risk of colon cancer in men. *Cancer Res.*, **54**, 2390–2397.
21. Lang, N.P., Butler, M.A., Massengill, J.P., Lawson, M., Stotts, R.C., Hauer-Jensen, M., and Kadlubar, F.F. (1994) Rapid metabolic phenotypes for acetyltransferase and cytochrome P4501A2 and putative exposure to food-borne heterocyclic amines increase the risk for colorectal cancer or polyps. *Cancer Epidemiol. Biomarkers Prev.*, **3**, 675–682.
22. Sinha, R., Chow, W.H., Kulldorff, M., Denobile, J., Butler, J., Garcia-Closas, M., Weil, R., Hoover, R.N., and Rothman, N. (1999) Well-done, grilled red meat increases the risk of colorectal adenomas. *Cancer Res.*, **59**, 4320–4324.
23. Sinha, R., Kulldorff, M., Chow, W.H., Denobile, J., and Rothman, N. (2001) Dietary intake of heterocyclic amines, meat-derived mutagenic activity, and risk of colorectal adenomas. *Cancer Epidemiol. Biomarkers Prev.*, **10**, 559–562.
24. Peters, U., DeMarini, D.M., Sinha, R., Brooks, L.R., Warren, S.H., Chatterjee, N., and Rothman, N. (2003) Urinary mutagenicity and colorectal adenoma risk. *Cancer Epidemiol. Biomarkers Prev.*, **12**, 1253–1256.
25. Sinha, R., Rothman, N., Brown, E.D., Mark, S.D., Hoover, R.N., Caporaso, N.E., Levander, O.A., Knize, M.G., Lang, N.P., and Kadlubar, F.F. (1994) Pan-fried meat containing high levels of heterocyclic aromatic amines but low levels of polycyclic aromatic hydrocarbons induces cytochrome P4501A2 activity in humans. *Cancer Res.*, **54**, 6154–6159.
26. Le Marchand, L., Hankin, J.H., Pierce, L.M., Sinha, R., Nerurkar, P.V., Franke, A.A., Wilkens, L.R., Kolonel, L.N., Donlon, T., Seifried, A., Custer, L.J., Lum-Jones, A.,

and Chang, W. (2002) Well-done red meat, metabolic phenotypes and colorectal cancer in Hawaii. *Mutat. Res.*, **506/507**, 205–214.
27. Dai, Q., Shu, X.O., Jin, F., Gao, Y.T., Ruan, Z.X., and Zheng, W. (2002) Consumption of animal foods, cooking methods, and risk of breast cancer. *Cancer Epidemiol. Biomarkers Prev.*, **11**, 801–808.
28. Sinha, R., Gustafson, D.R., Kulldorff, M., Wen, W.Q., Cerhan, J.R., and Zheng, W. (2000) 2-Amino-1-methyl-6-phenylimidazo[4,5-b]pyridine, a carcinogen in high-temperature-cooked meat, and breast cancer risk. *J. Natl. Cancer Inst.*, **92**, 1352–1354.
29. Zheng, W., Gustafson, D.R., Sinha, R., Cerhan, J.R., Moore, D., Hong, C.P., Anderson, K.E., Kushi, L.H., Sellers, T.A., and Folsom, A.R. (1998) Well-done meat intake and the risk of breast cancer. *J. Natl. Cancer Inst.*, **90**, 1724–1729.
30. Zheng, W., Xie, D., Cerhan, J.R., Sellers, T.A., Wen, W., and Folsom, A.R. (2001) Sulfotransferase 1A1 polymorphism, endogenous estrogen exposure, well-done meat intake, and breast cancer risk. *Cancer Epidemiol. Biomarkers Prev.*, **10**, 89–94.
31. Terry, P.D., Lagergren, J., Wolk, A., Steineck, G., and Nyren, O. (2003) Dietary intake of heterocyclic amines and cancers of the esophagus and gastric cardia. *Cancer Epidemiol. Biomarkers Prev.*, **12**, 940–944.
32. Sadrieh, N., Davis, C.D., and Snyderwine, E.G. (1996) *N*-Acetyltransferase expression and metabolic activation of the food-derived heterocyclic amines in the human mammary gland. *Cancer Res.*, **56**, 2683–2687.
33. Williams, J.A., Martin, F.L., Muir, G.H., Hewer, A., Grover, P.L., and Phillips, D.H. (2000) Metabolic activation of carcinogens and expression of various cytochromes P450 in human prostate tissue. *Carcinogenesis*, **21**, 1683–1689.
34. Williams, J.A. and Phillips, D.H. (2000) Mammary expression of xenobiotic metabolizing enzymes and their potential role in breast cancer. *Cancer Res.*, **60**, 4667–4677.
35. Williams, J.A., Stone, E.M., Millar, B.C., Hewer, A., and Phillips, D.H. (2000) Pathways of heterocyclic amine activation in the breast: DNA adducts of 2-amino-3-methylimidazo[4,5-f]quinoline (IQ) formed by peroxidases and in human mammary epithelial cells and fibroblasts. *Mutagenesis*, **15**, 149–154.
36. Sugimura, T. (1997) Overview of carcinogenic heterocyclic amines. *Mutat. Res.*, **376**, 211–219.
37. Sachse, C., Smith, G., Wilkie, M.J., Barrett, J.H., Waxman, R., Sullivan, F., Forman, D., Bishop, D.T., and Wolf, C.R. (2002) A pharmacogenetic study to investigate the role of dietary carcinogens in the etiology of colorectal cancer. *Carcinogenesis*, **23**, 1839–1849.
38. Sachse, C., Bhambra, U., Smith, G., Lightfoot, T.J., Barrett, J.H., Scollay, J., Garner, R.C., Boobis, A.R., Wolf, C.R., and Gooderham, N.J. (2003) Polymorphisms in the cytochrome P450 CYP1A2 gene (CYP1A2) in colorectal cancer patients and controls: allele frequencies, linkage disequilibrium and influence on caffeine metabolism. *Br. J. Clin. Pharmacol.*, **55**, 68–76.
39. Barrett, J.H., Smith, G., Waxman, R., Gooderham, N., Lightfoot, T., Garner, R.C., Augustsson, K., Wolf, C.R., Bishop, D.T., and Forman, D. (2003) Investigation of interaction between N-acetyltransferase 2 and heterocyclic amines as potential risk factors for colorectal cancer. *Carcinogenesis*, **24**, 275–282.
40. Le Marchand, L., Hankin, J.H., Wilkens, L.R., Pierce, L.M., Franke, A., Kolonel, L. N., Seifried, A., Custer, L.J., Chang, W., Lum-Jones, A., and Donlon, T. (2001) Combined effects of well-done red meat, smoking, and rapid N-acetyltransferase 2 and CYP1A2 phenotypes in increasing colorectal cancer risk. *Cancer Epidemiol. Biomarkers Prev.*, **10**, 1259–1266.
41. Cross, A.J. and Sinha, R. (2004) Meat-related mutagens/carcinogens in the etiology of colorectal cancer. *Environ. Mol. Mutagen.*, **44**, 44–55.
42. DeBruin, L.S. and Josephy, P.D. (2002) Perspectives on the chemical etiology of breast cancer. *Environ. Health Perspect.*, **110** Suppl 1, 119–128.
43. Hecht, S.S. (2002) Human urinary carcinogen metabolites: biomarkers for investigating tobacco and cancer. *Carcinogenesis*, **23**, 907–922.

44. Turesky, R. J., Garner, R. C., Welti, D. H., Richoz, J., Leveson, S. H., Dingley, K. H., Turteltaub, K. W., and Fay, L. B. (1998) Metabolism of the food-borne mutagen 2-amino-3,8-dimethylimidazo[4,5-*f*]quinoxaline in humans. *Chem. Res. Toxicol.*, **11**, 217–225.
45. Murray, S., Lake, B. G., Gray, S., Edwards, A. J., Springall, C., Bowey, E. A., Williamson, G., Boobis, A. R., and Gooderham, N. J. (2001) Effect of cruciferous vegetable consumption on heterocyclic aromatic amine metabolism in man. *Carcinogenesis*, **22**, 1413–1420.
46. Kulp, K. S., Knize, M. G., Fowler, N. D., Salmon, C. P., and Felton, J. S. (2004) PhIP metabolites in human urine after consumption of well-cooked chicken. *J. Chromatogr. B Analyt. Technol. Biomed. Life Sci.*, **802**, 143–153.
47. Skipper, P. L. and Tannenbaum, S. R. (1990) Protein adducts in the molecular dosimetry of chemical carcinogens. *Carcinogenesis*, **11**, 507–518.
48. Phillips, D. H. (1996) DNA adducts in human tissues: biomarkers of exposure to carcinogens in tobacco smoke. *Environ. Health Perspect.*, **104** Suppl 3, 453–458.
49. Farmer, P. B. (1999) Studies using specific biomarkers for human exposure assessment to exogenous and endogenous chemical agents. *Mutat. Res.*, **428**, 69–81.
50. Sinha, R., Rothman, N., Brown, E. D., Salmon, C. P., Knize, M. G., Swanson, C. S., Rossi, S. C., Mark, S. D., Levander, O. A., and Felton, J. S. (1995) High concentrations of the carcinogen 2-amino-1-methyl-6-phenylimidazo[4,5-*b*]pyridine (PhIP) occur in chicken but are dependent on the cooking method. *Cancer Res.*, **55**, 4516–4519.
51. Felton, J. S., Knize, M. G., Salmon, C. P., Malfatti, M. A., and Kulp, K. S. (2002) Human exposure to heterocyclic amine food mutagens/carcinogens: relevance to breast cancer. *Environ. Mol. Mutagen.*, **39**, 112–118.
52. Felton, J. S., Knize, M. G., Dolbeare, F. A., and Wu, R. (1994) Mutagenic activity of heterocyclic amines in cooked foods. *Environ. Health Perspect.*, **102** Suppl 6, 201–204.
53. Skog, K., Steineck, G., Augustsson, K., and Jagerstad, M. (1995) Effect of cooking temperature on the formation of heterocyclic amines in fried meat products and pan residues. *Carcinogenesis*, **16**, 861–867.
54. Sinha, R., Rothman, N., Salmon, C. P., Knize, M. G., Brown, E. D., Swanson, C. A., Rhodes, D., Rossi, S., Felton, J. S., and Levander, O. A. (1998) Heterocyclic amine content in beef cooked by different methods to varying degrees of doneness and gravy made from meat drippings. *Food Chem. Toxicol.*, **36**, 279–287.
55. Turesky, R. J., Taylor, J., Schnackenberg, L., Freeman, J. P., and Holland, R. D. (2005) Quantitation of carcinogenic heterocyclic aromatic amines and detection of novel heterocyclic aromatic amines in cooked meats and grill scrapings by HPLC/ESI-MS. *J. Agric. Food Chem.*, **53**, 3248–3258.
56. Jagerstad, M., Skog, K., Grivas, S., and Olsson, K. (1991) Formation of heterocyclic amines using model systems. *Mutat. Res.*, **259**, 219–233.
57. Skog, K. and Jagerstad, M. (1993) Incorporation of carbon atoms from glucose into the food mutagens MeIQx and 4,8-DiMeIQx using 14C-labeled glucose in a model system. *Carcinogenesis*, **14**, 2027–2031.
58. Matsumoto, T., Yoshida, D., and Tomita, H. (1981) Determination of mutagens, amino-alpha-carbolines in grilled foods and cigarette smoke condensate. *Cancer Lett.*, **12**, 105–110.
59. Sugimura, T. and Sato, S. (1983) Mutagens-carcinogens in foods. *Cancer Res.*, **43**, 2415s–2421s.
60. Gross, G. A. and Gruter, A. (1992) Quantitation of mutagenic/carcinogenic heterocyclic aromatic amines in food products. *J. Chromatogr.*, **592**, 271–278.
61. Gross, G. A., Turesky, R. J., Fay, L. B., Stillwell, W. G., Skipper, P. L., and Tannenbaum, S. R. (1993) Heterocyclic aromatic amine formation in grilled bacon, beef and fish and in grill scrapings. *Carcinogenesis*, **14**, 2313–2318.
62. Skog, K., Solyakov, A., Arvidsson, P., and Jagerstad, M. (1998) Analysis of nonpolar heterocyclic amines in cooked foods and meat extracts using gas chromatography-mass spectrometry. *J. Chromatogr. A*, **803**, 227–233.
63. Skog, K. and Solyakov, A. (2002) Heterocyclic amines in poultry products: a literature review. *Food Chem. Toxicol.*, **40**, 1213–1221.

64. Holder, C.L., Preece, S.W., Conway, S.C., Pu, Y.M., and Doerge, D.R. (1997) Quantification of heterocyclic amine carcinogens in cooked meats using isotope dilution liquid chromatography/atmospheric pressure chemical ionization tandem mass spectrometry. *Rapid Commun. Mass Spectrom.*, **11**, 1667–1672.
65. Sugimura, T. (1988) Successful use of short-term tests for academic purposes: Their use in identification of new environmental carcinogens with possible risk for humans. *Mutat. Res.*, **205**, 33–39.
66. Yamaizumi, Z., Kasai, H., Nishimura, S., Edmonds, C.G., and McCloskey, J.A. (1986) Stable isotope dilution quantification of mutagens in cooked foods by combined liquid chromatography-thermospray mass spectrometry. *Mutat. Res.*, **173**, 1–7.
67. Spingarn, N.E., Kasai, H., Vuolo, L.L., Nishimura, S., Yamaizumi, Z., Sugimura, T., Matsushima, T., and Weisburger, J.H. (1980) Formation of mutagens in cooked foods. III. Isolation of a potent mutagen from beef. *Cancer Lett.*, **9**, 177–183.
68. Turesky, R.J., Wishnok, J.S., Tannenbaum, S.R., Pfund, R.A., and Buchi, G.H. (1983) Qualitative and quantitative characterization of mutagens in commercial beef extract. *Carcinogenesis*, **4**, 863–866.
69. Hargraves, W.A. and Pariza, M.W. (1983) Purification and mass spectral characterization of bacterial mutagens from commercial beef extract. *Cancer Res.*, **43**, 1467–1472.
70. Holland, R.D., Gehring, T., Taylor, J., Lake, B.G., Gooderham, N.J., and Turesky, R.J. (2005) Formation of a mutagenic heterocyclic aromatic amine from creatinine in urine of meat eaters and vegetarians. *Chem. Res. Toxicol.*, **18**, 579–590.
71. Knize, M.G., Dolbeare, F.A., Carroll, K.L., Moore, D.H., and Felton, J.S. (1994) Effect of cooking time and temperature on the heterocyclic amine content of fried beef patties. *Food Chem. Toxicol.*, **32**, 595–603.
72. Busquets, R., Bordas, M., Toribio, F., Puignou, L., and Galceran, M.T. (2004) Occurrence of heterocyclic amines in several home-cooked meat dishes of the Spanish diet. *J. Chromatogr. B*, **802**, 79–86.
73. Holland, R.D., Taylor, J., Schoenbachler, L., Jones, R.C., Freeman, J.P., Miller, D.W., Lake, B.G., Gooderham, N.J., and Turesky, R.J. (2004) Rapid biomonitoring of heterocyclic aromatic amines in human urine by tandem solvent solid phase extraction liquid chromatography electrospray ionization mass spectrometry. *Chem. Res. Toxicol.*, **17**, 1121–1136.
74. Felton, J.S. and Knize, M.G. (1990) Heterocyclic amine mutagens/carcinogens in foods. In Cooper, C.S. and Grover, P.L. (eds.) *Handbook of Experimental Pharmacology.* Springer, Berlin Heidelberg, vol. 94/I, pp 471–502.
75. Yamazoe, Y. and Nagata, K. (2000) In vitro metabolism. In Sugimura, T. and Nagao, M. (eds.) *Food Borne Carcinogens Heterocyclic Amines.* John Wiley & Sons Ltd., Chichester, England, pp 74–89.
76. Alexander, J., Heidenreich, B., Reistad, R., and Holme, J.A. (1995) Metabolism of the food carcinogen 2-amino-1-methyl-6-phenylimidazo[4,5-*b*]pyridine (PhIP) in the rat and other rodents. In Adamson, R.H., Gustafsson, J.-A., Ito, N., Nagao, M., Sugimura, T., Wakabayashi, K., and Yamazoe, Y. (eds.) *Heterocyclic amines in cooked foods: Possible human carcinogens. 23rd Proceedings of the Princess Takamatusu Cancer Society.* Princeton Scientific Publishing Co., Inc., New Jersey, pp 59–68.
77. Snyderwine, E.G., Turesky, R.J., Turteltaub, K.W., Davis, C.D., Sadrieh, N., Schut, H.A., Nagao, M., Sugimura, T., Thorgeirsson, U.P., Adamson, R.H., and Thorgeirsson, S.S. (1997) Metabolism of food-derived heterocyclic amines in nonhuman primates. *Mutat. Res.*, **376**, 203–210.
78. King, R.S., Kadlubar, F.F., and Turesky, R.J. (2000) In vivo Metabolism of Heterocyclic Amines. In Nagao, M. and Sugimura, T. (eds.) *Food Borne Carcinogens: Heterocyclic Amines.* John Wiley & Sons, Ltd., Chichester, England, pp 90–111.
79. Shimada, T., Iwasaki, M., Martin, M.V., and Guengerich, F.P. (1989) Human liver microsomal cytochrome P-450 enzymes involved in the bioactivation of procarcinogens detected by *umu* gene response in *Salmonella typhimurium* TA 1535/pSK1002. *Cancer Res.*, **49**, 3218–3228.

80. Butler, M. A., Iwasaki, M., Guengerich, F. P., and Kadlubar, F. F. (1989) Human cytochrome P-450$_{PA}$ (P450IA2), the phenacetin O-deethylase, is primarily responsible for the hepatic 3-demethylation of caffeine and N-oxidation of carcinogenic arylamines. *Proc. Natl. Acad. Sci. USA*, **86**, 7696–7700.
81. Yamazoe, Y., Abu-Zeid, M., Manabe, S., Toyama, S., and Kato, R. (1988) Metabolic activation of a protein pyrolysate promutagen 2-amino-3,8-dimethylimidazo[4,5-*f*]quinoxaline by rat liver microsomes and purified cytochrome P-450. *Carcinogenesis*, **9**, 105–109.
82. Shimada, T. and Guengerich, F. P. (1991) Activation of amino-*α*-carboline, 2-amino-1-methyl-6-phenylimidazo[4,5-*b*]pyridine, and a copper phthalocyanine cellulose extract of cigarette smoke condensate by cytochrome P-450 enzymes in rat and human liver microsomes. *Cancer Res.*, **51**, 5284–5291.
83. Wallin, H., Mikalsen, A., Guengerich, F. P., Ingelman-Sundberg, I., Solberg, K. E., Rossland, O. J., and Alexander, J. (1990) Differential rates of metabolic activation and detoxification of the food mutagen 2-amino-1-methyl-6-phenylimidazo[4,5-*b*]pyridine by different cytochrome P450 enzymes. *Carcinogenesis*, **11**, 489–492.
84. Crofts, F. G., Strickland, P. T., Hayes, C. L., and Sutter, T. R. (1997) Metabolism of 2-amino-1-methyl-6-phenylimidazo[4,5-*b*]pyridine (PhIP) by human cytochrome P4501B1. *Carcinogenesis*, **18**, 1793–1798.
85. Crofts, F. G., Sutter, T. R., and Strickland, P. T. (1998) Metabolism of 2-amino-1-methyl-6-phenylimidazo[4,5-b]pyridine by human cytochrome P4501A1, P4501A2 and P4501B1. *Carcinogenesis*, **19**, 1969–1973.
86. Kadlubar, F. F. and Beland, F. A. (1985) Chemical properties of ultimate carcinogenic metabolites of arylamines and arylamides. In Harvey, R. G. (ed.) *Polycyclic Hydrocarbons and Carcinogenesis*. American Chemical Society, Washington, D.C., vol. 283, pp 332–370.
87. Hein, D. W. (2002) Molecular genetics and function of NAT1 and NAT2: role in aromatic amine metabolism and carcinogenesis. *Mutat. Res.*, **506/507**, 65–77.
88. Turesky, R. J. (1994) DNA adducts of heterocyclic aromatic amines, arylazides and 4-nitroquinoline 1-oxide. In Hemminki, K., Dipple, A., Shuker, D. E. G., Kadlubar, F. F., Segerbäck, D., and Bartsch, H. (eds.) *DNA Adducts: Identification and Biological Significance*. International Agency for Research on Cancer, Lyon, pp 217–228.
89. Schut, H. A. and Snyderwine, E. G. (1999) DNA adducts of heterocyclic amine food mutagens: implications for mutagenesis and carcinogenesis. *Carcinogenesis*, **20**, 353–368.
90. Turesky, R. J. and Vouros, P. (2004) Formation and analysis of heterocyclic aromatic amine-DNA adducts in vitro and in vivo. *J. Chromatogr. B*, **802**, 155–166.
91. Turesky, R. J., Rossi, S. C., Welti, D. H., O'Lay, Jr. J., and Kadlubar, F. F. (1992) Characterization of DNA adducts formed in vitro by reaction of *N*-hydroxy-2-amino-3-methylimidazo[4,5-*f*]quinoline and *N*-hydroxy-2-amino-3,8-dimethylimidazo[4,5-*f*]quinoxaline at the C-8 and N_2 atoms of guanine. *Chem. Res. Toxicol.*, **5**, 479–490.
92. Turesky, R. J., Gremaud, E., Markovic, J., and Snyderwine, E. G. (1996) DNA adduct formation of the food-derived mutagen 2-amino-3-methylimidazo[4,5-*f*]quinoline in nonhuman primates undergoing carcinogen bioassay. *Chem. Res. Toxicol.*, **9**, 403–408.
93. Turesky, R. J., Markovic, J., and Aeschlimann, J. M. (1996) Formation and differential removal of C-8 and N_2-guanine adducts of the food carcinogen 2-amino-3-methylimidazo[4,5-*f*]quinoline in the liver, kidney, and colorectum of the rat. *Chem. Res. Toxicol.*, **9**, 397–402.
94. Turesky, R. J., Box, R. M., Markovic, J., Gremaud, E., and Snyderwine, E. G. (1997) Formation and persistence of DNA adducts of 2-amino-3-methylimidazo[4,5-*f*]quinoline in the rat and nonhuman primates. *Mutat. Res.*, **376**, 235–241.
95. Ochiai, M., Nakagama, H., Turesky, R. J., Sugimura, T., and Nagao, M. (1999) A new modification of the ^{32}P-post-labeling method to recover IQ-DNA adducts as mononucleotides. *Mutagenesis*, **14**, 239–242.
96. Paehler, A., Richoz, J., Soglia, J., Vouros, P., and Turesky, R. J. (2002) Analysis and quantification of DNA adducts of 2-amino-3,8-dimethylimidazo[4,5-f]quinoxaline in liver of rats by liquid chromatography/electrospray tandem mass spectrometry. *Chem. Res. Toxicol.*, **15**, 551–561.

97. Turesky, R.J., Parisod, V., Huynh-Ba, T., Langouet, S., and Guengerich, F.P. (2001) Regioselective differences in C(8)- and N-oxidation of 2-amino-3,8-dimethylimidazo [4,5-*f*]quinoxaline by human and rat liver microsomes and cytochromes P450 1A2. *Chem. Res. Toxicol.*, **14**, 901–911.
98. Turesky, R.J., Constable, A., Fay, L.B., and Guengerich, F.P. (1999) Interspecies differences in metabolism of heterocyclic aromatic amines by rat and human P450 1A2. *Cancer Lett.*, **143**, 109–112.
99. Langouët, S., Welti, D.H., Kerriguy, N., Fay, L.B., Huynh-Ba, T., Markovic, J., Guengerich, F.P., Guillouzo, A., and Turesky, R.J. (2001) Metabolism of 2-amino-3,8-dimethylimidazo[4,5-*f*]quinoxaline in human hepatocytes: 2-amino-3-methylimidazo [4,5-*f*]quinoxaline-8-carboxylic acid is a major detoxification pathway catalyzed by cytochrome P450 1A2. *Chem. Res. Toxicol.*, **14**, 211–221.
100. Turesky, R.J., Guengerich, F.P., Guillouzo, A., and Langouet, S. (2002) Metabolism of heterocyclic aromatic amines by human hepatocytes and cytochrome P450 1A2. *Mutat. Res.*, **506/507**, 187–195.
101. Belloc, C., Baird, S., Cosme, J., Lecoeur, S., Gautier, J.-C., Challine, D., de Waziers, I., Flinois, J.-P., and Beaune, P.H. (1996) Human cytochrome P450 expressed in *Escherichia coli*: production of specific antibodies. *Toxicology*, **106**, 207–219.
102. Guengerich, F.P. (1997) Comparisons of catalytic selectivity of cytochrome P450 subfamily members from different species. *Chem.-Biol. Interact.*, **106**, 161–182.
103. Snyderwine, E.G., Sinha, R., Felton, J.S., and Ferguson, L.R. (2002) Highlights of the eighth international conference on carcinogenic/mutagenic N-substituted aryl compounds. *Mutat. Res.*, **506/507**, 1–8.
104. Totsuka, Y., Fukutome, K., Takahashi, M., Takashi, S., Tada, A., Sugimura, T., and Wakabayashi, K. (1996) Presence of N_2-(deoxyguanosin-8-yl)-2-amino-3,8-dimethylimidazo[4,5-*f*]quinoxaline (dG-C8-MeIQx) in human tissues. *Carcinogenesis*, **17**, 1029–1034.
105. Turteltaub, K.W. and Dingley, K.H. (1998) Application of accelerated mass spectrometry (AMS) in DNA adduct quantification and identification. *Toxicol Lett.*, **102/103**, 435–439.
106. Rindgen, D., Turesky, R.J., and Vouros, P. (1995) Determination of in vitro formed DNA adducts of 2-amino-1-methyl-6-phenylimidazo[4,5-*b*]pyridine using capillary liquid chromatography/electrospray ionization/tandem mass spectrometry. *Chem. Res. Toxicol.*, **8**, 1005–1013.
107. Soglia, J.R., Turesky, R.J., Paehler, A., and Vouros, P. (2001) Quantification of the heterocyclic aromatic amine DNA adduct N-(deoxyguanosin-8-yl)-2-amino-3-methylimidazo[4,5-f]quinoline in livers of rats using capillary liquid chromatography/microelectrospray mass spectrometry: a dose-response study. *Anal. Chem.*, **73**, 2819–2827.
108. Gangl, E.T., Turesky, R.J., and Vouros, P. (2001) Detection of in vivo formed DNA adducts at the part-per-billion level by capillary liquid chromatography/microelectrospray mass spectrometry. *Anal. Chem.*, **73**, 2397–2404.
109. Pfau, W., Schulze, C., Shirai, T., Hasegawa, R., and Brockstedt, U. (1997) Identification of the major hepatic DNA adduct formed by the food mutagen 2-amino-9H-pyrido[2,3-b]indole (A alpha C). *Chem. Res. Toxicol.*, **10**, 1192–1197.
110. Turteltaub, K.W., Mauthe, R.J., Dingley, K.H., Vogel, J.S., Frantz, C.E., Garner, R.C., and Shen, N. (1997) MeIQx-DNA adduct formation in rodent and human tissues at low doses. *Mutat. Res.*, **376**, 243–252.
111. Dingley, K.H., Curtis, K.D., Nowell, S., Felton, J.S., Lang, N.P., and Turteltaub, K.W. (1999) DNA and protein adduct formation in the colon and blood of humans after exposure to a dietary-relevant dose of 2-amino-1-methyl-6-phenylimidazo[4,5-*b*]pyridine. *Cancer Epidemiol. Biomarkers Prev.*, **8**, 507–512.
112. Magagnotti, C., Pastorelli, R., Pozzi, S., Andreoni, B., Fanelli, R., and Airoldi, L. (2003) Genetic polymorphisms and modulation of 2-amino-1-methyl-6-phenylimidazo[4,5-b]pyridine (PhIP)-DNA adducts in human lymphocytes. *Int. J. Cancer*, **107**, 878–884.
113. Turteltaub, K.W., Dingley, K.H., Curtis, K.D., Malfatti, M.A., Turesky, R.J., Garner, R.C., Felton, J.S., and Lang, N.P. (1999) Macromolecular adduct formation and me-

tabolism of heterocyclic amines in humans and rodents at low doses. *Cancer Lett.*, **143**, 149–155.
114. Lightfoot, T.J., Coxhead, J.M., Cupid, B.C., Nicholson, S., and Garner, R.C. (2000) Analysis of DNA adducts by accelerator mass spectrometry in human breast tissue after administration of 2-amino-1-methyl-6-phenylimidazo[4,5-b]pyridine and benzo [a]pyrene. *Mutat. Res.*, **472**, 119–127.
115. Friesen, M.D., Kaderlik, K., Lin, D., Garren, L., Bartsch, H., Lang, N.P., and Kadlubar, F. F. (1994) Analysis of DNA adducts of 2-amino-1-methyl-6-phenylimidazo[4,5-*b*]pyridine in rat and human tissues by alkaline hydrolysis and gas chromatography/electron capture mass spectrometry: validation by comparison with ^{32}P-postlabeling. *Chem. Res. Toxicol.*, **7**, 733–739.
116. Zhu, J., Chang, P., Bondy, M.L., Sahin, A.A., Singletary, S.E., Takahashi, S., Shirai, T., and Li, D. (2003) Detection of 2-amino-1-methyl-6-phenylimidazo[4,5-b]-pyridine-DNA adducts in normal breast tissues and risk of breast cancer. *Cancer Epidemiol. Biomarkers Prev.*, **12**, 830–837.
117. Turesky, R.J., Skipper, P.L., and Tannenbaum, S.R. (1987) Binding of 2-amino-3-methylimidazo[4,5-*f*]quinoline to hemoglobin and alcumin in the rat. Identification of an adduct suitable for dosimetry. *Carcinogenesis*, **8**, 1537–1542.
118. Watkins, B.E., Suzuki, M., Wallin, H., Wakabayashi, K., Alexander, J., Vanderlaan, M., Sugimura, T., and Esumi, H. (1991) The effect of dose and enzyme inducers on the metabolism of 2-amino-1-methyl-6-phenylimidazo[4,5-*b*]pyridine (PhIP) in rats. *Carcinogenesis*, **12**, 2291–2295.
119. Lynch, A.M., Murray, S., Boobis, A.R., Davies, D.S., and Gooderham, N.J. (1991) The measurement of MeIQx adducts with mouse haemoglobin in vitro and in vivo: implications for human dosimetry. *Carcinogenesis*, **12**, 1067–1072.
120. Dingley, K.H., Freeman, S.P., Nelson, D.O., Garner, R.C., and Turteltaub, K.W. (1998) Covalent binding of 2-amino-3,8-dimethylimidazo[4,5-*f*]quinoxaline to albumin and hemoglobin at environmentally relevant doses. Comparison of human subjects and F344 rats. *Drug Metab. Dispos.*, **26**, 825–828.
121. Garner, R.C., Lightfoot, T.J., Cupid, B.C., Russell, D., Coxhead, J.M., Kutschera, W., Priller, A., Rom, W., Steier, P., Alexander, D.J., Leveson, S.H., Dingley, K.H., Mauthe, R.J., and Turteltaub, K.W. (1999) Comparative biotransformation studies of MeIQx and PhIP in animal models and humans. *Cancer Lett.*, **143**, 161–165.
122. Chepanoske, C.L., Brown, K., Turteltaub, K.W., and Dingley, K.H. (2004) Characterization of a peptide adduct formed by N-acetoxy-2-amino-1-methyl-6-phenylimidazo[4,5-b]pyridine (PhIP), a reactive intermediate of the food carcinogen PhIP. *Food Chem. Toxicol.*, **42**, 1367–1372.
123. Magagnotti, C., Orsi, F., Bagnati, R., Celli, N., Rotilio, D., Fanelli, R., and Airoldi, L. (2000) Effect of diet on serum albumin and hemoglobin adducts of 2-amino-1-methyl-6-phenylimidazo[4,5-b]pyridine (PhIP) in humans. *Int. J. Cancer*, **88**, 1–6.
124. Kelada, S.N., Eaton, D.L., Wang, S.S., Rothman, N.R., and Khoury, M.J. (2003) The role of genetic polymorphisms in environmental health. *Environ. Health Perspect.*, **111**, 1055–1064.
125. Lynch, A.M., Knize, M.G., Boobis, A.R., Gooderham, N., Davies, D.S., and Murray, S. (1992) Intra- and interindividual variability in systemic exposure in humans to 2-amino-3,8-dimethylimidazo[4,5-*f*]quinoxaline and 2-amino-1-methyl-6-phenylimidazo[4,5-*b*]pyridine, carcinogens present in food. *Cancer Res.*, **52**, 6216–6223.
126. Boobis, A.R., Lynch, A.M., Murray, S., de la Torre, R., Solans, A., Farré, M., Segura, J., Gooderham, N.J., and Davies, D.S. (1994) CYP1A2-catalyzed conversion of dietary heterocyclic amines to their proximate carcinogens is their major route of metabolism in humans. *Cancer Res.*, **54**, 89–94.
127. Sinha, R., Rothman, N., Mark, S.D., Murray, S., Brown, E.D., Levander, O.A., Davies, D.S., Lang, N.P., Kadlubar, F.F., and Hoover, R.N. (1995) Lower levels of urinary 2-amino-3,8-dimethylimidazo[4,5-f]-quinoxaline (MeIQx) in humans with higher CYP1A2 activity. *Carcinogenesis*, **16**, 2859–2861.

128. Ushiyama, H., Wakabayashi, K., Hirose, M., Ito, N., Sugimura, T., and Nagao, M. (1991) Presence of carcinogenic heterocyclic amines in urine of healthy volunteers eating normal diet, but not of inpatients receiving parenteral alimentation. *Carcinogenesis*, **12**, 1417–1422.
129. Murray, S., Lynch, A.M., Knize, M.G., and Gooderham, M.J. (1993) Quantification of the carcinogens 2-amino-3,8-dimethyl- and 2-amino-3,4,8-trimethylimidazo[4,5-*f*]-quinoxaline and 2-amino-1-methyl-6-phenylimidazo[4,5-*b*]pyridine in food using a combined assay based on gas chromatography-negative ion mass spectrometry. *J. Chromatogr.*, **616**, 211–219.
130. Reistad, R., Rossland, O.J., Latva-Kala, K.J., Rasmussen, T., Vikse, R., Becher, G., and Alexander, J. (1997) Heterocyclic aromatic amines in human urine following a fried meat meal. *Food Chem. Toxicol.*, **35**, 945–955.
131. Stillwell, W.G., Kidd, L.-C.K.S.-B., Wishnok, J.W., Tannenbaum, S.R., and Sinha, R. (1997) Urinary excretion of unmetabolized and phase II conjugates of 2-amino-1-methyl-6-phenylimidazo[4,5-*b*]pyridine and 2-amino-3,8-dimethylimidazo[4,5-*f*]quinoxaline in humans: Relationship to cytochrome P450 1A2 and *N*-acetyltransferase activity. *Cancer Res.*, **57**, 3457–3464.
132. Strickland, P.T., Qian, Z., Friesen, M.D., Rothman, N., and Sinha, R. (2002) Metabolites of 2-amino-1-methyl-6-phenylimidazo[4,5-*b*]pyridine (PhIP) in human urine after consumption of charbroiled or fried beef. *Mutat. Res.*, **506/507**, 163–173.
133. Stillwell, W.G., Turesky, R.J., Gross, G.A., Skipper, P.L., and Tannenbaum, S.R. (1994) Human urinary excretion of sulfamate and glucuronide conjugates of 2-amino-3,8-dimethylimidazo[4,5-*f*]quinoxaline (MeIQx). *Cancer Epidemiol. Biomarkers Prev.*, **3**, 399–405.

9 Molecular Epidemiology of Food Pyrolysis Products in Relation to Colon, Breast, and Prostate Cancer

Susan A. Nowell *, Rashmi Sinha **, Luke Ratnasinghe ***, Nicholas P. Lang ****, and Fred F. Kadlubar ***

Abstract

Certain food pyrolysis products are known mutagens that cause DNA damage, including both nucleotide alterations and chromosomal aberrations that are initiated by the formation of carcinogen-DNA adducts. However, adduct formation can be modified by host factors that affect the ultimate exposure of DNA to a food mutagen, such as variations in low-penetrance genes responsible for metabolic activation/detoxification (e.g., CYPs, NATs, SULTs, GSTs) of the putative mutagen or in genes responsible for cellular response to damage. The best characterized of these products are the heterocyclic amines (HCAs) and polycyclic aromatic hydrocarbons (PAHs). Both PhIP- and benzo[a]pyrene-(BP)-DNA adducts have now been detected in human colon and breast, and PhIP in human prostate. Epidemiological studies have indicated that components of the Western-type diet, particularly high fat and meat consumption, are closely associated with the risk of colorectal cancer; prospective studies addressing meat consumption, particularly red meat, and colorectal cancer have shown a fairly consistent association. Over thirty case-control and cohort studies have examined the relationship between HCA exposure, genetic variation in metabolic enzymes, and colorectal cancer risk and many have shown a gene–diet interaction. In one study, metabolic phenotypes for CYP1A2 and NAT2, in combination with lifestyle variables showed that rapid phenotypes for these enzymes in conjunction with a preference for well-done meat conferred a substantial increased risk ($OR = 6.45$). Subsequent studies examining meat intake and metabolic genotypes and phenotypes in relation to both colorectal cancer and the occurrence of colorectal adenomas have not always provided consistent results, but this discrepancy could be due to differences in study design, and the fact that doneness of meat and the dietary content of individual HCAs and PAHs were not usually addressed. In two studies that provided study participants with photographs of meat cooked to varying degrees of doneness, there was a

* Correspondence to: Susan Nowell, Department of Environmental and Occupational Health, College of Public Health, University of Arkansas for Medical Sciences, 4301 W. Markham St., #820, Little Rock AR 72205, USA.
** National Cancer Institute, Rockville, MD 20852, USA.
*** National Center for Toxicological Research, Jefferson, AR 72079, USA.
**** Central Arkansas Veterans' Healthcare System, Little Rock, AR 72205, USA.

Thermal Processing of Food: Potential Health Benefits and Risks
DFG, Deutsche Forschungsgemeinschaft

ISBN: 978-3-527-31909-1

significant association between doneness and the occurrence of colorectal adenomas and of colon cancer risk. Moreover, this was correlated with the specific content of MeIQx, PhIP, and BP. High consumption of meat has also been associated with an increased risk of breast and prostate cancer. However, estimated exposure to PhIP (but not MeIQx) and BP was more strongly associated with breast and prostate cancer than intake of red meat alone. In a recent study, we have shown that the rapid SULT1A1 genotype and phenotype presumed to be responsible for activating *N*-OH-PhIP to form a DNA adduct in human prostate was associated with increased prostate cancer risk, particularly in African-Americans ($OR = 5.0$ and 9.6, respectively).

9.1 Introduction

Cancer etiology is complex, and many factors, both environmental and genetic, contribute to susceptibility to this disease. Humans are exposed to compounds from dietary sources that can elevate the risk of cancer occurrence and the link between diet and cancer risk has long been recognized. Migration studies have lent support to the hypothesis that diet contributes to cancer risk; migrant populations who move from low to high cancer incidence areas experience an increase in cancer rates, with risk approaching that of the newer region within a few generations. Nutritional epidemiological studies have identified numerous dietary components that are associated with increased cancer risk, but the factors identified are not consistent across all studies and only explain a portion of disease variability in populations. Therefore, one might ask, "Why can we not explain more of cancer etiology in nutritional epidemiologic studies, and what is the basis of inconsistencies in observations?" Several different factors could account for discrepancies in findings from nutritional epidemiological studies. These include biases inherent to epidemiologic and, particularly, case-control studies (recall, selection and misclassification bias which could influence associations) to the heterogeneity of study populations and the fact that most nutritional epidemiological studies of diet and cancer have not taken into account the contribution of individual biochemistry to risk.

Humans have a complex system of enzymes and enzyme families that function to maintain cellular homeostasis. This includes enzymes that metabolize exogenous environmental compounds and promutagens ingested in food. Metabolism allows the utilization of nutrients and the subsequent detoxification and excretion of potentially harmful compounds and metabolites. The genes encoding metabolic enzymes are polymorphically expressed in humans; molecular biology and enzymology studies have shown that there are many polymorphisms that have a functional consequence for the expressed protein. Therefore, the interaction of genetic polymorphisms with consumed nutrients or with food-borne promutagens could serve to modulate diet-influenced cancer etiology.

In the early 1980's, researchers demonstrated that specific heterocyclic amines (HCAs) produced by pyrolysis of meats cooked at high temperatures were highly mutagenic [1]. In animal studies, exposure to HCAs results in the development of colon tumors in rodents [2]. However, HCAs are considered to be promutagens in that they require metabolic activation in order to achieve their full mutagenic potency [3–5]. Bioactivation of HCAs to carcinogenic species *in vivo* is initiated primarily by *N*-oxidation of the compound. This reaction occurs mainly in the liver and, in humans, is catalyzed by cytochrome P4501A2 (CYP1A2). Subsequent acetylation or sulfation of the *N*-hydroxy-HCA, catalyzed by acetyltransferases (NAT) and sulfotransferases (SULT), generates *N*-acetoxy and *N*-sulfonyloxy esters, electrophiles that are much more reactive with DNA [6, 7]. Polycyclic aromatic hydrocarbons (PAHs) are also produced during grilling of meat and can additionally be detected in preserved or smoked meat or other foodstuffs [8, 9]. PAHs exist in many forms and more than 100 different PAH compounds have been identified. However, the best characterized PAH is benzo(a)pyrene (BaP). Extensive animal studies have demonstrated the carcinogenic potency of BaP, with formation of tumors particularly in the stomach and esophagus [10]. BaP is highly lipophilic and requires metabolism before it can be excreted. It is during this process that reactive species such as benzo(a)pyrene-7,8-dihydrodiol epoxide can be formed primarily by the actions of CYP1A1 and CYP1B1, which can be expressed in target tissues [11] and have been shown to be polymorphic in humans. This metabolite is also a substrate for glutathione conjugation catalyzed by a variety of polymorphic members of the glutathione-*S*-transferase family. However, assessment of the dietary exposure of humans to BaP and other PAHs is complicated by the fact that these compounds are fairly ubiquitous and human exposures can come from air pollution, cigarette smoking, contaminated vegetables and other burnt foods besides meats [12, 13]. Hence, assessment of the contribution of PAH exposure and metabolic variation to cancer risk is problematic due to difficulties in quantifying those exposures.

9.2 Meat Consumption and Colorectal Cancer

Case-control studies have examined the relationship between HCA exposure, genetic variation in metabolic enzymes and colorectal cancer risk and these studies were the subject of a recent review article by Pisani and Mitton [14]. Briefly, in one study, metabolic phenotypes for CYP1A2 and NAT2, in combination with lifestyle variables collected in a patient questionnaire were evaluated [15]. Rapid phenotypes for these enzymes in conjunction with a preference for well-done meat conferred an increased risk of colorectal cancer ($OR=6.45$). Subsequent studies examining meat intake and metabolic genotypes and phenotypes in relation to both colorectal cancer and the occurrence of colorectal adenomas have not provided consistent results, but this discrepancy could be due to differences in study design, and the fact that dietary content of individual HCAs was not addressed in most instances. A majority of studies have not addressed the issue of the doneness of meat, which can substantially affect the amount of carcinogens in the meat [16–19]. In two studies that provided study participants with photographs of meat cooked to varying degrees of doneness, there was a significant association between doneness and the occurrence of colorectal adenomas (precursor lesions to colorectal cancer) [20] and of colon cancer risk [21], presumably due to increased exposure to HCAs. A recent study by Gunter et al. demonstrated an association between increasing BaP exposure and the occurrence of large adenomas [22]. However, the study population was modest compared to a recent study by Sinha et al., which examined the risk of colorectal adenomas and food mutagen exposure (including BaP) in 3696 patients with left-sided adenomas and 34817 endoscopy-negative controls from the Prostate, Lung, Colon and Ovarian cancer screening trial [23]. The results of these studies indicated that within the highest quintile of BaP exposure, there was increased risk of both non-advanced adenomas ($OR = 1.18$, 95% CI 1.02–1.37) and the occurrence of adenomas within the colon ($OR = 1.18$, 95% CI 1.02–1.35). A subsequent molecular epidemiological study examining the association between the CYP1A2*1F polymorphism and the occurrence of colorectal adenomas in 94 individuals found that this polymorphism was positively associated with adenoma [24].

In a large molecular epidemiologic study of colorectal cancer, meat consumption, and NAT2 genotype and CYP1A2 phenotype, LeMarchand and colleagues [25] showed that risk was greatest for those who were rapid for both NAT2 and CYP1A2, and were consumers of well-done meat ($OR = 8.8$, CI 1.7–44.9). SULT1A1 and SULT1E1 have been demonstrated to be involved in the activation of HCAs, but studies have not been able to demonstrate a convincing relationship between exposure to HCAs, genetic polymorphisms in these enzymes and risk of colorectal cancer.

While oxidation, followed by acetylation or sulfation are activation pathways for HCAs, several detoxification pathways do exist. Of particular importance is the reduction of reactive electrophilic HCA metabolites by glutathione. PhIP is the most mass-abundant HCA and human GSTA1 is the most effective member of the GST family that is capable of detoxifying the reactive PhIP metabolite *N*-acetoxy-PhIP [26]. A polymorphism in the promoter region of the *hGSTA1* gene has been shown to predict the levels of hepatic expression of both hGSTA1 and hGSTA2 and, perhaps more importantly, the ratio of hGSTA1/GSTA2 expression [26]. When the effect of *hGSTA1* genotype on susceptibility to colorectal cancer was examined [26], the allele that produces the lowest amount of hGSTA1 expression was significantly associated with risk of colorectal cancer.

Glucuronidation of HCAs by the UDP-glucuronosyltransferase (UGT) family of Phase II enzymes provides an additional detoxification pathway. Studies have identified several UGTs that are capable of catalyzing the glucuronidation of *N*-OH-PhIP. Research with recombinant enzymes has indicated that UGT1A1 plays a major role in the detoxification of *N*-OH-PhIP [27], although no studies of dietary exposures, UGT1A1 genotype and colorectal cancer risk have been performed to date. However, UGT1A7 has been demonstrated to be involved in the detoxification of both HCAs and PAHs. In a recent study of the joint effects of UGT1A7 genotype and exposure to HCA/PAHs on risk of colon cancer, greater than additive effects were observed for UGT1A7 low activity genotypes and greater HCA exposure [28].

9.3 Meat Consumption and Breast Cancer

High consumption of meat has also been associated with an increased risk of breast cancer. Several prospective studies have also identified a link between meat consumption and breast cancer risk (reviewed in [29]). Case-control studies examining NAT2 genotype, meat consumption and breast cancer risk have not been entirely in agreement in their conclusions. A case-control study by Ambrosone et al. found no association [30], while a later study by Dietz et al. demonstrated that NAT2 rapid acetylators who consumed well-done meat were at increased risk of breast cancer [31]. A subsequent study in a Turkish population, however, did not find an association between NAT2 polymorphisms and breast cancer risk. Recent studies by Sinha et al. demonstrated an association between PhIP exposure and breast cancer risk [32]. Using a HCA database, the authors were able to estimate exposure to specific HCAs from food-frequency questionnaires, and found that PhIP was more strongly associated with breast cancer than intake of red meat alone.

Since most of the initial metabolism of HCAs from the diet occurs in the liver, the question of exposure of the breast target tissue to HCAs arises. A recent study by Zhu et al. [33] demonstrated the presence of PhIP-DNA adducts in the normal breast tissue obtained from breast cancer patients and from normal tissue of women undergoing reduction mammoplasty. These investigators found that significantly more adducts were detected in the tissue of breast cancer patients compared to tissue from healthy women. A potential interaction between *NAT2* genotype, well-done meat consumption and levels of PhIP-DNA adducts was also identified. PhIP has also been detected in epithelial cells isolated from the breast milk of healthy women [34] and PhIP-DNA adducts have been detected in human breast tissues by accelerator mass spectrometry after administration of ^{14}C-labeled PhIP in patients undergoing breast surgery [35], and in exfoliated epithelial cells obtained from human breast milk [36]. Hence, it is biologically plausible that HCA exposure and genetic variability could be involved in the etiology of breast cancer.

A genetic polymorphism in *SULT1A1* results in an amino acid change (Arg to His, designated *SULT1A1*1* and *SULT1A1*2*) at the conserved residue 213. In a predominantly Caucasian population, the frequency of the polymorphic allele has been reported to be 0.674 and 0.313 for *SULT1A1*1* and *SULT1A1*2*, respectively. The *SULT1A1*2* allele is associated with reduced sulfotransferase activity and thermostability in platelets, although the relationship to activity in human liver cytosol is not clear [37]. This polymorphism has been investigated in relation to breast cancer in several studies, with conflicting results, but only one study thus far has examined SULT1A1 polymorphism, consumption of well-done cooked meat and breast cancer risk. In this study, individuals who possessed the high activity *SULT1A1* allele were at increased risk of breast cancer when they consumed large amounts of well-done cooked meat [38].

Recently, two other pathways of HCA activation/detoxification have been identified. Gorlewska-Roberts et al. reported that the lactoperoxidase enzyme, which is secreted from the mammary gland, is capable of activating several HCAs to DNA binding species [39]. In contrast to this potential activation pathway, another study has demonstrated that the breast cancer resistance protein 1 (Bcrp1), which is highly expressed in lactating mammary glands in several species (including humans), is capable of transporting HCAs into milk in mice exposed to these compounds [40]. Since polymorphisms have been identified in both of these genes, it will be interesting to examine the effect of genetic variability within these pathways, HCA exposure and risk of breast cancer in future studies.

9.4 Meat Consumption and Prostate Cancer

Prostate cancer incidence varies substantially by ethnicity and geographical location. Chinese men living in China have the lowest rates of prostate cancer; African-American men have the highest. Consumption of meat has also been suggested to be involved in prostate cancer risk. PhIP exposure produces prostate tumors in rodents and human prostate tissue has been shown to metabolically activate *N*-hydroxy-PhIP to DNA-binding species [41, 42]. Additionally, when transplantation of human prostate into athymic mice is followed by exposure to PhIP PhIP-DNA adducts are detectable in approximately 95% of samples [43]. Finally, evidence has been presented that human prostate DNA contains the same PhIP-DNA adduct formed by metabolic activation of *N*-hydroxy-PhIP [42].

Several studies have examined the potential interaction of meat consumption and HCA exposure in relation to prostate cancer risk. A large study in New Zealand did not find a clear association between HCA intake and prostate cancer risk, but an association between well-done beefsteak and the HCA, 2-amino-1,6-dimethylfuro[3,2-*e*]imidazo[4,5-*b*]pyridine, and risk of prostate cancer was suggested [44]. A recent study in the United States, which examined NAT1 and NAT2 genotypes, HCA exposure and prostate cancer risk found a trend toward increased risk of prostate cancer in patients who were exposed to high levels of MeIQx and who had alleles for rapid NAT2 [45]. Another study of a southern European population found that the NAT2 slow acetylator genotype was protective against prostate cancer, although dietary exposures to

HCAs were not considered [46]. A recent pilot study of an Indian population found an association between the rapid acetylator NAT2 genotype and prostate cancer risk in smokers [47].

While these studies examined the NAT pathway of HCA-activation, heterologous expression of human NAT and SULT enzymes in *Salmonella typhimurium* and mutagenicity testing of heterocyclic amines indicated that *N*-OH-PhIP was activated specifically by SULT1A1 in humans [48]. *SULT1A1* transcript and SULT1A1 protein have been detected in human prostate, so a role for SULT1A1 in PhIP-induced prostate carcinogenesis is biologically plausible. The *SULT1A1*1/*2* polymorphism was investigated in relation to prostate cancer risk in two studies. While one report found no association of *SULT1A1* genotype with prostate cancer risk [49], another study found that both *SULT1A1* genotype and particularly SULT1A1 enzymatic activity was significantly associated with risk of this disease [50]. The difference could be contributed to the size of the respective study populations. The study population from the latter study consisted of 403 prostate cancer patients and 450 control individuals, while that of the former consisted of 134 patients and 184 control individuals, all of Caucasian origin. Although their results did not achieve statistical significance, Steiner [49] also reported a lower frequency of the *SULT1A1*2/*2* genotype in prostate cancer patients compared to control individuals, which is consistent with the study by Nowell et al. [50].

9.5 Summary and Conclusions

Cancer is a multi-factorial disease, with both genetic and environmental factors contributing to its occurrence. Understanding of the interaction between exposures from dietary sources to compounds that are either potentially harmful or potentially beneficial and genetic polymorphisms in the xenobiotics metabolizing enzymes involved in their biotransformation is critically important to the individual in making informed lifestyle choices. It is also possible that positive or negative associations found by traditional epidemiological methods will be modified when genetic variation is taken into account. For this reason, large scale, multi-disciplinary studies are required to elucidate the relative importance of these interactions. This information will be invaluable in making recommendations for individualized dietary changes that could greatly enhance the field of cancer prevention.

References

1. Nagao M, Fujita Y, Wakabayashi K, Sugimura T. Ultimate forms of mutagenic and carcinogenic heterocyclic amines produced by pyrolysis. *Biochem. Biophys. Res. Comm.* 1983, **114**(2): 626–631.
2. Wakabayashi K, Nagao M, Esumi H, Sugimura T. Food-derived mutagens and carcinogens. *Cancer Res.* 1992, **52**(7 Suppl): 2092s–2098s.
3. Becher G, Knize MG, Nes IF, Felton JS: Isolation and identification of mutagens from a fried Norwegian meat product. *Carcinogenesis* 1988, **9**(2): 247–253.
4. Hatch FT, Knize MG, Healy SK, Slezak T, Felton JS: Cooked-Food Mutagen Reference List and Index. *Environ. Mol. Mutagenesis* 1988, **12** Suppl 14: 1–85.
5. Knize MG, Shen NH, Felton JS. A comparison of mutagen production in fried ground chicken and beef: effect of supplemental creatine. Mutagenesis 1988, **3**(6): 503–508.
6. Minchin RF, Reeves PT, Teitel CH, McManus ME, Mojarrabi B, Ilett KF, Kadlubar FF. N- and O-acetylation of aromatic and heterocyclic amine carcinogens by human monomorphic and polymorphic acetyltransferases expressed in COS-1 cells. *Biochem. Biophys. Res. Comm.* 1992, **185**(3): 839–844.
7. Chou HC, Lang NP, Kadlubar FF. Metabolic activation of N-hydroxy arylamines and N-hydroxy heterocyclic amines by human sulfotransferase(s). *Cancer Res.* 1995, **55**(3): 525–529.
8. Kazerouni N, Sinha R, Hsu CH, Greenberg A, Rothman N. Analysis of 200 food items for benzo[a]pyrene and estimation of its intake in an epidemiologic study. *Food Chem. Toxicol.* 2001, **39**: 423–436.
9. Phillips DH. Polycyclic aromatic hydrocarbons in the diet. *Mutat. Res.* 1999, **443**: 139–147.
10. Culp SJ, Gaylor DW, Sheldon WG, Goldstein LS, Beland FA. A comparison of the tumors induced by coal tar and benzo[a]pyrene in a 2-year bioassay. *Carcinogenesis* 1998, **19**: 117–124.
11. Keshava C, Divi RL, Whipkey DL, Frye BL, McCanlies E, Kuo M, Poirier MC, Weston A. Induction of CYP1A1 and CYP1B1 and formation of carcinogen-DNA adducts in normal mammary epithelial cells treated with benzo[a]pyrene. *Cancer Lett.* 2005, **221**(2): 213–224.
12. Guillen MD, Sopelana P, Partearroyo MA. Food as a source of polycyclic aromatic carcinogens. *Rev. Environ. Health.* 1997, **12**: 133–146.
13. Rothman N, Correa-Villasenor A, Ford DP, Poirier MC, Haas R, Hansen JA, O'Toole T, Strickland PT. Contribution of occupation and diet to white blood cell polycyclic aromatic hydrocarbon-DNA adducts in wildland firefighters. *Cancer Epidemiol. Biomarkers Prev.* 1993, **2**: 341–347.
14. Pisani P, Mitton N. Cooking methods, metabolic polymorphisms and colorectal cancer. *Eur. J. Cancer Prev.* 2002, **11**: 75–84.
15. Lang NP, Butler MA, Massengill J, Lawson M, Stotts RC, Hauer-Jensen M, Kadlubar FF. Rapid metabolic phenotypes for acetyltransferase and cytochrome P4501A2 and putative exposure to food-borne heterocyclic amines increase the risk for colorectal cancer or polyps. *Cancer Epidemiol. Biomarkers Prev.* 1994, **3**(8): 675–682.
16. Sinha R, Rothman N, Salmon CP, Knize MG, Brown ED, Swanson CA, Rhodes D, Rossi S, Felton JS, LevanderOA. Heterocyclic amine content in beef cooked by different methods to varying degrees of doneness and gravy made from meat drippings. *Food Chem. Toxicol.* 1998, **36**: 279–287.
17. Knize MG, Salmon CP, Pais P, Felton JS. Food heating and the formation of heterocyclic aromatic amine and polycyclic aromatic hydrocarbon mutagens/carcinogens. *Adv. Exp. Med. Biol.* 1999, **459**: 179–193.
18. Knize MG, Kulp KS, Salmon CP, Keating GA, Felton JS. Factors affecting human heterocyclic amine intake and the metabolism of PhIP. *Mutat. Res.* 2002, **506/507**: 153–162.

19. Sinha R, Rothman N. Exposure assessment of heterocyclic amines (HCAs) in epidemiologic studies. *Mutat. Res.* 1997, **376**: 195–202.
20. Sinha R, Kulldorff M, Chow WH, Denobile J, Rothman N. Dietary intake of heterocyclic amines, meat-derived mutagenic activity, and risk of colorectal adenomas. *Cancer Epidemiol. Biomarkers Prev.* 2001, **10**: 559–562.
21. Nowell S, Coles BF, Sinha R, MacLeod S, Ratnasinghe L, Stotts C, Kadlubar FF, Ambrosone CB, Lang NP. Analysis of total meat intake and exposure to individual heterocyclic amines in a case-control study of colorectal cancer: contribution of metabolic variation to risk. *Mutat. Res.* 2002, **506/507**(C): 175–185.
22. Gunter MJ, Probst-Hensch NM, Cortessis VK, Kulldorff M, Haile RW, Sinha R. Meat intake, cooking-related mutagens and risk of colorectal adenoma in a sigmoidoscopy-based case-control study. *Carcinogenesis* 2005, 26(3): 637–642.
23. Sinha R, Peters U, Cross AJ, Kulldorff M, Weissfeld JL, Pinsky PF, Rothman N, Hayes RB, and Prostate, Lung, Colorectal and Ovarian Cancer Project Team. Meat, meat cooking methods and preservation, and risk for colorectal adenoma. *Cancer Res.* 2005, **65**(17): 8034–8041.
24. Moonen H, Engels L, Kleinjans J, Kok T. The CYP1A2-164A>C polymorphism (CYP 1A2*1F) is associated with the risk for colorectal adenomas in humans. *Cancer Lett.* 2005, **229**(1): 25–31.
25. Le Marchand L, Hankin JH, Wilkens LR, Pierce LM, Franke A, Kolonel LN, Seifried A, Custer LJ, Chang W, Lum-Jones A, Donlon T. Combined effects of well-done red meat, smoking, and rapid N-acetyltransferase 2 and CYP1A2 phenotypes in increasing colorectal cancer risk. *Cancer Epidemiol. Biomarkers Prev.* 2001, **10**: 1259–1266.
26. Coles B, Nowell SA, MacLeod SL, Sweeney C, Lang NP, Kadlubar FF. The role of human glutathione S-transferases (hGSTs) in the detoxification of the food-derived carcinogen metabolite N-acetoxy-PhIP, and the effect of a polymorphism in hGSTA1 on colorectal cancer risk. *Mutat. Res.* 2001, **482**: 3–10.
27. Malfatti MA, Felton JS. N-glucuronidation of 2-amino-1-methyl-6-phenylimidazo[4,5-b]pyridine (PhIP) and N-hydroxy-PhIP by specific human UDP-glucuronosyltransferases. *Carcinogenesis* 2001, **22**: 1087–1093.
28. Butler LM, Duguay Y, Millikan RC, Sinha R, Gagne JF, Sandler RS, Guillemette C. Joint effects between UDP-glucuronosyltransferase 1A7 genotype and dietary carcinogen exposure on risk of colon cancer. *Cancer Epidemiol. Biomarkers Prev.* 2005, **14**(7): 1626–1632.
29. Nowell SA, Ahn J, Ambrosone CB. Gene-nutrient interactions in cancer etiology. *Nutr. Rev.* 2004, **62**: 427–438.
30. Ambrosone CB, Freudenheim JL, Sinha R, Graham S, Marshall JR, Vena JE, Laughlin R, Nemoto T, Shields PG. Breast cancer risk, meat consumption and N-acetyltransferase (NAT2) genetic polymorphisms. *Int. J. Cancer* 1998, **75**(6): 825–830.
31. Dietz AC, Zheng W, Leff MA, Gross M, Wen WQ, Doll MA, Xiao GH, Folsom AR, Hein DW. N-Acetyltransferase-2 genetic polymorphism, well-done meat intake, and breast cancer risk among postmenopausal women. *Cancer Epidemiol. Biomarkers Prev.* 2000, **9**(9):905–910.
32. Sinha R, Gustafson DR, Kulldorff M, Wen WQ, Cerhan JR, Zheng W. 2-Amino-1-methyl-6-phenylimidazo[4,5-b]pyridine, a carcinogen in high-temperature-cooked meat, and breast cancer risk. *J. Natl. Cancer Inst.* 2000, **92**: 1352–1354.
33. Zhu J, Chang P, Bondy ML, Sahin AA, Singletary SE, Takahashi S, Shirai T, Li D. Detection of 2-amino-1-methyl-6-phenylimidazo[4,5-b]-pyridine-DNA adducts in normal breast tissues and risk of breast cancer. *Cancer Epidemiol. Biomarkers Prev.* 2003, **12**: 830–837.
34. DeBruin LS, Martos PA, Josephy PD. Detection of PhIP (2-amino-1-methyl-6-phenylimidazo[4,5-b]pyridine) in the milk of healthy women. *Chem. Res. Toxicol.* 2001, **14**: 1523–1528.
35. Lightfoot TJ, Coxhead JM, Cupid BC, Nicholson S, Garner RC. Analysis of DNA adducts by accelerator mass spectrometry in human breast tissue after administration of

2-amino-1-methyl-6-phenylimidazo[4,5-b]pyridine and benzo[a]pyrene. *Mutat. Res.* 2000, **472**: 119–127.
36. Gorlewska-Roberts K, Green B, Fares M, Ambrosone CB, Kadlubar FF. Carcinogen-DNA adducts in human breast epithelial cells. *Environ. Mol. Mutagen.* 2002, **39**: 184–192.
37. MacLeod SL, Nowell S, Lang NP. Genetic Polymorphisms. *In* Food Borne Carcinogens: Heterocyclic Amines. Nagao M, Sigumura T (eds) John Wiley & Sons, West Sussex, England, 2000, pp 112–130.
38. Zheng W, Xie D, Cerhan JR, Sellers TA, Wen W, Folsom AR. Sulfotransferase 1A1 polymorphism, endogenous estrogen exposure, well-done meat intake, and breast cancer risk. *Cancer Epidemiol. Biomarkers Prev.* 2001, **10**: 89–94.
39. Gorlewska-Roberts KM, Teitel CH, Lay JO, Roberts DW, Kadlubar FF. Lactoperoxidase-catalyzed activation of carcinogenic aromatic and heterocyclic amines. *Chem. Res. Toxicol.* 2004, **17**(12): 1659–1666.
40. van Herwaarden AE, Wagenaar E, Karnekamp B, Merino G, Jonker JW, Schinkel AH. Breast cancer resistance protein (Brcp1/Abcg2) reduces systemic exposure of the dietary carcinogens aflatoxin B1, IQ and Trp-P-1 but also mediates their secretion into breast milk. *Carcinogenesis*, 2005, epub ahead of print Jul 18.
41. Williams J, Stone E, Fakis G, Johnson N, Meinl W, Glatt H, Sim E, Phillips D. Human mammary NAT and SULT enzymes metabolically activate N-hydroxylated heterocyclic amines, but NAT enzyme activity is not influenced by NAT genotype. *In* American Association for Cancer Research, San Francisco, CA. 2000.
42. DiPaolo O, Teitel C, Nowell S, Coles B, Kadlubar FF. Expression of cytochromes P450 and glutathione S-transferases in human prostate, and the potential for activation of heterocyclic amine carcinogens via acetyltransferase-, PAPS- and ATP-dependent pathways. *Int. J. Cancer*, in press, 2005.
43. Cui L, Takahashi S, Tada M, Kato K, Yamada Y, Kohri K, Shirai T. Immunohistochemical detection of carcinogen-DNA adducts in normal human prostate tissues transplanted into the subcutis of athymic nude mice: results with 2-amino-1-methyl-6-phenylimidazo[4,5-b]pyridine (PhIP) and 3,2'-dimethyl-4-aminobiphenyl (DMAB) and relation to cytochrome P450s and N-acetyltransferase activity. *Jpn. J. Cancer Res.* 2000, **91**: 52–58.
44. Norrish AE, Ferguson LR, Knize MG, Felton JS, Sharpe SJ, Jackson RT. Heterocyclic amine content of cooked meat and risk of prostate cancer. *J. Natl. Cancer Inst.* 1999, **91**(23): 2038–2044.
45. Rovito PM, Morse PD, Spinek K, Newman N, Jones RF, Wang CY, Haas GP. Heterocyclic amines and genotype of N-acetyltransferases as risk factors for prostate cancer. *Prostate Cancer Prostatic Dis.* 2005, **8**(1): 69–74.
46. Costa S, Pinto D, Morais A, Vasconcelos A, Oliveira J, Lopes C, Medeiros R. Acetylation genotype and the genetic susceptibility to prostate cancer in a southern European population. *Prostate* 2005, **64**(3): 246–282.
47. Srivastava DS, Mittal RD. Genetic polymorphism of the N-acetyltransferase 2 gene, and susceptibility to prostate cancer: a pilot study in north Indian population. *BMC Urol.* 2005, Aug 6; **5**: 12.
48. Muckel E, Frandsen H, Glatt HR: Heterologous expression of human N-acetyltransferases 1 and 2 and sulfotransferase 1A1 in Salmonella typhimurium for mutagenicity testing of heterocyclic amines. *Food Chem. Toxicol.* 2002, **40**(8): 1063–1068.
49. Steiner M, Bastian M, Schulz WA, Pulte T, Franke KH, Rohring A, Wolff JM, Seiter H, Schuff-Werner P. Phenol sulphotransferase SULT1A1 polymorphism in prostate cancer: lack of association. *Arch. Toxicol.* 2000, **74**(4/5): 222–225.
50. Nowell S, Ratnasinghe DL, Ambrosone CB, Williams S, Teague-Ross T, Trimble L, Runnels G, Carrol A, Green B, Stone A *et al.* Association of SULT1A1 phenotype and genotype with prostate cancer risk in African-Americans and Caucasians. *Cancer Epidemiol. Biomarkers Prev.* 2004, **13**(2): 270–276.

10 The Formation of 3-Monochloropropane-1,2-diol (3-MCPD) in Food and Potential Measures of Control

Richard H. Stadler *, Viviane Theurillat *, Alfred Studer **, Francis Scanlan **, and Walburga Seefelder **

Abstract

The findings of 3-MCPD in foods such as roasted cereals, bread, toasted bread, non-HVP seasonings, cheese, cooked meat, etc. have attracted significant attention in the past years. Despite the widespread occurrence of the food-borne contaminant 3-MCPD, the mechanisms of its formation have not been studied in much detail in foods other than acid-hydrolysed vegetable protein (acid-HVP). Several reports focused mainly on model systems attempt to describe the underlying reaction mechanisms of 3-MCPD formation in the different foods, and essentially two very basic pathways can be described, i.e. thermally driven and enzyme catalyzed reactions. The latter involves the action of lipases upon "bound" or esterified MCPD, also termed chloroesters. Lipases can originate from different sources (e.g. microbial) and catalyze the hydrolysis of MCPD esters and subsequently release 3-MCPD. This pathway may be valid in products that contain ingredients with lipase activity together with fat which may contain low amounts of the intermediate chloroesters. On the other hand, heat-driven mechanisms seem predominant in roasted cereals, bread crust and toast, and are correlated to the degree of browning of the product. They are, however, also dependent on the availability of the direct precursors, i.e. mainly glycerol and chloride. The rate of formation of 3-MCPD in simple dough systems has been shown to be dependent on several salient parameters such as the moisture content, temperature/time and pH. Based on the recent studies related to better understanding the formation of 3-MCPD in different foods, several options have been proposed to control or lower 3-MCPD. However, most of the mitigation research that has been conducted in non-acid HVP products is at laboratory or pilot scale using rather simple models. This report summarizes the main reaction pathways to 3-MCPD in the key food categories, highlighting potential options for reduction. Clearly, more research into the mechanisms of formation of 3-MCPD and related compounds in food systems is warranted, and any reduction strategies must also take into account the potential impact on the formation of other "undesired" compounds without compromising on the quality and safety of food.

* Correspondence to: Richard H. Stadler, Nestlé Product Technology Centre, CH-1350 Orbe, Switzerland.
** Nestlé Research Centre, Lausanne, Switzerland.

Thermal Processing of Food: Potential Health Benefits and Risks
DFG, Deutsche Forschungsgemeinschaft

ISBN: 978-3-527-31909-1

10.1 Introduction

3-Monochloropropane-1,2-diol (3-MCPD) was originally identified in 1981 as a processing contaminant of acid-hydrolysed vegetable proteins (HVP), which are frequently used as ingredients of savoury foods such as soups, snacks, gravy mixes and bouillon cubes [1]. 3-Monochloropropane-1,2-diol is generally the most widely occurring of the chloropropanols in food [2], and exists as a racemic mixture of (R)- and (S)-isomers. Based on toxicological studies, the EU SCF and the Joint WHO/FAO Expert Committee on Food Additives (JECFA) have recommended a provisional maximum Tolerable Daily Intake (PTDI) of 2 μg/kg bodyweight/day for 3-MCPD [3, 4]. In view of the exposure and toxicity of 3-MCPD, the European Commission has adopted a regulatory limit of 0.02 mg/kg for 3-MCPD in liquid soy sauce and HVP [5].

However, 3-MCPD has also been found in several foods subject to thermal treatment or fermented, such as grilled cheese, fermented cheese, roasted cereals, meats (ham, salami) as well as barley during the production of malt or domestically toasted bread [6, 7]. The formation of 3-MCPD was also found to be dependent on the cooking process [8]. Several mechanisms of formation of chloropropanols and their potential precursors in foods have been proposed and discussed, albeit most related to model systems and savoury foods [9]. Heat-induced reactions leading to the formation of chloropropanols include reactions of hydrochloric acid with glycerol and lipids. All reactions require prolonged heating temperatures above 100 °C. Collier et al. [10] initially proposed a mechanism explaining the heat-induced formation of chloropropanols from triacylglycerols under acidic conditions. The key step involves the nucleophilic substitution of the acyl group by the chloride anion at positions activated by neighbouring ester groups. The resulting intermediate is a chloropropanediol-diester that under hydrolytic conditions leads to the formation of chloropropanol.

The unexplained traces of 3-MCPD in several heat-treated foods have stimulated fundamental model studies to assess the possible involvement of hydrolytic enzymes in the formation of chloropropanols. Consequently, an alternative pathway involving lipase and operating at relatively low temperatures has been proposed [11], pointing to a more important role of esterified 3-MCPD (mono- or diesters of fatty acids, also termed 3-MCPD-esters). These fatty acid mono- and diesters of 3-MCPD can be chemically formed from triacylglycerols in the presence of hydrochloric acid and heat [12]. Chloroesters have also been described in a wide range of different foods, including goat milk and milk fat [13, 14], cereal-based foods and coffee [15–17]. In general, where chloroesters are present in foods their concentration significantly exceeds the concentration of free 3-MCPD and as such could provide an additional source of 3-MCPD in a low temperature reaction, e.g. through the action of commercial lipases used in food manufacture or lipases present in the foods [11].

The rather diverse mechanisms of formation of 3-MCPD in food will inevitably necessitate different strategies to achieve a reduction of exposure to this food-borne contaminant. This report summarizes the progress made to date in understanding the formation pathways of chloropropanols in the key food categories (savoury, cereal-based foods including bread and coffee), highlighting possible strategies of mitigation.

10.2 Materials and Methods

10.2.1 Determination of Free and Bound 3-MCPD in Roasted Barley

Sample Preparation

Free 3-MCPD: A test portion (2 g) was weighed into a 150 mL beaker and 20 μL of d_5-3-MCPD spiking solution (10 ng/μL) were added. Then, a sodium chloride (5 M) solution was added to give a total weight of 20 g. The mixture was placed in an ultrasonic bath for 10 min.

Bound 3-MCPD: 3-MCPD esters were analyzed according to the method of Hamlet and Sadd [15]. Samples (4 g) were dispersed in 50 mL of 0.1 M phosphate buffer (pH 7.0). LIPOPAN 50BG (activity: 50 kilo lipase units per gram) was added and the pH adjusted to 7 with NaOH (0.1 N). The sample was incubated at room temperature with slight agitation for 24 h. After lyophilisation, d_5-3-MCPD (100 μL) spiking solution and 15 mL of a sodium chloride (5 M) solution were added and the samples extracted as described below.

Extraction and Cleanup

Extrelut™ was added to the prepared test sample before it was mixed thoroughly with a spatula and left standing at room temperature for 15–20 min. Glass wool was inserted in the bottom of a chromatography column (450×22 mm i.d.) fitted with a stopcock. The column was filled with 20 g of anhydrous sodium sulfate, followed by ca. 15 g of sea sand. The sample mixture was added and finally the column was topped with some glass wool. Polar and mid polar compounds were rinsed out with 50 mL of diethylether/hexane (10+90) and afterwards with a mixture of 100 mL diethylether/hexane (40+60). After the washing step, 2-MCPD and 3-MCPD were eluted with 250 mL of diethyl ether over a 20 mL column containing ca. 10 g of deactivated Florisil. The extract was concentrated to 2 mL *in vacuo* (40 °C) and subsequently concentrated to 200 μL under a stream of nitrogen. The concentrated extract was transferred to a 5 mL volumetric flask with ethyl acetate and diluted to volume with ethyl acetate. An aliquot of 2.5 mL was taken, evaporated to about 100 μL under a nitrogen stream and diluted with 900 μL of iso-octane.

Derivatization with Heptafluorobutyrylimidazole (HFBI)

Derivatization of extracted 3-MCPD with heptafluorobutyrylimidazole (HFBI) was performed as described in the literature [18]. The reconstituted extracts and the calibration standards were mixed with HFBI (50 μL) and the vials tightly closed. These were then vortexed for a few seconds and heated at 70 °C for 20 min in a convection oven. The mixtures were allowed to cool to room temperature, and the solutions were then transferred to clean vials. Distilled water (4 mL) was added and the mixture vortexed for a few seconds. The vials were centrifuged at 2500 rpm for 5 min. The lower aqueous phase was discarded and distilled water (4 mL) added. The sample was again agitated for a few seconds and centrifuged as described before. The upper organic phase was transferred to a sample vial and ready for injection into GC-MS system.

Calibration Standards

Standards of 3-MCPD were prepared as described in Brereton et al. [19] at final concentrations of 3-MCPD of 0, 5, 10, 25, 50, 100, 150, and 200 μg/L. The final concentration of d_5-MCPD in each calibration solution was 100 μg/L. Pooled calibration standard solutions (1 mL of 3-MCPD) were transferred to a set of eight 5 mL conical vials, and derivatised as described above.

Gas Chromatography-Mass Spectrometry (GC-MS)

Analyses were performed on a Hewlett-Packard 6890 series MS (Wilmington, DE, USA) combined with a Hewlett-Packard 6890 series GC. The whole system was controlled by HP 6890 Series ChemStation software. Chromatography was performed on a DB-XLB ITD column (length 30 m, i.d. 0.25 mm, film 0.25 μm) from J & W Scientific, Agilent Technologies (Basel, Switzerland). The carrier gas was helium at a constant flow of 1.1 mL/min. The oven temperature gradient was as follows: from 0 to 1.0 min, hold at 50 °C, then ramp to 90 °C at 6 °C/min, ramp to 325 °C at 50 °C/min, and keep the temperature constant at 325 °C for 5 min. The transfer line temperature for the MS was 270 °C.

The MS system was operated in the selected ion monitoring mode (SIM). The ions monitored for 3-MCPD were *m/z* 75, 253, 289 and 453. The ion *m/z* 257 was recorded for the deuterated internal standard. Retention times (min.) for 3-MCPD and the deuterated standard were 7.40 and 7.32, respectively.

Quantitation

All data evaluation was normalized to the area response ratio of the analyte to the internal standard, and the amount of the compounds in the samples was extrapolated from the respective linear regression equation. For the final quantitation, the median of all ions in a given sample was taken into account. 3-MCPD is expressed as the amount formed per kilogram of sample.

10.3 Results and Discussion

10.3.1 3-MCPD in Savoury Food Products

Conventional acid-hydrolysed vegetable protein (acid-HVP) is produced using strong hydrochloric acid. Chloropropanols are formed as a result of the chlorination of glycerol at elevated temperatures present in fats and oils in the crude protein starting materials.

Surveys on the occurrence of chloropropanols have revealed that 3-MCPD is found in several foodstuffs that are fermented and/or not subjected to thermal treatment, such as meats (ham, salami), fish and fermented cheese [7, 9]. Recently, we reported a non-thermal process involving lipase as catalyst of 3-MCPD formation and/or release in model systems comprised of vegetable oil, water and sodium chloride [11]. Lipases catalyze the hydrolysis of acyl esters and most enzymes display a regioselectivity toward the sn-1 or sn-3 position of the prochiral substrate. Further, we proposed that chloropropanediol esters may be formed by lipase-catalyzed transesterification of triacylglycerols. Similar observations have recently been made by Hamlet and Sadd [15], i.e. the formation of both 3-MCPD and 2-MCPD over time (48 h) from purified tributyrin, albeit at the same rate with or without added chloride. This mechanism may account for the spurious amounts of MCPD's detected in some dairy products, that have not been subjected to heat treatment [13, 14]. In dried savoury foods that contain salts and lipids, lipase or esterase activity – originating either from the food ingredients or microbiological load – will be important to monitor as it may reflect the potential formation of MCPD over time. A study encompassing several savoury foods indeed showed a correlation of 3-MCPD formation (r^2=0.791) over time and corresponding lipase activity (Fig. 10.1).

A survey of the occurrence of 3-MCPD in seasonings of Asian convenience products (instant noodle dishes) has shown that 3-MCPD is present in these products at rather low amounts (30–50 µg/kg). However, the authors reported two products with elevated levels (average 200 µg/kg), one of which did not contain acid-HVP nor soy sauce as ingredients [20]. Onion or garlic extracts were proposed as the source of 3-MCPD, possibly generated via allylalcohol. Our observations, however, indicate that the most plausible source of 3-MCPD

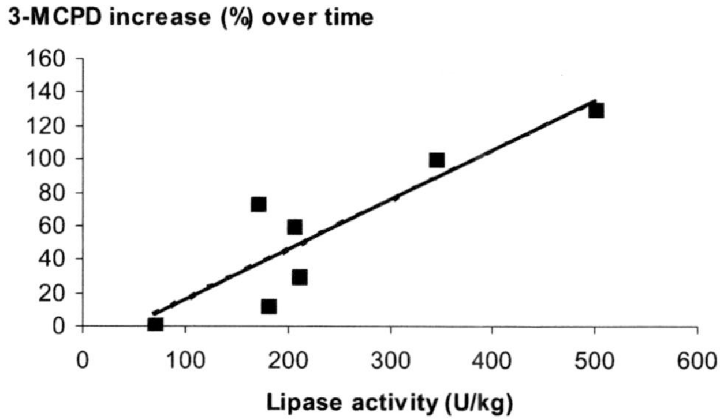

Figure 10.1: Change of 3-MCPD concentration over time (2–6 months) in selected savoury food products and correlation to lipase activity.

in such products is lipase activity and release of 3-MCPD by enzyme-catalyzed hydrolysis of the esterified intermediates (Fig. 10.1).

The minimization of 3-MCPD in soy sauce and acid-HVP over the past decade has been achieved by the food industry through good manufacturing practice (GMP). GMP entails careful control of the acid hydrolysis step (temperature, time, acidity), and subsequent neutralization of the slurry by alkaline (e.g. NaOH) treatment. Notably, the (R)- and (S)-isomers of 3-MCPD are destroyed equally effectively by this process. In fact, some manufacturers have introduced enzyme or fermentation (yeast) based processes (also termed biohydrolysates), the final products of which (either liquid or powders) tend to show concentrations well below 20 μg/kg. Ingredients such as spices (e.g. pepper) or dried onions, used in the manufacture of culinary products may contain lipase activity, consequently leading to the release of low amounts of 3-MCPD from chloroesters over time, depending on the water activity and storage temperature. Mitigation options to avoid formation during storage could include a heat treatment to inactivate enzymes and/or reduce the microbial load, or the replacement of native spices with spice extracts.

10.3.2 3-MCPD in Cereal Products

10.3.2.1 Roasted Barley

Roasted cereals such as barley and rye are important constituents of coffee surrogates, and in such products amounts of 3-MCPD ranging from 40–800 μg/kg have recently been reported [21]. In the same study, two bio-products manufactured from malted barley showed levels of 377 and 483 μg/kg. This is in line with information from the UK brewing and malt industries that indicate 3-MCPD levels of up to 400 μg/kg in roasted cereals and dark specialty malts which are used to impart colour and flavour to most dark beers and some lagers. The concentration of 3-MCPD has been found to parallel the degree of roasting given to barley, the darker malts having the greatest concentration. However, within this context the formation of 3-MCPD esters has never been investigated.

Roasting of barley is also a common process applied in the manufacture of cereal-based beverages. Therefore, we determined the impact of thermal treatment of barley over time on the formation of 3-MCPD and its corresponding esters using the analytical method described in [15]. Furthermore, we also investigated the effect of water extraction and subsequent drying of the barley on the concentrations of MCPD esters. Chloroesters are essentially halo-substituted lipids and therefore share their hydrophobic properties. It can thus be envisaged that only a small fraction of the chloroesters are actually extracted during brewing.

Barley was heated over time (5, 15, 25, and 35 min) at different temperatures (220, 230, and 240 °C) in a laboratory oven. Esters of 3-MCPD were determined as the amount of 3-MCPD released after enzymatic (lipase) treatment. Samples were also analysed for their concentrations of free 3-MCPD to compare ratios of free 3-MCPD and chloroesters. As depicted in Figure 10.2, the concentrations of free 3-MCPD increased exponentially up to 150 μg/kg within the first 15 to 20 minutes of the roasting cycle, reaching an apparent maximum in the samples roasted at 220 °C and 230 °C. At higher temperatures (240 °C), a clear decline is already observed after 20 min, probably due to a faster rate of decomposition versus formation of 3-MCPD.

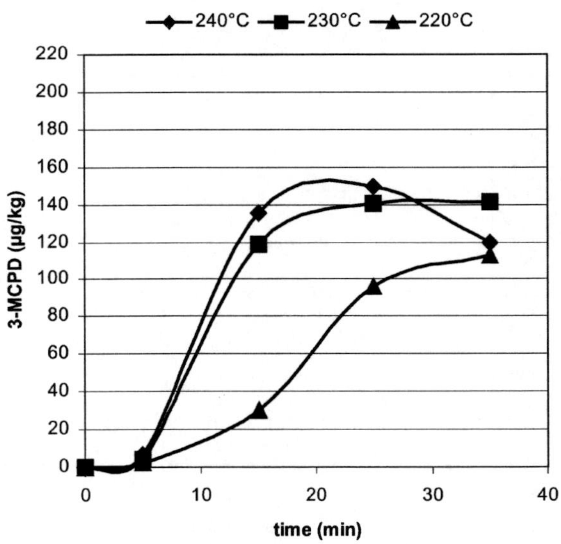

Figure 10.2: Formation of free 3-MCPD during roasting of barley at temperatures of 220, 230 and 240 °C over time (5–35 min). Each value represents the mean value of two independent determinations.

The pH-values of the individual samples were also determined and shown to decrease (from pH 6.4 to pH 5.3–5.4) within the first 15 minutes of roasting, and stabilizing at these values toward the end of the cycle (data not shown). Acidity is attributed to the degradation products of carbohydrates (e.g. organic acids) and possibly also lipids. Chloroesters were also measured in the same samples and ranged from 557–1896 μg/kg, i.e. exceeding the amounts of free 3-MCPD 4–14-fold after a roasting period of 35 min. The formation of esterified 3-MCPD was correlated to both time and temperature, with the highest amounts of 3-MCPD bound in esters (approx. 1900 μg/kg) determined in samples roasted at 240 °C over 35 minutes (Fig. 10.3). The ratio of free 3-MCPD to the chloroesters was in accordance to those reported in other processed foods [15, 16], where bound 3-MCPD may exceed the levels of free 3-MCPD 5–400-fold.

It is important to remember that roasted barley is not consumed directly after the roasting process. During domestic preparation of cereal-based beverages, roasted cereals are extracted with water and afterwards the extract is normally filtered before consumption. Roasted cereals are manufactured into cereal-based beverages analogously, but followed by a drying step to attain the final soluble powder. In the second part of our study, we investigated how water extraction impacts the final content of chloroesters. As 3-MCPD bound in esters are structurally related to lipids it is likely that they are poorly extracted with water due to their hydrophobic properties. Therefore, we analyzed a sample of a 100% roasted barley directly after the roasting step, and then after extraction and evaporation to afford the soluble product. As depicted in Figure 10.4, the level of bound 3-

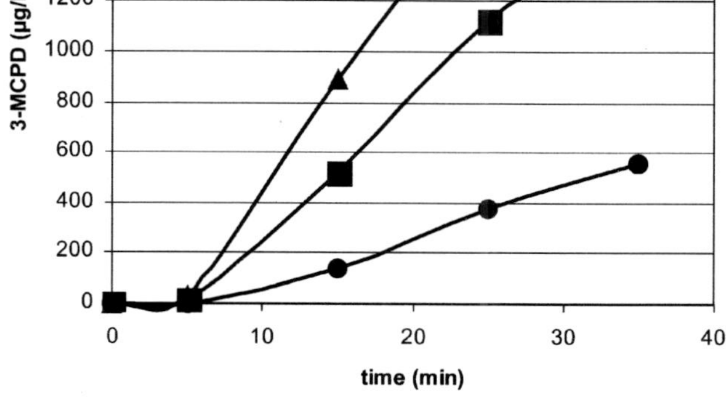

Figure 10.3: Formation of 3-MCPD esters during roasting of barley at temperatures of 220, 230 and 240 °C over time (5–35 min). Each data point represents a mean value of two independent determinations.

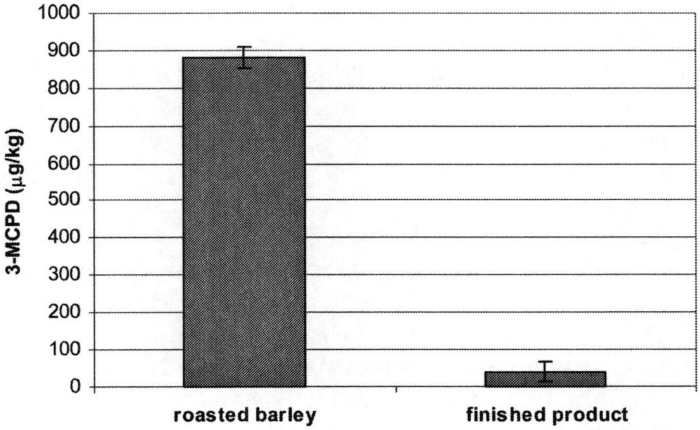

Figure 10.4: Impact of water extraction of roasted cereals on the amount of chloroesters in the final product.

MCPD in the soluble product was 95% lower than in the roasted cereal. The results clearly show that the role of 3-MCPD-esters in terms of overall MCPD exposure in instant cereal-based beverages is only a very minor one.

10.3.2.2 Bread and Toast

A model dough comprised simply of wheat flour, salt and water is sufficient to rapidly generate 3-MCPD upon baking, theoretically following zero order kinetics as both precursors (lipids, chloride ions) are well in excess [22]. The factors that contribute to the rate of formation of 3-MCPD are temperature, time, moisture content, and pH. The moisture content of the dough influences the reaction, and below approx. 15% the moisture content becomes negligible and temperature is the main factor that determines the amount of 3-MCPD in the baked product [22]. Other determinants that may impact 3-MCPD are dough additives such as fats, emulsifiers and sugars [23], and are discussed in more detail later on in this section. Kinetic models to predict the formation and decay of 3-MCPD in leavened and unleavened dough have recently been established [24, 25]. In leavened wheat dough, glycerol is a major precursor of MCPD's, formed primarily by the yeast during proving. Glycerol is a well established precursor of 3-MCPD [10] and can react with chloride ions from added cooking salt to afford 3-MCPD in the dough when baked. In fact, glycerol could account for up to 68% of the MCPD formed in proved dough [24]. On the other hand, model systems comprised of unleavened dough indicate that the contribution of glycerol to MCPD levels is relatively small [25]. Precursors in flour (monoacylglycerols, lysophospholipids and phosphatidylglycerols) as well as constituents of commercial flour improvers (e.g. diacetyl tartaric acid esters and monoacylglycerols) play a more important role in unleavened dough. The addition of monoacylglycerols to model dough afforded a first order rate response in MCPD formation. The mechanism proposed describing the formation of MCPD from 1- and 2-monoacylglycerols is analogous to that initially described by Collier et al. [10], i.e. opening of the cyclic acyloxonium ion intermediate by chloride anions.

3-MCPD is not very stable in foods and any net amount is the result of the mass balance of formation and decomposition. The stability of 3-MCPD in a wheat flour dough model is sensitive to temperature, pH and moisture content [22]. In fact, 3-MCPD may decompose via glycidol and the intermediate formation of 2-MCPD. Kinetic models that take into account the salient process parameters (pH, moisture, time, temperature) could be of use in devising potential strategies of reduction, i.e. setting parameters so as to drive the reaction toward favouring rapid "decay", and avoiding formation of 3-MCPD via reverse reactions (e.g. from glycerol and liberated chloride).

The use of microorganisms, cell extracts, or purified enzymes with dehalogenating properties is a further option to reduce the levels of 3-MCPD. Facile models using suspensions of whole cells bakers yeast in water, with or without glucose as a carbon source, have been proposed and have shown to effectively degrade 3-MCPD, with an apparently higher preference for the (S)-isomer [26]. Cell-free extracts of *Pseudomonas* sp. display esterase activity towards esterified dichloropropanols as well as dehalogenating properties of 2,3-dichloropropanediol [27]. However, this or similar approaches do have limitations as the studies are often conducted with non-food grade microorganisms or enzymes derived thereof. An additional constraint is that certain models are over-simplified and cannot be extrapolated to a food system, exemplified by the study of Bel-Rhlid *et al.* [26], that fails to consider the formation of glycerol, a natural by-product of yeast fermentation, that in the presence of chloride may in a reverse reaction furnish 3-MCPD after baking.

Analogous to the roasted barley experiments reported above, the toasting of bread leads to measurable amounts of MCPD, the concentration of which is dependent on the degree of toast. Hence, darker toasted bread will contain more 3-MCPD [20] that can reach amounts up to 0.275 mg/kg (Fig. 10.5). Studies by Hamlet and Sadd [15] support these observations showing that MCPD is formed predominantly in the crust region of bread.

Additives such as baking agents and fats added to the dough of bread intended for toasting can increase the amount of 3-MCPD in the final product [23]. In fact, a proportional increase of 3-MCPD was shown with increasing amount of baking agent added to the dough [23]. Sucrose, a major ingredient (29 g/100 g) of the baking agent, was suggested as the determinant factor. However, no explanation of the possible reaction mechanism or role of sucrose was provided in the study, but other authors have indicated that simple organic acids – derived from the breakdown of sugars during thermal treatment of the dough [29] – may favour the formation of 3-MCPD from glycerol [10, 30]. These acids may also contribute to lowering the pH of the cooked dough, which in turn increases the stability of 3-MCPD [22].

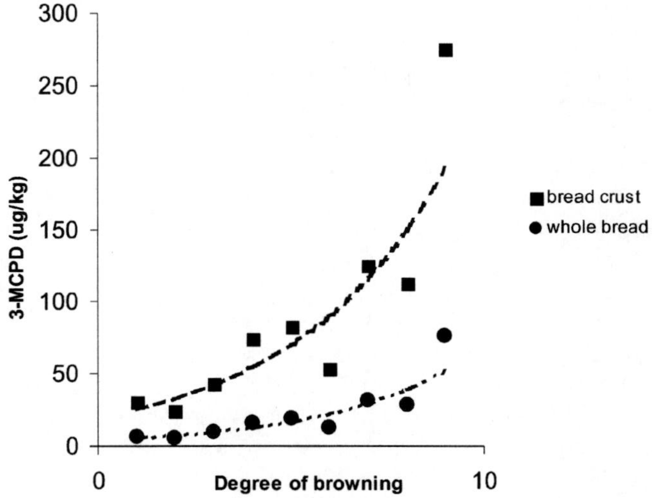

Figure 10.5: Correlation of the degree of browning (increasing tendency across the numerical scale) of the crust and 3-MCPD content in a white bread (data extracted from [20]).

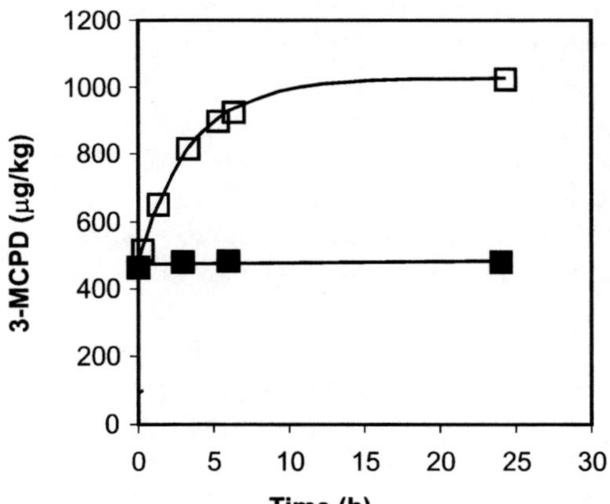

Figure 10.6: Release of bound 3-MCPD (open squares) by treatment of bread crust with lipase over time (23 °C) versus free 3-MCPD (shaded squares). Diagram adapted from [15].

The possible role of 3-MCPD esters as a potential source of MCPD in toasted bread has recently been addressed [15]. In this study, higher concentrations of bound MCPD esters (>1 mg/kg) versus free 3-MCPD (approx. 0.5 mg/kg) were detected. The authors employed a lipase from *Aspergillus oryzae* to hydrolyze the chloroesters and subsequently release the bound MCPD's (Fig. 10.6).

Hence, intermediate chloroesters formed during the baking process may be important to consider when assessing the overall contribution of the food to 3-MCPD exposure.

10.3.3 3-MCPD in Coffee

Green coffee beans do not contain chloropropanols and no 3-MCPD [21], or only spurious amounts of have been detected in roast and ground coffees and soluble coffee [17]. The concentrations reported in the literature range from <3 µg/kg to 18.5 µg/kg on a powder basis [17], i.e. considerably lower than the concentrations reported in roasted barley products and bread. Analogous to the positive correlation of colour to 3-MCPD in toast, higher concentrations were measured in darker roasted beans, that is contrary to the behaviour of the food-borne contaminant acrylamide in roasted coffee [28]. However, the same study [17] has revealed that roasted and soluble coffees (commercial samples and prepared in the laboratory) may also contain low amounts of 3-MCPD esters, isolated from the coffee lipid fraction. In roast and ground coffee powder up to 390 µg/kg 3-MCPD bound in esters have been measured, whereas in the soluble instant coffee powder only 6 µg/kg were found due to the considerably lower amount of lipids in soluble coffee [17]. However, coffee brews were not analyzed. The results of the instant coffees indicate clearly that MCPD-esters are not extracted during the water extraction process.

Conclusively, it can be stated that the role of MCPD-esters in brewed coffee is a minor one. Similar to other food products, the chloroester levels significantly exceeded those of free 3-MCPD. Furthermore, statistically significant differences were determined in Arabica versus Robusta beans, in that the latter apparently harbours relatively lower amounts of bound MCPD.

Attempts to correlate MCPD formation with other coffee constituents is a difficult task, considering the higher temperatures of roast of coffee than most other foods and the chemical complexity of the product. Interestingly, the free MCPD content seems to correlate positively to the diacyl- and triacylglycerol concentrations, and MCPD esters decrease in concentration with increase degree of roast [17]. On a comparative basis, however, if at all roasted coffee contains only spurious amounts of free 3-MCPD and as also in the case of roasted barley, essentially no chloroesters are extracted into the brew (valid also for soluble coffee).

10.4 Conclusion

Progress has been made in the past years in understanding the key precursors and intermediates that lead to the formation of 3-MCPD in different foods. Certain catalysts have been identified (e.g. enzymes, organic acids) and kinetic models have been established to predict the formation in staple products such as bread. In the case of acid-hydrolysed vegetable proteins (HVP) and soy sauce, manufacturers have already more than a decade ago made modifications to their processes to ensure that 3-MCPD in the products resides within acceptable levels.

Based on process knowledge and the results of fundamental mechanistic research, several options can be identified to control or lower 3-MCPD in foods. However, their efficacy in terms of reduction has in most cases only been assessed at the laboratory or pilot stage, and some general approaches are listed below:

- raising the pH (e.g. neutralization of acid-HVP by addition of NaOH)
- adjustment of *T/t* profiles
- use of colour as an endpoint (e.g. toasting bread to a "golden yellow" colour)
- lowering the chloride (salt) content
- limiting the amount of glycerol in the food produced during preparation and storage
- avoiding the use of partial glycerides as additives in heat processed foods
- inactivation of lipases/esterases
- use of spice extracts in place of native spices, or reducing the microbial load via thermal treatment
- screening foods, food ingredients and additives for the presence of chloroester intermediates
- screening food contact materials for 3-MCPD precursors.

Clearly, more research into the key parameters that determine the formation of 3-MCPD in different foods is warranted, and the contribution of the purported intermediates and role of catalysts is not adequately understood. Furthermore, mitigation strategies can be recommended but with certain caveats, e.g. practicality, impact on overall quality and safety, consideration of other "undesired" processing compounds, e.g. acrylamide. This also raises questions of how to set priorities for single chemicals and how to position these within a risk/risk analysis, taking into account that humans have been exposed to such molecules for thousands of years through the normal diet.

References

1. Velíšek, J., Davidek, J., Hajslova, J., Kubelka, V., Janicek, G., Mankova, B. *Z. Lebensm. Unters. Forsch.* 1978, **167**, 241–244.
2. Velíšek, J., Davídek, J., Kubelka, V., Pánek, J. *Sbor. ÚVTIS-Potr. Vìdy* 1985, **3**, 84–94.
3. SCF 2001, Opinion of the Scientific Committee on Food on 3-monochloro-propane-1,2-diol (3-MCPD).
4. JECFA 2001, Joint FAO/WHO Expert Committee on Food Additives Fifty-seventh meeting Rome, 5–14 June 2001. Summary and Conclusions.
5. European Commission, 2001, Commission Regulation (EC) No 466/2001. Setting maximum levels for certain contaminants in foodstuffs. *Official Journal of the European Communities.* 8th March 2001.

6. Hamlet, C.G., Jayaratne, S.M., Matthews, W. *Food Addit. Contam.* 2002, **19**, 15–21.
7. Crews, C., Hough, P., Brereton, P., Harvey, D., Macarthur, R., Matthews, W. *Food Addit. Contam.* 2002, **19**, 22–27.
8. Crews, C., Brereton, P. and Davies, A. *Food Addit. Contam.* 2001, **18**, 271–280.
9. Hamlet, C.G., Sadd, P.A., Crews, C., Velíšek, J., Baxter, D.E. *Food Addit. Contam.* 2002, **19**, 619–631.
10. Collier, P.D., Cromie, D.D.O., Davies, A.P., *J. Amer. Oil Chem. Soc.* 1991, **68**, 785–790.
11. Robert, M.C., Oberson, J.M., Stadler, R.H. *J. Agric. Food Chem.* 2004, **52**, 5102–5108.
12. Davídek, J., Velíšek, J., Kubelka, V., Janicek, G., Simicova, Z. Z. *Lebensm. Unters. Forsch.* 1980, **171**, 14–17.
13. Kuksis, A., Marai, L., Myher, J.J., Cerbulis, J., Farrell, H.M., Jr. *Lipids* 1986, **21**, 183–190.
14. Myher, J.J., Kuksis, A., Marai, L., Cerbulis, J. *Lipids* 1986, **21**, 309–314.
15. Hamlet, C.G., Sadd, P.A. *Czech. J. Food Sci.* 2004, **22**, 259–262.
16. Svejkovska, B., Novotny, O., Divinova, V., Reblova, Z., Dolezal, M., Velíšek, J. *Czech. J. Food Sci.* 2004, **22**, 190–196.
17. Dolezal, M., Chaloupska, M., Divinova, V., Svejkovska, B., Velíšek, J. *Eur. Food Res. Technol.* 2005, **221**, 221–225.
18. Hamlet, C.G. *Food Add. Contam.* 1998, **15**, 451–465.
19. Brereton, P., Kelly, J., Crews, C., Honour, S., Wood, R., Davies, A. *J. AOAC Int.* 2001, **84**, 455–465.
20. Breitling-Utzmann, C.M., Köbler, H., Herbolzheimer, D., Maier, A.D. *Lebensm. Rund.* 2003, **99**, 280–285.
21. Kurzrock, T., Speer, K. Determination of 3-monochloropropane-1,2-diol in coffee and coffee surrogates. Symposium of the DFG-Senate Commission on Food Safety, 25th–27th September 2005, Kaiserslautern, Germany.
22. Hamlet, C.G., Sadd, P.A., Gray, D.A. *Eur. Food Res. Technol.* 2003, **216**, 122–128.
23. Breitling-Utzmann, C.M., Hrenn, H., Haase, N.U., Unbehend, G.M. *Food Addit. Contam.* 2005, **22**, 97–103.
24. Hamlet, C.G., Sadd, P.A., Gray D.A., *J. Agric. Food Chem.* 2004, **52**, 2059–2066.
25. Hamlet, C.G., Sadd, P.A., Gray, D.A. *J. Agric. Food Chem.* 2004, **52**, 2067–2072.
26. Bel-Rhlid, R., Talmon, P., Fay, L.B., Juillerat, M.A. *J. Agric Food Chem.* 2004, **52**, 6165–6169.
27. Kasai, N., Suzuki, T., Idogaki, H. *LWT* 2004, **39**, 86–90.
28. Stadler, R.H., Scholz, G. *Nutrition Reviews* 2004, **62**, 449–467.
29. Hamlet, C.G., Sadd, P.A. *Food Addit. Contam.* 2005, **22**, 616–623.
30. Gassenmeier, K., Goeke, A. *Lebensmittelchemie* 1999, **53**, 29–46.

11 Minimization Strategies: Acrylamide

Norbert U. Haase *

Abstract

Acrylamide findings in several foodstuffs of plant origin are of major concern. Total daily intake is shared to different food categories and minimization efforts should cover all of them, because consumer preferences are incalculable. Acrylamide formation of semi-finished food takes place at the stage of final preparation in restaurants or in the domestic sector. To reduce the acrylamide level in these products, an education campaign with modified cooking advices is necessary. Industrial food of plant origin with a ready to eat character may be under control of national food authorities.

The concrete acrylamide content in a given product depends upon several aspects. The raw material contains the precursors of the formation. Furthermore a recipe may enhance the precursors concentration, but specific pretreatments reduce it. Final heating determines the local formation of acrylamide, and food size determines the total acrylamide concentration.

In potato food the formation mainly depends on the amount of reducing sugars. Several factors were identified to have an influence on that, and as a result each potato lot has its individual acrylamide formation potential. Among others, concentration of precursors can be reduced by an additional leaching or blanching step. Also a changed thermal input (lower temperature over a prolonged cooking time) may be successful. Vacuum frying reduces the formation, whereas high pressure frying may be helpful in specific circumstances (e.g. prefabricated crisps) only. Furthermore, in prefabricated products an enzymatic asn reduction at dough stage is under design.

In cereal based food, free asparagine limits the acrylamide formation. On principal, the same tools as in potato food are relevant, concerning a reduced heat treatment and a precursor reduction. The acrylamide formation in wheat is lower than in rye products. Light flours contain less free asparagine than whole meal flours (e.g. type 405 vs. 1050). Also the shape to volume ratio is of relevance, and half-dry or dry products (e.g. crisp bread) are higher in acrylamide than wet ones. Furthermore, several ingredients may have an influence upon the formation potential. Oven geometry and temperature distribution inside the oven are important aspects at baking itself. Dough rising by yeast may reduce

* Federal Research Centre for Nutrition and Food, Institute for Cereal, Potato and Starch Technology, Schuetzenberg 12, 32756 Detmold, Germany.

Thermal Processing of Food: Potential Health Benefits and Risks
DFG, Deutsche Forschungsgemeinschaft

ISBN: 978-3-527-31909-1

the free asn level. An industrial application of such an enzymatic reduction of asn is under development. Ammonia as a rising agent should be substituted.

Until now the identified reduction potential in coffee is low, whereas coffee extract and coffee substitutes may offer some possibilities including heat treatment and modified ingredients.

Summarizing the list of acrylamide reduction potentialities, the best procedure will be a combination of several single steps. Thereby, the remaining nutritional and organoleptic product quality must be controlled to fit the consumers demand. In this context the European food industry (CIAA) has described a toolbox system for different food categories covering all relevant aspects.

11.1 Introduction

Acrylamide findings in several foodstuffs are of major concern, because of a potential health risk. Since 2002, a lot of research activities were initiated to reduce levels. Next to heated potato food also cereal food and coffee have come into focus. A national German project on these subjects was finished in June 2005, sponsored by both government and industry [1]. Next to food specific formation aspects and perspectives to reduce levels, also analytical and toxicological details were included. Several other world-wide activities are finished too or still are under progress. Also a EU sponsored project within the 6th framework (HEATOX) will offer data soon. A database of all registered European activities is available on the web [2]. Furthermore, the EU authorities have organized stakeholder meetings to promote exchange of latest acrylamide research [3]. New knowledge of fundamental and applied research was collected on acrylamide in a relative short period, to allow an updated view upon potato products combining the overall product quality with health aspects.

A risk assessment has to overcome a couple of incalculable problems. The individual food choice and food uptake differ widely. Furthermore, the concrete acrylamide concentration in the meal is unknown in most cases. Nevertheless some data are available, indicating a mean daily uptake between 0.2 and 2 μg/kg body weight [4–6]. One of these studies has estimated the acrylamide uptake by children and adolescents, not detecting any differences compared with adults [5].

The upcoming public discussion about acrylamide findings in foods of plant origin focussed mainly on potato food, but the latest FAO/WHO report of the expert committee on food additives [6] indicated several food categories to be involved in the acrylamide exposition (Fig. 11.1). Interestingly, coffee rates next to French fries and crisps, a noticeable part of the daily acrylamide uptake.

With respect to the unsolved uncertainties of the acrylamide discussion consumers are asked to continue their individual diets. Furthermore, the acrylamide concentration in the food should be lowered to reduce the overall risk.

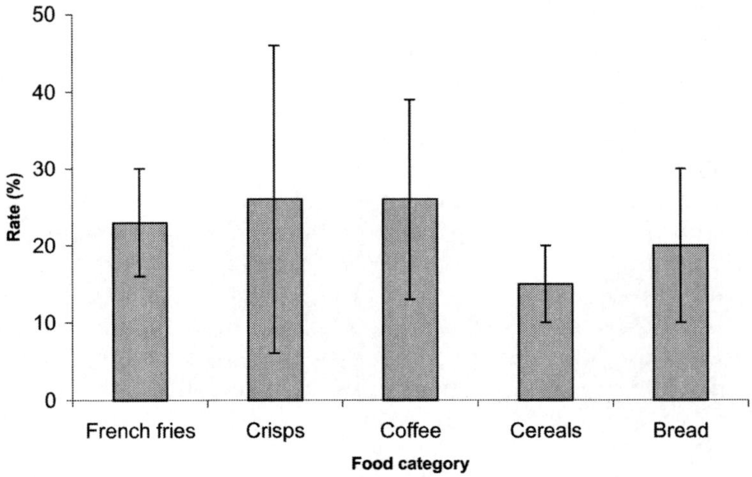

Figure 11.1: Share of several food categories in the human acrylamide exposition.

This report gives an overview upon minimization tools in acrylamide formation. Concrete data are with potato and cereal food in particular.

11.2 Formation Aspects

Acrylamide in fried and roasted products is formed by several chemical reactions, e.g. lipid degradation via acrolein [7], and an enzymatic modification of the amino acid asparagine [8, 9]. The principal reaction includes the free amino acid asparagine and carbonyl groups, derived from carbohydrates and especially from reducing sugars (e.g. glucose, fructose) [10–12]. Thus the reaction pathway is linked with the early Maillard reaction, covering a decarboxylation of the Schiff base, an rearrangement to Amadori products, and a final β-elimination reaction to release acrylamide [12].

With view to the concentration of the precursors, the overall reaction efficiency is low [13]. Further on, Biedermann et al. have shown by addition of deuterated acrylamide before heat treatment, that the final acrylamide concentration in food is a result of concurrent formation and elimination reactions [14]. Model studies with different food matrices have demonstrated specific factors to determine the acrylamide formation in the different food categories [14–17].

A reduction of any acrylamide potential has to focus on two aspects, a reduction of the precursors concentration in the food to be processed and an optimised heat treatment. In any case, both potentialities have to be considered carefully, because a reduction of acrylamide should maintain the overall organoleptic and nutritional quality of the products as far as possible.

11.3 Minimization Potentialities in Potato Food

The consumers demand of fried, baked, and roasted potato products increased significantly during the last thirty years. Especially the additional taste from the frying medium has promoted the economic meaning of that products, which actually exceeds fresh potato trading in several countries. Next to staple articles like French fries or potato crisps, the processing companies have diversified their offerings with a number of semi-finished and finished products.

Potatoes are known to have a relative high level of free asparagine [18, 19]. The second precursor, the sum of different reducing sugars, has a much lower concentration, but it varies in a large range, depending of several factors to be discussed later on. Consequently, the reducing sugars will limit the acrylamide formation in potatoes and have to be controlled in particular when trying to optimise the raw material.

11.3.1 The Raw Material

The quality of the raw material is an important tool within potato processing, because a high variability of the relevant constituents between single lots guides to large variations of the quality of the final products. Basis of that behaviour is the plant origin: Potato tubers (*Solanum tuberosum* ssp. *tuberosum* L.) are vegetative parts of the plant, grown subterranean, and therefore be influenced by a couple of environmental factors.

Asparagine as the most relevant amino acid for acrylamide formation [20] was found to be 39% of total free amino acids [21]. In a set of nine cultivars, grown at two locations in 2003, concentration of free asparagine (asn) ranged between 58.7 and 143 mmol asn/kg dry matter (DM). The corresponding values of reducing sugars (RS) were between 3.64 and 13.8 mmol RS/kg DM [22].

Since many years, reducing sugars (RS), defined as sum of glucose and fructose and regardless of other reducing sugars, are discussed critically in view of heated potato products which undergo the Maillard reaction, because colour formation has been found to be closely connected with reducing sugars. A recently published review is from Kumar et al. [23]. In a stepwise regression analysis, around 90% of the colour variation could be explained by that constituents [24], while free asparagine was not correlated with Maillard derived colour [25].

Continuous monitoring of sugar level in tubers during storage will be useful in determining when the sugar concentration rises to critical concentrations. Consequently, different strategies were developed by the industry to control sugar level by laboratory techniques or by rapid test procedures. Next to the content of reducing sugars, defined as the sum of glucose and fructose, also glucose and sucrose levels were used as a criterion for acceptance or refusal of the concrete potato lot. In connection with the acrylamide discussion this approach has received increased attention by the different stakeholders.

To get an impression on that variability, samples of ten cultivars, grown at five locations, were analyzed according to the concentration of relevant acrylamide precursors (reducing sugars, RS; free asparagine, asn). The reducing sugars concentration varied significantly ($p<0.05$) between cultivars, but on a non-significant level between growing locations. Variation between locations was not the same for all cultivars, as to be shown by the minimum-maximum range of data. Asparagine concentration differed between cultivars and growing locations on a significant level ($p<0.05$), but again a cultivar dependent variation could be observed (Fig. 11.2).

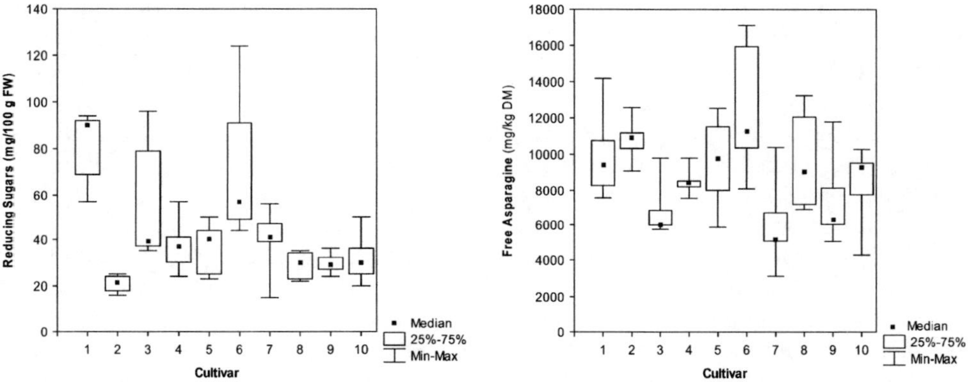

Figure 11.2: Reducing sugars and free asparagine in ten cultivars, grown at five locations.

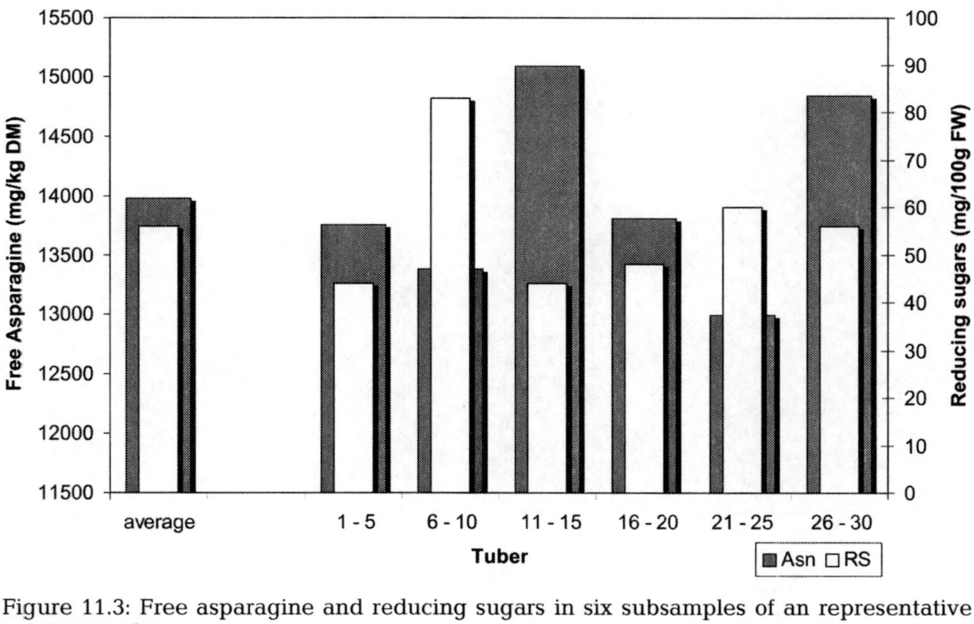

Figure 11.3: Free asparagine and reducing sugars in six subsamples of an representative potato sample.

Acrylamide data of industrial potato processing indicate a high variability even within one cultivar [26]. To demonstrate such inhomogeneities, a representative aliquot of 30 tubers of a single lot was randomly divided into six subsamples of five tubers each and analyzed according to the main acrylamide precursors (Fig. 11.3).

The results confirmed the above mentioned variability. The level of reducing sugars was between 44 and 83 mg/100 g FW, whereas the content of free asparagine was between 13000 and 15000 mg/kg DM. A correlation of both criteria according to the subsample was not possible.

In developing a long term perspective of the acrylamide potential of the raw material an retrospect of several years points out year-dependencies. Four cultivars, grown at three different locations, were analyzed according to their content of reducing sugars, both after harvest and after a half year of storage at 8 °C (Fig. 11.4).

Next to year dependent RS concentrations at harvest time, also increases after a long term storage were different. 2003 was noticeable as a growing season with lower RS values after storage than at harvest time.

Absolutely, the concentration of free asparagine exceeded RS values in all investigated samples, indicating the relevance of reducing sugars in minimization concepts for acrylamide. Since many years, reducing sugars are discussed critically with heated potato products, because colour formation within the Maillard reaction has been found to be closely connected with reducing sugars.

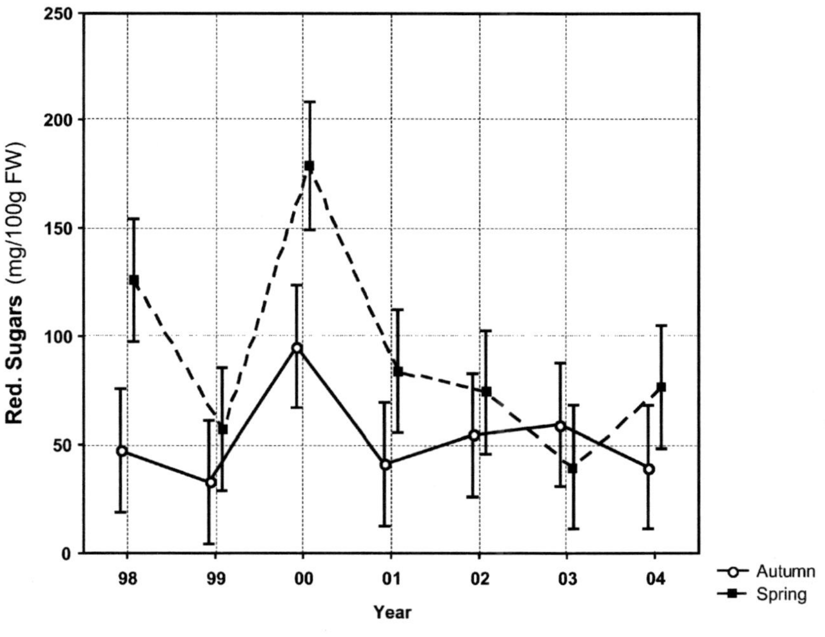

Figure 11.4: Reducing sugars in four cultivars, suitable for processing, between 1998 and 2004 (data of harvest and storage time); vertical bars indicate a 0.95 confidence interval.

Variability of RS data is confirmed in relation to cultivars [27], soil type, weather [28–30], specific growing conditions with individual plant treatments [15, 30], harvest [31], and storage [32]. Summarizing these aspects, several reduction possibilities are offered, but in each concrete situation it has to be proven which factor is of relevance.

Asparagine as the most relevant amino acid for acrylamide formation was found to be 39% of total free amino acids [21]. On the other hand, around 90% of the colour variation of fried potato food could be explained by reducing sugars [24]. In another experiment, sucrose explained 3% and RS explained only 29% of crisp colour variation at harvest time. After a five months storage at 8 °C, the values were 23% and 38%, respectively [33]. Free asparagine was not correlated with the Maillard derived colour [25]. With respect to the dimension of asn values, the effect of an reduction is still on debate. Heuser et al. [34] have shown some principal aspects in relation to fertilisation, which have to be proven in relation to the overall processing suitability of those potato tubers.

11.3.2 Potato Processing

Potato products include a lot of different preparation stages. Thereby, an acrylamide potential is noticeable both at industrial cooking and at domestic preparation. In the latter case, the concrete acrylamide formation depends on the individual heat intensity and may contribute a substantial part of the overall daily acrylamide intake.

On the other hand, industrially finished products are under control of national food control authorities offering the challenge of a reduced acrylamide content in food, e.g. by guidelines of processing. The German dynamic minimisation concept aims such a gradual reduction of acrylamide contents, agreed among the German authorities and food industry [35].

11.3.2.1 Semi-finished Potato Food (French Fries)

Industrially produced par-fried French fries contain only traces of acrylamide. Nevertheless, it seems to be possible to reduce the overall potential by a prolonged leaching period, especially in case of potato lots with high sugar level. An experiment with samples from two cultivars with an additional leaching step after the typical blanching procedure reduced the concentration of reducing sugars. As a consequence the acrylamide formation at final preparation dropped drastically (Tab. 11.1).

Cooking recommendations for French fries finishing were changed to reduce the acrylamide formation [26, 36]. As data have shown (Tab. 11.1), the overall potential may also be reduced by an intensified leaching at par-frying.

Table 11.1: Acrylamide concentration in French fries as well as RS and asn concentration in the raw material; results of a 50 min additional leaching of two cultivars, grown at three locations, autumn and spring data.

Leaching	N	Acrylamide [µg/kg]		Red. Sugars [mg/100 g FW]		Free asn [mg/kg DM]	
		average	standard deviation	average	standard deviation	average	standard deviation
All	24	110	67.1	54.5	54.2	9958	5035
No	12	138	82.1	67.5	64.2	12495	5616
Yes	12	81.9	31.3	41.5	40.7	7420	2724
	Autumn						
No	6	103	26.9	30.8	19.3	8574	3149
Yes	6	68.7	23.2	19.2	9.7	5772	2073
	Spring						
No	6	172	106	104	73.9	16416	4751
Yes	6	95.0	34.7	63.8	48.4	9069	2345

The finished French fries had an acceptable organoleptic quality with less browning and with a bit firm texture.

Otherwise also a simple renouncement of fine cuts (6×6 mm, 8×8 mm) and a preference of normal or thick cuts also reduced the acrylamide values significantly [37] because of the changed surface to volume ratio.

11.3.2.2 Finished Potato Food (Potato Crisps)

Processing of potato crisps represents a dehydration of thin sliced potatoes in hot oil. An acceptable level of crispiness needs an residual moisture level below 5%. Under the guideline of acrylamide minimisation tools, industry has increased the moisture level. To follow that procedure, a set of potato samples was crisped under standardized conditions. As a result, the final moisture level was not correlated with the acrylamide concentration (Fig. 11.5), indicating that rather a change of the individual heat treatment will have an influence upon the concrete acrylamide formation. On the same subject, Biedermann et al. [14] have reported higher acrylamide levels in case of prolonged heating but without any change in residual moisture.

A reduction of low molecular weight components including sugars and free asparagine in crisps potatoes by leaching or blanching is possible [38]. Therefore, this effect was tested in relation to a potential acrylamide reduction. Calcium ions were added to the water (2% $CaCl_2$), because some negative textural changes of the pectin network are reported [39]. Also the influence of a simple pH-lowering by citric acid (0.1%) was tested at different temperature regimes (30 °C, 50 °C, 70 °C and 90 °C, respectively; 2.5 min each) (Tab. 11.2).

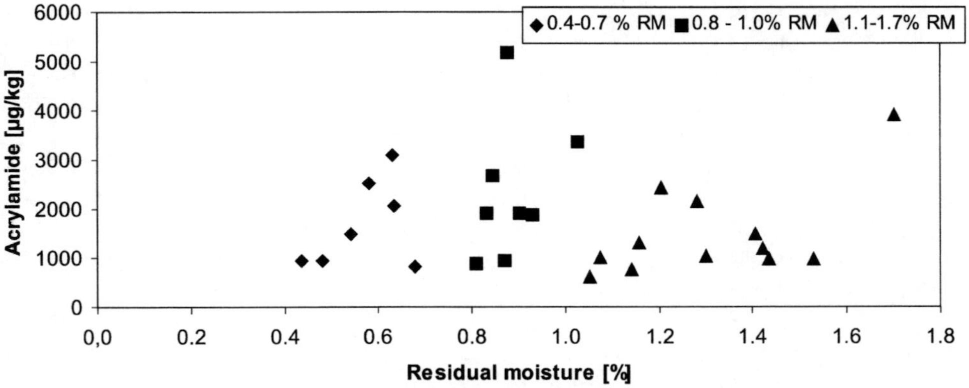

Figure 11.5: Relation between residual moisture of crisps and acrylamide level.

Table 11.2: Efficiency of different leaching conditions (water, calcium chloride (2% w/v), citric acid (0.1% w/v), temperature) on several quality aspects.

Variant	Temp. [°C]	RS [mg/100 g FW]	Free asn [mg/kg DM]	Colour [L-value]	Texture [N]	Acrylamide [µg/kg]
Water	30	73	11335	64.2	2.56	n.d.
	50	45	14642	63.9	2.3	1770
	70	34	8405	63.5	2.09	627
	90	31	6425	66.7	2.55	371
$CaCl_2$	30	73	11753	59.4	3.78	n.d.
	50	53	5322	58.1	3.82	1260
	70	55	6766	59.4	2.76	54
	90	31	6555	58.1	3.09	46
Citric acid	30	67	12263	61.0	2.29	n.d.
	50	50	11583	62.8	2.60	1030
	70	36	8015	61.5	2.2	979
	90	26	7639	65.3	2.35	558

n.d.: not determined.

A short water contact at 50 °C caused a noticeable decrease in sugars, even continued above the gelatinization range of potato starch (55–65 °C). The same effect could be observed for free asparagine. Pedreschi et al. [40] reported a 76% reduction of glucose and a 68% reduction of asparagine by blanching of sliced potatoes. Colour of the final product became brighter, crispiness was not effected, and acrylamide concentration dropped down drastically, reported also by Haase et al. [41]. The addition of calcium chloride to the water produced no further reduction of low molecular weight compounds, but breaking force was a bit higher in relation to water leached samples. As a main result, the acrylamide concentration was almost zero at water temperatures beyond the gelatinization temperature. Lowering the pH of the water to 4.2, no additional effect of the pH was detected. On the same subject, Pedreschi et al. [40] described a reduced acrylamide formation by almost 70% against a control for slices fried at 150 °C. Surprisingly this effect was not present at higher frying temperatures.

The crisps samples of the calcium chloride variants had a bitter taste. Therefore, the experiment was repeated with a reduced calcium chloride concentration (0.1–1.0% w/v). A separate washing step after leaving the blancheur was implemented to clean the surface of the slices (Tab. 11.3).

The additional cleaning of the blanched slices removed most of the gelatinized starch from the surface, and the concentration of water soluble compounds, e.g. RS, was reduced. As a result, the acrylamide formation declined. A low calcium chloride supplement to the blanching medium (up to 0.2% w/v) had no further reduction potential for acrylamide compared with water treatments. Also a substantial influence upon crisps hardness (texture) could not be detected. A negative bitter taste of the crisps was not described below a 0.5% $CaCl_2$ supplement. This indicates a small effective concentration range of an acrylamide reduction potential in combination with a most perfect organoleptic behaviour.

The formation of acrylamide in relation to the frying temperature itself was investigated in a temperature range between 140 °C and 220 °C. Frying time was adjusted to get crispy potato slices (Fig. 11.6).

The increase in acrylamide followed a non-linear progression line with a correlation coefficient of $r = 0.92$ against frying temperature. The highest value was measured at highest temperature. At rising temperatures the calculated temperature load (temp.×time) went down from 52,920 (°C s) at 140 °C to

Table 11.3: Salt effect at blanching (80 °C, 2 min) upon several quality parameters with and without additional washing.

Variant	Salt	Washing	RS [mg/100 g FW]	Texture [N]	Sensoric quality [Score]	Acrylamide [µg/kg]
Control		No	108	2.88	4.4	1874
Water	0	No	51	1.97	5.0	867
		Yes	42	2.30	4.4	446
$CaCl_2$	0.1	No	60	2.39	4.8	1185
		Yes	65	2.44	4.5	676
	0.2	No	54	2.61	4.5	751
		Yes	41	3.07	4.5	681
	0.5	No	43	3.14	4.0	260
		Yes	34	2.84	3.8	441
	1.0	No	53	2.78	3.5	158
		Yes	39	2.45	3.8	214

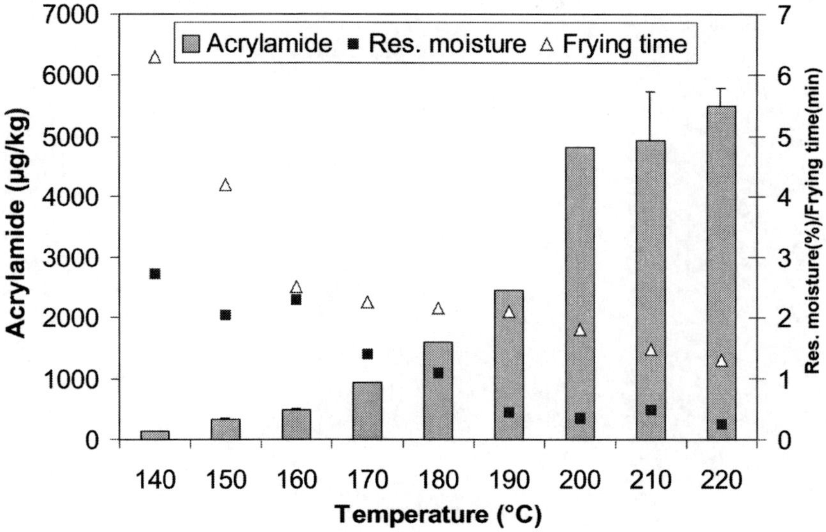

Figure 11.6: Acrylamide formation in relation to the frying temperature.

17,160 (°C s) at 220 °C. The topmost sample still required a relatively long frying time of 80 s. As a consequence the present advice toward moderate frying temperatures [26] was verified. A temperature decrease from 180 °C to 165 °C reduced the acrylamide content in potato crisps by 51% during traditional frying [42].

Another subject related to an overfrying of the samples, being present especially in inhomogeneous potato lots with different dry matter contents. A first sample set was heated until no large bubbles were formed anymore, and a second set of samples was heated for one more minute (Fig. 11.7).

At frying temperatures above 170 °C, a prolonged frying resulted in increased acrylamide levels. This would indicate a need for specified frying times, if the actual preparation advice of moderate frying [26] would not exist. The fat content was the same at all three temperatures independent of the different frying times, but increased substantially at prolonged frying.

The frying experiments described above were conducted with a most stable temperature of the oil bath. In the following experiment, temperature was lowered after a first frying at 170 °C (Fig. 11.8).

Temperature was changed after 1.5 min initial frying at 170 °C. At that time, the water content was about 26%, and the fat content was about 37%. An acrylamide formation was already present (300 µg/kg). The gradient (Δ 10 °C to 40 °C) had no effect upon colour and texture, whereas fat concentration increased up to 60%. The acrylamide formation was reduced substantially, especially at an intensified temperature decline.

A clear temperature decline resulted in quality losses (fat uptake). To overcome these negative product qualities at low frying temperatures, samples

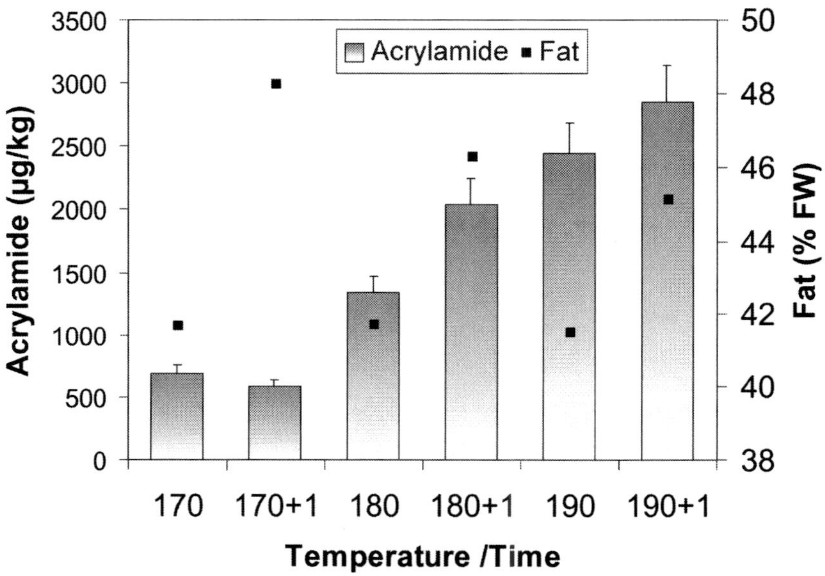

Figure 11.7: Effect of a prolonged frying time upon acrylamide concentration and fat uptake.

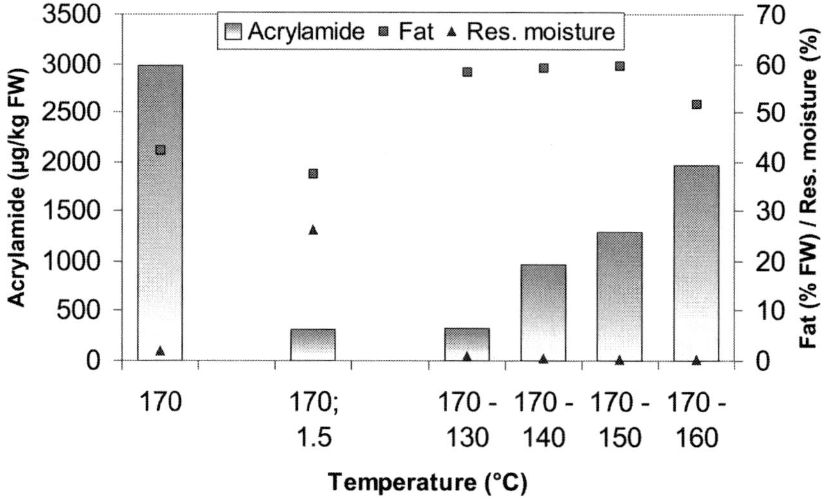

Figure 11.8: Effect of a temperature gradient during frying upon acrylamide and fat concentration.

were finished fried at reduced atmospheric pressure in a special fryer of BMA Florigo B.V. (Woerden, NL). Initial frying was with the atmospheric fryer of BFEL prior to switching over to the closed system (at that time sample moisture was between 5 and 10%) (Tab. 11.4).

Table 11.4: Effect of vacuum frying upon potato crisps quality.

Atmospheric		Vacuum		Product			
Temp. [°C]	Res. moisture [%]	Pressure [mbar]	Temp. [°C]	Res. moisture [%]	Fat [% air-dry]	Colour [L-value]	Acrylamide [µg/kg]
160	2.5	–	–	2.5	36	67	2140
160	9	200	100	2.7	43.8	64	610
160	4.5	100	90	3.2	44.9	66	486
160	4.5	100 →500	90	4.9	47.1	63	468
160	9	100	100	3.3	41.7	66	279

A significant decrease of the frying temperature at the final frying stage resulted in a considerable reduction of acrylamide. In this connection Granda et al. [42] reported an acrylamide reduction of 63% when the temperature decreased from 140 °C to 125 °C in vacuum frying. The overall product quality was different with respect to specific processing parameters (e.g. pressure) indicating the need for further optimization of chemical and organoleptic quality of the product. Other experiments in this test bed have shown that the organoleptic quality of the vacuum fried crisps was not significantly different in relation to the frying method for texture and flavour characteristics (mouth feeling flavour). Otherwise colour was significantly different between frying methods [42].

The inverse technique with a pressurised frying system is used in some special cases to reduce frying time (especially for fabricated crisps) [26]. In further experiments it has to be examined whether the anticipated reduction in temperature load will break the inverse relationship with acrylamide (see above).

11.4 Minimization Potentialities in Cereals

The average daily uptake of cereal based food exceeds all other plant food on a dry weight basis. In comparison with potato food the diversity of products is much higher. Next to the number of raw materials also the range of products is most complex. Therefore, it seems to be almost impossible to state a relationship toward the concrete acrylamide uptake.

11.4.1 The Raw Material

Wheat and rye are the most significant sources for cereal food. In comparison with potatoes, the concentration of free asparagine is much lower, wheat lower than rye (400 and 1000 mg/kg DM respectively). The more nitrogen fertilizer used, the more protein tends to be made by the grain and more asparagine is available [43]. Following the acrylamide formation, free asparagine limits the reaction in cereal processing [44]. On the other hand sugars are even added to some recipes. As a typical processing tool of cereals, the kernels are ground and used as a flour. A partial exception is the use as a breakfast cereal. Milling results in different flour qualities. Next to a wholemeal flour several flour types with different components of the kernel are available. Figure 11.9 indicates the different concentrations of free asparagine in relation to the mineral concentration, following a typical mill stream.

Bright flours with a low mineral concentration also have a low asparagine concentration, indicating a low acrylamide formation potential. On the other hand, whole meal flours with health benefits contain more asparagine and a relatively high acrylamide potential. To overcome this negative aspect of whole meal flours, an exemplary test was carried out to lower the asparagine level but to maintain the character of a wholemeal flour. A rye sample was abrasive peeled with mass losses between 1.9 and 5%. The concentration of free asparagine dropped from 1400 to 1250 and 900 mg/kg DM, respectively [33]. This pre-treatment before milling seems to be qualified to reduce the acrylamide potential and to keep the nutritional benefits of the flour.

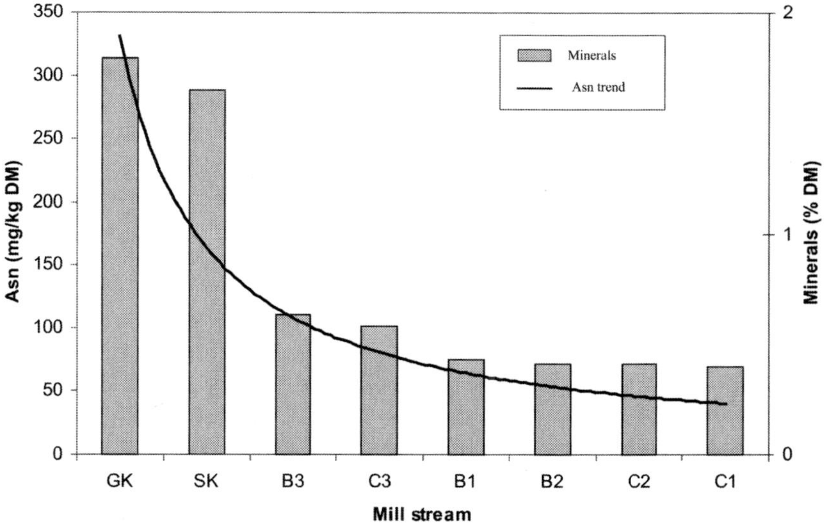

Figure 11.9: Concentration of minerals and free asparagine in wheat flour following the mill stream, cultivar Drifter.

11.4.2 Processing

The high diversity of products leads to incalculable human exposure. Also regional or national specialities contribute to this. German consumers choose between two hundred types of bread, whereas in other countries a more uniform and small assortment exists. Furthermore the dominant flour type varies between wholemeal and bright flours. For breakfast cereal soft wheat with a low protein content is used often, whereas bread making requires higher protein concentrations for high quality products. Some ingredients of the recipe were identified to contain acrylamide. Also a shift of the reaction efficiency is possible [26]. Bakery products for diabetics often contain high fructose levels suitable for an increased acrylamide formation.

With respect to the overall acrylamide formation principles, the predominantly formation takes place in the outermost crust layer, whereas the crumb contains small amounts only. In the base crust the levels were found to be higher than in the top crust [43]. Bread toasting resulted in an acrylamide increase, but absolute values remained on a low level (<100 μg/kg) [43]. Intensively baked breads (e.g. traditional wholemeal flour breads) tended to higher acrylamide concentrations than soft baked white breads. Low volume breads were higher in acrylamide content than large volume bread loafs. A reduced temperature profile resulted in a brighter crust and in a lower acrylamide concentration despite of a prolonged baking time [45]. Another possibility to lower the acrylamide content is the partial or full elimination of free asparagine in the dough. The addition of pure asparaginase was successful in a baking model [33], but so was the use of a baker's yeast in combination with an extended reaction time [46]. A typical short fermentation of a dough did not reduce the asparagine level significantly, but longer fermentation did. The analysis of the applied baker's yeast resulted in a mix of *Saccharomyces cerevisiae* and *Lactobacillus curvatus* and *Lactobacillus para-para casei*, respectively. Isolates of the yeast were able to reduce asparagine of about 99% [33].

Following the principles of acrylamide formation, dry or most dry products, e.g. biscuits and crisp bread, are more sensitive than bread loafs to high acrylamide concentrations, manufacturing of wafers with two types of wheat flour indicated that (Tab. 11.5). Addition of sodium hydrogencarbonate (an ingredient of baking powder) had no influence on acrylamide formation.

An experiment with gingerbread has shown an over-proportional acrylamide increase by the raising agent ammonium hydrogencarbonate [47]. The reason for this is still unclear, as ^{15}N-labelled ammonium hydrogencarbonate did not lead to ^{15}N-acrylamide in the product. A simple exchange of the raising agent by baking powder eliminated most of the acrylamide, but the typical organoleptic profile of gingerbread changed significantly [48]. An optimization procedure of the raising agent with focus on organoleptic aspects reduced the acrylamide concentration of at least 30% (Tab. 11.6).

A likewise strong problem outlines the inconsistency of the baking process. Springer et al. [49] have shown at crisp bread production a high variability within and between industrial scale ovens. Therefore, a conclusion from a single unit analysis toward a consumer exposition seems to be almost impossible.

Table 11.5: Acrylamide concentration in wafers produced with two different types of wheat flour and with two recipes (from [45]).

Variant	V1	V2	V3	V4
Flour type	550	1050	550	1050
Recipe				
Flour [g]	1000	1000	1000	1000
Peanut oil [mL]	40	40	40	40
Sugar [g]	25	25	25	25
Salt [g]	2	2	2	2
Sodium hydrogencarbonate [g]	6	6	–	–
Result				
Acrylamide [μg/kg]	993	2423	995	2329

Table 11.6: Recipe variation and quality data of gingerbread, from [45] (SAPP: sodium acidified pyro-phosphate).

Variant	1	2	3	4
Recipe				
Potash	+	+	+	+
Ammonium	100	–	–	33
Baking powder	–	100	SAPP optimized	SAPP optimized
Quality data				
Pores	Coarse	Medium	Fine	Medium
Taste	Most perfect	Affected	Perfect	Most perfect
Acrylamide	1283	139	187	854

11.5 Minimization Potentialities in Coffee

Initial studies on coffee have shown acrylamide formation in the early stages of roasting and a sharply declination until the end point of roasting [50]. A comparison between coffea arabica (Arabica) and coffea canephora robusta (Robusta) indicated a slightly higher acrylamide level of the roasted Robusta beans [26]. Labelled acrylamide experiments have shown a most complete degradation during the process. Only a small percentage of the free asparagine is converted to acrylamide [26]. Furthermore, acrylamide is not stable in the finished product, pointing to nonthermal interactions [51, 52].

11.6 Minimization Potentialities in Vegetables

Acrylamide formation is associated with plant materials rich in starch, even though the starch itself is not implemented in the reaction. Focussing on the two relevant precursors, reducing sugars and free asparagine, it could be shown that vegetables comprise an acrylamide potential. Roasting broccoli, peppers and onions at three intensity levels led to remarkable acrylamide concentrations in broccoli in particular (Fig. 11.10).

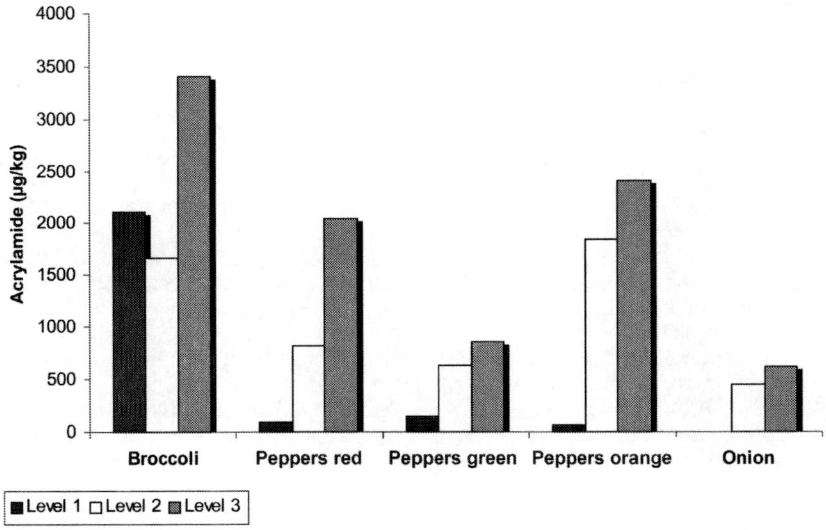

Figure 11.10: Acrylamide concentration in vegetables roasted with three levels.

11.7 Conclusion

Several research activities have identified minimization tools for a systematic and permanent reduction of acrylamide in food. Raw material and processing offer successful starting points. Variation of single steps may be effective, but in view of a high organoleptic and chemical product quality a combination of several modifications with slight changes each will be the best strategy. Excluding coffee, a moderate heat treatment will lower the acrylamide formation. Education of the consumers is necessary to change the overall thermal load of domestic food preparation, still an unsolved task of the minimization concept.

Acknowledgments

Potato data mentioned were generated within a research project, supported by the FEI (Forschungskreis der Ernährungsindustrie e.V.), the AiF, and the Ministry of Economics and Technology, Project 108 ZBG.

References

1. Bund für Lebensmittelrecht und Lebensmittelkunde e.V. (BLL) 2005, Entwicklung von neuen Prozesstechniken zur Vermeidung des Acrylamidgehaltes in Lebensmitteln, http://www.ilu-ev.de/acrylamid/projekt.htm
2. European Commission Joint Research Centre Institute for Reference Materials and Measurements 2004, Monitoring database on acrylamide levels in food, http://www.irmm.jrc.be/html/activities/acrylamide/acrylamidemonitoringdatabase_status sDecember04.xls
3. European Commission DG Health and Consumer Protection Food Contaminants 2005, Acrylamide, http://europa.eu.int/comm/food/food/chemicalsafety/contaminants/acrylamide_en.htm
4. Madle, S., Broschinski, L., Mosbach-Schulz, O., Schöning, G. and Schulte, A. Zur aktuellen Risikobewertung von Acrylamid in Lebensmitteln, *Bundesgesundheitsbl. Gesundheitsforsch. Gesundheitsschutz* 2003, **46**, 405–415.
5. Hilbig, A., Freidank, N., Kersting, M., Wilhelm, M. and Wittsiepe, J. Estimation of the dietary intake of acrylamide by German infants, children and adolescents as calculated from dietary records and available data on acrylamide levels in food groups, *International Journal Hygiene Environmental Health* 2004, **207**, 463–471.
6. FAO/WHO Joint FAO/WHO expert committee on food additives 64th meeting, 8–17 Feb 2005, Rome, http://www.who.int/ipcs/food/jecfa/summaries/en/summary_report _64_final.pdf
7. Yasuhara, A., Tanaka, Y., Hengel, M. and Shibamoto, T. Gas chromatographic investigation of acrylamide formation in browning model systems, *J. Agric. Food Chem.* 2003, **51**, 3999–4003.
8. Zyzak, D.V., Sanders, R.A., Stojanovic, M., Tallmadge, D.H., Eberhart, B.L. and Ewald, D.K. Acrylamide formation mechanism in heated foods, *J. Agric. Food Chem.* 2003, **51**, 4782–4787.
9. Granvogel, M., Jezussek, M., Köhler, P. and Schieberle, P. Quantitation of 3-aminopropionamide in potatoes – a minor but potent precursor in acrylamide formation, *J. Agric. Food Chem.* 2004, **52**, 4751–4757.
10. Mottram, D.S., Wedzicha, B.L. and Dodson, A.T. Acrylamide is formed in the Maillard reaction, *nature* 2002, **419**, 448–449.
11. Stadler, R.H., Blank, I., Varga, N., Robert, F., Hau, J., Guy, P.A., Robert, M.C. and Riediker, S. Acrylamide from maillard reaction products, *nature* 2002, **419**, 449–450.
12. Yaylayan, V.A., Wnorowski, A. and Locas, C.P. Why asparagine needs carbohydrates to generate acrylamide, *J. Agric. Food Chem.* 2003, **51**, 1753–1757.
13. Stadler, R.H., Robert, F., Riediker, S., Varga, N., Davidek, T., Devaud, S., Goldmann, T. and Ha, J. In-depth mechanistic study on the formation of acrylamide and other vinylogous compounds by the Maillard reaction, *J. Agric. Food Chem.* 2004, **52**, 5550–5558.

14. Biedermann, M., Biedermann-Brem, S., Noti, A. and Grob, K. Methods for determining the potential of acrylamide formation and its elimination in raw materials for food preparation, such as potatoes, *Mitt. Geb. Lebensmittelunters. Hyg.* 2002, **93**, 653–667.
15. Amrein, T.M., Bachmann, S., Noti, A., Biedermann, M., Barbosa, M.F., Biedermann-Brem, S., Grob, K., Keiser, A., Realini, P., Escher, F. and Amado, R. Potential of acrylamide formation, sugars, and free asparagine in potatoes: A comparison of cultivars and farming systems, *J. Agric. Food Chem.* 2003, **51**, 5556–5560.
16. Becalski, A., Lau, B.P.Y., Lewis, D. and Seaman, S.W. Acrylamide in foods: occurrence, sources, and modelling, *J. Agric. Food Chem.* 2003, **51**, 802–808.
17. Biedermann, M., Noti, A., Biedermann-Brem, S., Mozzetti, V. and Grob, K. Experiments on acrylamide formation and possibilities to decrease the potential of acrylamide formation in potatoes, *Mitt. Geb. Lebensmittelunters. Hyg.* 2002, **93**, 668–687.
18. Brierley, E.R., Bonner, P.L.R. and Cobb, A.H. Aspects of amino acid metabolism in stored potato tubers (cv Pentland Dell), *Plant Sci.* 1997, **127**, 17–24.
19. Mack, G. and Schjoerring, J.K. Effect of nitrite supply on nitrogen metabolism of potato plants (Solanum tuberosum L.) with special focus on the tubers, *Plant, Cell and Environment* 2002, **25**, 999–1009.
20. Surdyk, N., Rosen, J., Andersson, R. and Aman, P. Effects of asparagine, fructose, and baking conditions on acrylamide content in yeast-leavened wheat bread, *J. Agric. Food Chem.* 2004, **52**, 2047–2051.
21. Gerendas, J., Heuser, F. and Sattelmacher, B. Influence of nutrient supply on contents of acrylamide precursors in potato and of acrylamide in French fries. In: S.-E. Jacobsen, C.R. Jensen, and J.R. Porter (Eds.): VIII ESA Congress: Book of Proceedings, Copenhagen, Denmark, KVL, 2004, pp 511–512.
22. Haase, N.U., Matthäus, B. and Vosmann, K. Aspects of acrylamide formation in potato crisps, *J. of Applied Botany and Food Quality* 2004, **78**, 144–147.
23. Kumar, D., Singh, B.P. and Kumar, P. An overview of the factors affecting sugar content of potatoes, *Ann. appl. Biol.* 2004, **145**, 247–256.
24. Roe, M.A., Faulks, R.M. and Belsten, J.L. Role of reducing sugars and amino acids in fry colour of chips from potatoes grown under different nitrogen regimes, *J. Sci. Food Agric.* 1990, **52**, 207–214.
25. Rodriguez-Saona, L.E. and Wrolstad, R.E. Influence of potato composition on chip color quality, *American Potato Journal* 1997, **74**, 87–106.
26. Taeymans, D., Andersson, A., Ashby, P., Blank, I., Gonde, P., Eijck van, P., Faivre, V., Lalljie, S.P.D., Lingnert, H., Lindblom, M., Matissek, R., Müller, D., Stadler, R.H., Studer, A., Silvani, D., Tallmadge, D., Thompson, G., Whitmore, T., Wood, J. and Zyzak, D. Acrylamide: Update on selected research activities conducted by the European Food and Drink Industry, *J. AOAC Int.* 2005, **88**, 234–241.
27. Haase, N.U. and Weber, L. Variability of sugar content in potato varieties suitable for processing, *Food, Agriculture and Environment* 2003, **1**, 80–81.
28. Grob, K., Biedermann, M., Biedermann-Brem, S., Noti, A., Imhof, D., Amrein, T. and Pfefferle, A. French fries with less than 100 µg/kg acrylamide. A collaboration between cooks and analysts, *Eur. Food Res. Technol.* 2003, **217**, 185–194.
29. Ezekiel, R., Verma, S.C., Sukumaran, N.P. and Shekhawat, G.S. A guide to potato processors in India, Technical Bulletin No. 48, Shimla, India, Central Potato Research Institute, 1999, pp 14–16.
30. Kolbe, H. Einflußfaktoren auf die Inhaltsstoffe der Kartoffel. Teil 2: Zucker, *Kartoffelbau* 1996, **47**, 35–39.
31. Pritchard, M.K. and Adam, L.R. Preconditioning and storage of chemically immature Russet Burbank and Shepody potatoes, *American Potato Journal* 1992, **69**, 805–815.
32. Mazza, G. and Siemens, A.J. Carbon dioxide concentration in commercial potato storage and its effect on quality of tubers for processing, *American Potato Journal* 1990, **67**, 121–132.
33. Haase, N.U., *unpublished data* 2005.

34. Heuser, F., Gerendás, J. and Sattelmacher, B. Einfluss der N- und K-Düngung auf die Gehalte an reduzierenden Zuckern und freien Aminosäuren, *Kartoffelbau* 2005, **56**, 308–313.
35. Bundesamt für Verbraucherschutz und Lebensmittelsicherheit – BVL 2004, German dynamic minimisation concept, *http://www.bvl.bund.de/acrylamid/*
36. aid Infodienst Verbraucherschutz Ernährung Landwirtschaft Acrylamid (Flyer) 2002, *www.was-wir-essen.de/download/acrylamid.pdf* (25-05-2005).
37. Matthäus, B., Haase, N.U. and Vosmann, K. Factors effecting the concentration of acrylamide during deep-fat frying of potatoes, *Eur. J. Lipid Sci. Technol.* 2004, **106**, 793–801.
38. Gorun, E.G. and Potapow, W.D. Produkcja koncentratow sniadaniowych ze zboz i ziemniakow 1974, Warsaw, WNT.
39. Haase, N.U. Die Bedeutung der Zellwand für die Verarbeitungsqualität von Kartoffeln, *Kartoffelbau* 2001, **52**, 351–355.
40. Pedreschi, F., Kaack, K. and Granby, K. Reduction of acrylamide formation in potato slices during frying, *Lebensm.-Wiss. u. -Technol.* 2004, **37**, 679–685.
41. Haase, N.U., Matthäus, B. and Vosmann, K. Minimierungsansätze zur Acrylamid-Bildung in pflanzlichen Lebensmitteln – aufgezeigt am Beispiel von Kartoffelchips, *Deutsche Lebensmittel-Rundschau* 2003, **99**, 87–90.
42. Granda, C., Moreira, R.G. and Tichy, S.E. Reduction of acrylamide formation in potato chips by low-temperature vacuum frying, *J. Food Science* 2004, **69**, E405–E411.
43. Confederation of the food and drink industries of the EU (CIAA) Acrylamide status report december 2004. www.ciaa.be/documents/positions/Acrylamide-Status-Report-December-2004.pdf
44. Biedermann, M. and Grob, K. Model studies on acrylamide formation in potato, wheat flour and corn starch; ways to reduce acrylamide contents in bakery ware, *Mitt. Lebensmittelunters. und Hygiene* 2003, **94**, 5406–5422.
45. Haase, N.U., Matthäus, B. and Vosmann, K. Acrylamid in Backwaren – ein Sachstandsbericht, *Getreide Mehl und Brot* 2003, **57**, 180–184.
46. Fredriksson, H., Tallving, J., Rosen, J. and Aman, P. Fermentation reduces free asparagine in dough and acrylamide content in bread, *Cereal Chem.* 2004, **81**, 650–653.
47. Amrein, T.M., Schönbächler, B., Escher, F. and Amado, R. Acrylamide in Gingerbread: critical factors for formation and possible ways for reduction, *J. Agricult. Food Chem.* 2004, **52**, 4282–4288.
48. Grothe, K.-H., Unbehend, G.M., Haase, N.U., Ludewig, H.-G., Matthäus, B. and Vosmann, K. Einfluss von Backtriebmitteln auf die Acrylamidgehalte von Braunen Lebkuchen und Mürbkeksen, *Getreidetechnologie* 2005, **59**, 163–167.
49. Springer, M., Fischer, T., Lehrack, A. and Freund, W. Acrylamidbildung in Backwaren, *Getreide, Mehl und Brot* 2003, **57**, 274–278.
50. Taeymans, D., Wood, J., Ashby, P., Blank, I., Studer, A., Stadler, R.H., Gondé, P., van Eijck, P., Lalljie, S., Lingnert, H., Lindblom, M., Matissek, R., Müller, D., Tallmadge, D., O'Brien, J., Thompson, S., Silvani, D., and Whitmore, T. A review of acrylamide: An industry perspective on research, analysis, formation, and control, *Critical Reviews in Food Science and Nutrition* 2004, **44**, 323–347.
51. Andrzejewski, D., Roach, J.A.G., Gay, M.L. and Musser, S.M. Analysis of coffee for the presence of acrylamide by LC-MS/MS, *J. Agricult. Food Chem.* 2004, **52**, 71996–72002.
52. Delatour, T., Perisset, A., Goldmann, T., Riediker, S. and Stadler, R.H. Improved sample preparation to determine acrylamide in difficult matrixes such as chocolate powder, cocoa, and coffee by liquid chromatography tandem mass spectroscopy, *J. Agricult. Food Chem.* 2004, **52**, 154625–154631.

12 Deep-fat Frying as Food Heating Process: Product Quality, Safety and Process Control

Knut Franke * and Ernst H. Reimerdes

Abstract

Frying, in particular deep-fat frying, is one of the most important food heating processes. Typical sensory quality parameters of fried products expected by the consumers are:

- browning and flavor (determined by Maillard reaction components and fat intake),
- crispness and, especially for French fries, a soft kernel.

During deep-fat frying numerous exchange reactions occur at the interface between product surface and fat as heating medium. Water is released very quickly from outer layers of the product through the frying fat. Additionally, intake of fat and exchange of triglycerides in the case of par-fried products contribute to the final sensory and nutritional quality of fried foods.

On the other hand, the acrylamide discussion for heated foods has been focused on safety of fried products. The numerous projects and efforts of science and industry are naturally related to reduced acrylamide contents in this food category. Substantial progress in the case of fried products has been achieved by controlling the precursor situation (reducing sugars and free asparagine) but also by influencing surface processes by raw material pre-treatment and control of frying conditions.

However, current results also indicate the extraordinary complexity with respect to the variables of raw material, product quality and processing conditions. Therefore, sustainable manufacturing of high quality and safe fried foods will only be achieved if an integrated scientific approach considering all relevant variables is applied. This includes:

- optimization of raw material composition with respect to quality and acrylamide formation,
- specific pre-treatments influencing the kinetics of quality parameters positively and reducing the acrylamide formation,

* Deutsches Institut für Lebensmitteltechnik e.V., Prof.-von-Klitzing-Str. 7, 49610 Quakenbrück, Germany.

Thermal Processing of Food: Potential Health Benefits and Risks
DFG, Deutsche Forschungsgemeinschaft

ISBN: 978-3-527-31909-1

- adaptation/development of related processing technology also considering new processing equipment.

Based on these results new research projects are proposed which will contribute to a distinct reduction of acrylamide exposure of consumers by the transfer of new techniques and product concepts into industrial scale and by the implementation of new concepts for household fryer control and for restructured potato products.

12.1 Deep-fat Frying as Food Heating Process

Heating or cooking of food is one of the basic processes to produce high quality foods well-known for several ten thousand years. Examples of these techniques are baking, grilling or deep-fat frying. In all these processes temperatures above 100 °C are reached locally causing Maillard reactions which contribute to browning and taste. From the engineering point of view, the processes differ with respect to type of heat transfer medium, e.g. gas or liquid, and temperature range of heat transfer medium. This results in a broad range of heated foods, each with special taste and appearance.

The unique feature of deep-fat frying processes compared to other food heating processes is the double function of frying fat. It acts as a very effective heat transfer medium leading to short processing times, for example 3 to 5 min compared to baking with 20 to 50 min. On the other hand, the fat itself becomes a part of the fried product and therefore determines its quality. In contrast to pan frying the fat stays in the fryer for a longer time (days or weeks) often at temperatures of more than 150 °C. This leads to alterations in the fat regarding its chemical composition and also its sensory properties. Formation of polar components, e.g. di- and monoglycerides, oxidized and polymerized triglycerides, which modifies the surface properties of the fat, is of technological importance. These fat alteration processes cause generation of typical taste and aroma of deep-fat fried products which can not be substituted by other heating processes, e.g. baking. The quality of deep-fat fried foods can be characterized by following parameters:

- sensory properties
  - texture (firmness, crispness, thickness of crust)
  - flavor (frying flavor, fat distribution, off-flavor)
- color (browning, graying, black spots)
- storage stability, handling
  - residual moisture, surface fat
- nutritional properties
  - fat content, fat composition, toxicological aspects (acrylamide).

These quality parameters are influenced by several variables which can be divided in three main groups: raw material, equipment and process. All three groups of variables have an effect both on quality and formation of potential risk substances, e.g. acrylamide. To control both quality and formation of acrylamide in fried foods, it is necessary to investigate the influence of all relevant variables. The objective of all these efforts shall be the minimization of acrylamide formation (safety) whereas maintaining the expected product quality. Due to acrylamide formation being linked with Maillard reactions, it is necessary to decouple these reaction pathways to achieve the desired browning with lower acrylamide contents [1].

A special role in deep-fat frying can be attributed to the processes at the product surface which acts as an interface between product and frying fat. During frying the water in the outer product layers being in contact with the hot fat evaporates very fast. The reduction of moisture results in an increase of local temperatures to values above 100 °C. Then the brown crust is formed. The evaporation zone moves to the inner product layers and frying fat permeates into crust pores which are formed due to water evaporation [2]. The rate of temperature increase in the crust and the evaporation rate are determined by the heat transfer coefficient which is influenced by the content of polar components [3–5]. Additionally to the heat transfer, exchange of substances, e.g. triglycerides, aroma components and tocopherols, takes place through the product surface (interface). Water, water soluble compounds and fats (originating from par-frying) come into contact.

12.2 Role of Product Surface Processes and Frying Fat

As mentioned above, ready-to-eat French fries consist of a brown crispy crust at the outer product layers and a soft yellow kernel of cooked potato tissue. These two different taste impressions are the unique attribute of this product. Due to the fact that only in crust region the temperatures rise higher than 100 °C and local moisture content is reduced after evaporation, it can be assumed that this part is critical with respect to the formation of the health relevant substance acrylamide. Therefore, a distinct spatial distribution of acrylamide can be assumed in French fries resulting from internal differences in structure and temperature load. To evaluate these effects a model potato food possessing a composition similar to French fries but with a larger size (2×2×10 cm^3) was prepared and fried for 12 min to generate a brown crust. After frying thin layers from the crust of the model fries were truncated and analyzed with respect to contents of acrylamide, water and fat. The acrylamide content was determined by a mass spectroscopic method after liquid chromatographic separation from residual matrix and sample preparation. Figure 12.1 shows the contents of acrylamide, water and fat in the different layers of the model potato

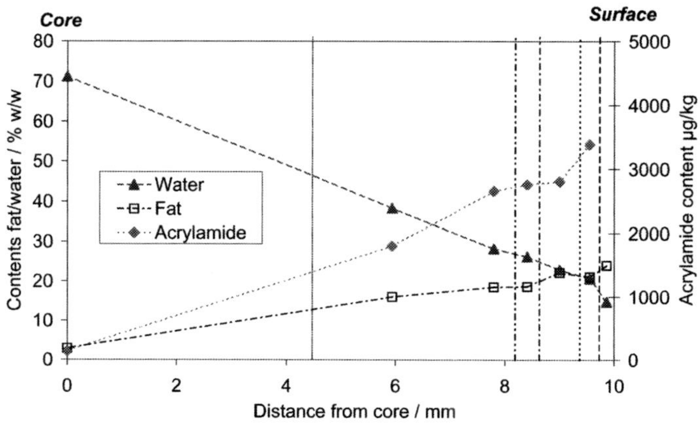

Figure 12.1: Spatial distribution of acrylamide, water and fat in model French fries.

food beginning from product core (left) to surface (right). Vertical lines show the thickness of truncated crust layers. Measured values are drawn in the middle of every layer. The total water content of the fried model system was about 49% and slightly higher compared to ready-to-eat French fries. This is due to the larger size of the model food possessing a higher part of moist core. The fat content was 11% which is in the range of French fries.

As expected, the water content decreases from core (approx. 70%) to surface (approx. 15%). On the other hand, fat content increases from core to surface. However, the largest changes are found in the acrylamide content. The content in the core is relatively low (about 150 µg/kg) but detectable. The reason for this finding may be the long frying time and the fact that acrylamide formed in the outer layers may diffuse into the soft core. The content at the outer layers is very high with about 3500 µg/kg. This is about 20 times higher than the core content and confirms the importance of surface processes for the acrylamide formation. As can be observed in Figure 12.1 the layers high in acrylamide also contain remarkable contents of frying fat. This fact supports the idea of influencing acrylamide formation by the frying process and by the frying fat itself.

To validate the supposition of influencing acrylamide formation by frying fat, different additives were added to the fat before the frying of par-fried French fries was finished. Especially addition of w/o-emulsions has been described as a way to influence frying processes with respect to heat transfer and oil deterioration [4]. A reactant in the water phase of such an emulsion is carried over to product surface by the frying fat and may influence Maillard reaction pathways during heating. Emulsions containing 20% inner aqueous phase and PGPR as emulsifier were added in a level of 1% to the frying fat. The inner phase consisted of an aqueous solution of amino acid glutamine. Par-fried French fries were finished in this fat using different frying times (3 to 6 min) to achieve a broad range of quality with respect to browning (brightness). The interrelation of surface browning of finished French fries shown as brightness (L^*) and their acrylamide contents is presented in Figure 12.2.

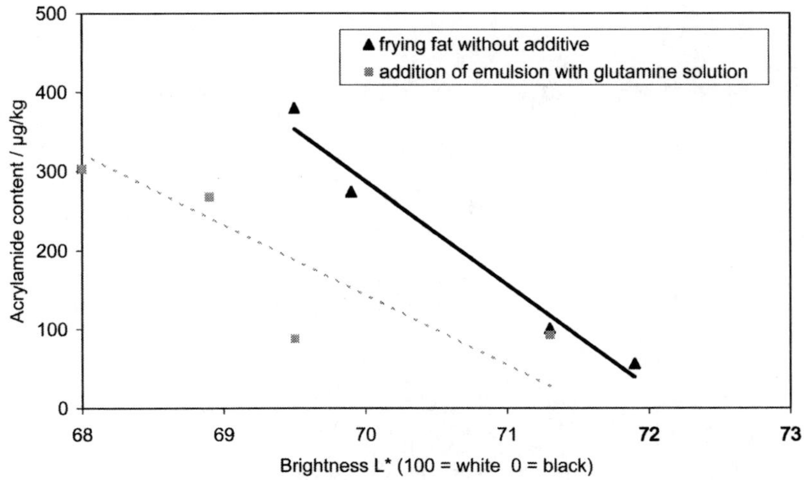

Figure 12.2: Connection between acrylamide content and brightness of French fries fried for different times in fat with and without an added w/o-emulsion (1% w/w) with a glutamine solution in the inner aqueous phase (20% w/w).

The straight line for French fries fried without added emulsion (pure fat) shows a clear increase of acrylamide with decreasing brightness i.e. darker French fries. The correlation confirms the findings that the darker the fries (lower brightness) the higher the acrylamide contents. This fact is well-known and integrated into the recommendations for frying operations [7]. If an emulsion with glutamine solution in the inner phase is added (dotted line), comparable results are achieved. However, the acrylamide contents of products fried in fats with added emulsions are lower. This means that the same level of browning or quality of the final product can be obtained at lower acrylamide levels. Glutamine is one of the precursors for browning in potatoes [8] and seems to be able to shift the Maillard reactions on product surfaces away from acrylamide formation. Therefore, providing of aqueous reaction components through the fat during frying influences the reaction mechanisms and can contribute to lower acrylamide contents in French fries.

12.3 Role of Frying Equipment and Frying Process

Additionally to the frying fat, the frying process itself and the frying equipment have an influence on surface processes and, therefore, on quality development and also acrylamide formation during frying. Relevant variables which have to be considered are frying temperature and its alteration during frying, size and location of the heating device, temperature distribution in the frying fat and others.

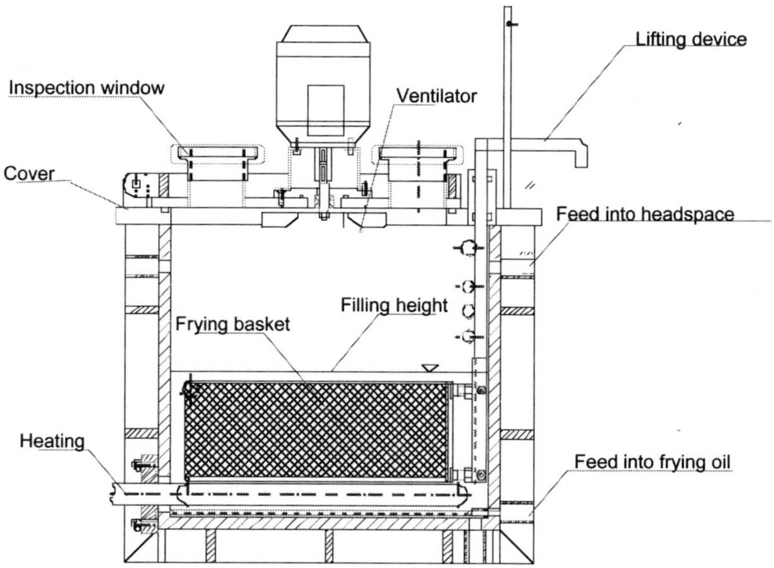

Figure 12.3: Multi-functional fryer for a well defined frying process.

However, in most of the practiced frying operations only few of these variables can be controlled or registered, due to limited measurement opportunities in conventional fryers. On the other side, earlier investigations demonstrated distinct effects of temperature distribution in the fryer on quality and acrylamide formation in French fries [9].

In order to enable a more defined frying process including measurement of important process variables, a new multi-functional frying apparatus was designed and constructed to enable a well defined frying process (Fig. 12.3). This apparatus allows frying in a closed system with a control of headspace conditions above the frying fat. Applying these features especially water evaporation from product surface can be controlled in a broad range of temperatures using different absolute pressures in the headspace above frying bath.

As one example, the multi-functional fryer was used to investigate the influence of headspace conditions (pressure variation) on sensory quality (browning) and acrylamide content of French fries. The pressure in the head space above the frying bath shifts the evaporation temperature of water and, therefore, the temperature course in French fries surface during frying. Using this approach, water evaporation and local temperature increase during frying can be modified.

For this purpose, par-fried French fries from one batch were fried for different times to produce the desired broader range of quality levels. The pressure in the headspace was varied from ambient pressure (1 bar absolute) to 4 bar absolute. This corresponds to water evaporation temperatures from 100 °C to about 140 °C. The resulting quality as brightness (browning) of the fries and acrylamide concentrations are shown in Figure 12.4.

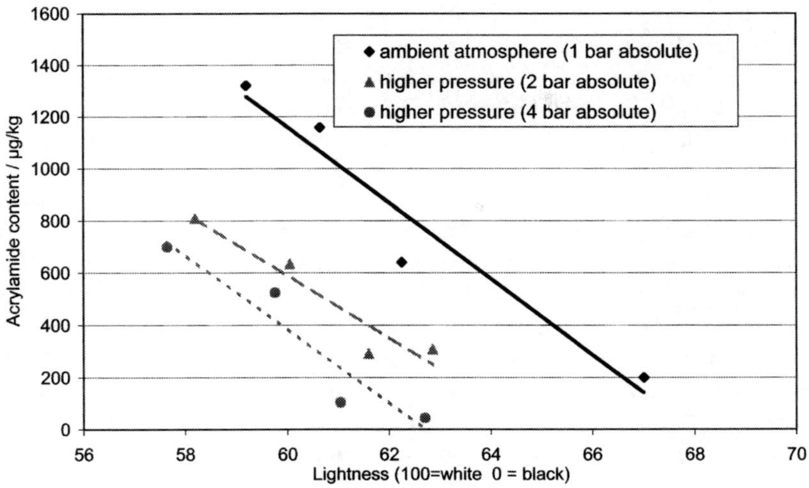

Figure 12.4: Acrylamide contents vs. color (brightness) of French fries fried under different headspace conditions.

A good correlation between the browning (brightness) and the acrylamide contents exists for all frying conditions. However, the course of the correlations differs depending on frying conditions. Frying under higher pressures in the headspace at 2 and 4 bar absolute results in similar slopes compared to frying at ambient conditions (open fryer bath), but the lines for pressure frying run below the line for traditional frying. This means that frying with application of higher pressure leads to lower acrylamide contents for the same browning level. Therefore, frying at higher pressure levels is one opportunity to control frying processes by frying equipment with respect to acrylamide minimization.

This selected example demonstrates the influence of frying process parameters on acrylamide formation and confirms that considering frying time and temperature alone is not sufficient to achieve comparable results with respect to acrylamide contents in ready-to-eat French fries. On the other side, the application of suitable and controlled frying conditions is an opportunity to reduce the formation of undesired substances in fried products while maintaining the quality expected by the consumer.

12.4 Pre-treatment of Raw Material

The integrated approach for investigating frying processes also includes the consideration of raw material influence and its control. Similar to modification of process conditions to realize a variation of water evaporation and product surface moisture, also pre-treatment of raw materials can be used to control surface processes. Possible methods to influence these processes based on raw material are coating of the product surface and pre-drying before frying to control water evaporation during frying. The coating of product surfaces with salt solutions or acids (lower pH-value) to reduce acrylamide formation during heating has been already described [10].

To investigate the influence of such a raw material, par-fried French fries were coated with salt solutions with concentrations between 3 and 15% NaCl in the aqueous phase for 5 min followed by dripping off for 5 min and freezing. After storage for 24 h at –20 °C the frozen potato sticks were pre-dried to a mass loss of 7 or 21%, respectively, using a tray dryer. Salt-coated and pre-dried French fries were finished in a catering fryer at 168 °C. The frying time was adjusted to 3 min for the non-dried French fries (standard conditions) and was adapted to 2 min or 90 s for pre-dried French fries, respectively, to obtain a similar browning and final water content compared to those fried at standard conditions.

As shown in Figure 12.5 coating with a salt solution of 6% NaCl and more leads to a significant reduction in acrylamide contents of French fries (gray columns). The higher salt contents in the potato surface during frying lower the local water activity and increase the ionic strength. Both influence the chemical reactions during heating, the precise influencing mechanism is not fully understood although similar results were found for baked goods [11]. Much more effective with respect to acrylamide reduction in French fries is pre-drying combined with an adapted frying time. This method reduces acrylamide contents in ready-to-eat French fries distinctly. Up to 70% reduction can be achieved. For

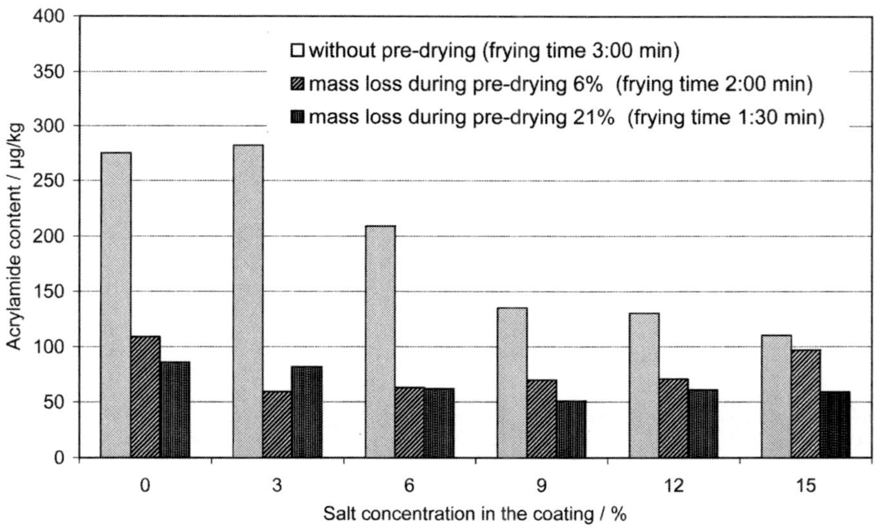

Figure 12.5: Acrylamide contents of par-fried French fries coated with salt solutions (NaCl) with different concentrations and pre-dried to different mass losses before frying.

salt contents in the range between 3 and 9% synergistic effects of salting and pre-drying leading to further reduction can be observed.

These results illustrate the opportunities of raw material pre-treatment with respect to minimizing acrylamide contents in French fries and maintaining the traditional quality considering browning.

12.5 Conclusion and Outlook

Deep-fat frying is an important and versatile food heating process enabling the manufacturing of high quality products. The processes at product surface and interface to frying fat determine development of quality (e.g. browning) and also formation of acrylamide as could be shown by the extremely high contents of acrylamide in outer product layers of model French fries.

Several approaches to influence these processes at the product surface can be applied to reduce acrylamide contents in ready-to-eat French fries. Examples tested positive were addition of reactants to frying fat and pre-treatment of raw material surfaces before frying to influence water evaporation and surface moisture development during frying. It could be verified that a pre-drying in combination with a sensory acceptable coating with salt solutions reduces acrylamide contents in ready-to-eat French fries up to 70%. Also modification of frying conditions, e.g. frying at higher pressure or also at very low pressure can contribute to lower acrylamide contents in French fries. A multifunctional frying apparatus was designed and built enabling such a control of headspace conditions with respect to pressure, e.g. vacuum or pressure frying, together with an extended control of process variables.

Next steps which have to be carried out in this technologically oriented context of frying processes are the implementations of reduction methods for fried potato products validated in laboratory scale into the industrial scale. Additionally, the use of recombination technologies, e.g. for French fries, to enable a better control of precursors for acrylamide formation on molecular level for example by enzymatic treatment, is another challenge to produce high quality products with negligible acrylamide contents. Another aspect which should be considered in future is the preparation of fried products in household and catering where a distinct part of dietary acrylamide intake is generated. In this field innovative equipment control concepts which are able to capture and process product data and to calculate automatically adapted frying conditions (temperature and time) should be developed.

All these steps will contribute to a distinct lowering of dietary acrylamide intake of consumers and to a higher level of safety for fried products.

References

1. Stadler, R. H.; Robert, F.; Riediker, S.; Varga, N.; Davidek, T.; Devaud, S.; Goldmann, T.; Hau, J.; Blank, I. (2004) In-depth mechanistic study on the formation of acrylamide and other vinylogous compounds by the Maillard reaction, *Journal of Agricultural and Food Chemistry* **52**, 5550–5558.
2. Krokida, M. K.; Oreopolou, V.; Maroulis, Z. B. (2000) Water loss and oil uptake as a function of frying time, *Journal of Food Engineering* **44**, 39–46.
3. Blumenthal, M. M. (1991) A new look at the chemistry and physics of deep-fat frying, *Food Technology* **45**, 68–71.
4. Gertz, C. (2004) Optimising the frying and baking process using oil improving agents, 4th International Symposium Deep-Frying, Hagen.
5. Moreira, R. G.; Castell-Perez, M. E.; Barrufet, M. A. (1999) Deep Fat Frying: Fundamentals and Applications, Aspen Publishers, Gaithersburg.
6. Whitfield, F. B. (1992) Volatiles from interactions of Maillard reactions and lipids, *Critical Reviews in Food Science and Nutrition* **31**, 1–58.
7. Dybing, E.; Farmer, P. B.; Andersen, M.; Fennell, T. R.; Lalljie, S. P. D.; Müller, D. J. G.; Olin, S.; Petersen, B. J.; Schlatter, J.; Scholz, G.; Scimeca, J. A.; Slimani, N.; Törnqvist, M.; Tuijtelaars, S.; Verger, P. (2005) Human exposure and internal dose assessment of acrylamide in food, *Food and Chemical Toxicology* **43**, 365–410.
8. Khanbari, O. S.; Thompson, A. K. (1993) Effects of amino acids and glucose on the fry colour of potato crisps, *Potato Research* **36**, 359–364.
9. Franke, K.; Reimerdes, E. H. (2004) Possibilities in simulating frying processes with respect to minimizing acrylamide contents, 4th International Symposium on Deep Frying, Hagen.
10. Jung, M. Y.; Choi, D. S.; Ju, J. W. (2003) A novel technique for limitation of acrylamide formation in fried and baked corn chips and in French fries, *Journal of Food Science* **68**, 1287–1290.
11. Levine, R. A.; Smith, R. E. (2005) Sources of variability of acrylamide levels in a cracker model, *Journal of Agricultural and Food Chemistry* **53**, 4410–4416.

13 Thermal Processing of Food: Allergenicity

Lars K. Poulsen *

Abstract

The allergenicity or allergenic potential to food may be understood in two ways: 1) as the ability to induce an IgE-response, to sensitize the patient; and 2) as the capability to induce a reaction leading to symptomatic food allergy in an already sensitized patient. We have still only limited knowledge of the former process, and the latter will be the primary focus of this paper. Thermal processing of foods may induce a plethora of biochemical reactions that can be characterized by traditionally physico-chemical methods such as gel electrophoresis, circular dichroism etc., but in order to evaluate the allergenicity it is necessary to employ food allergological methods based on patient material from well characterized food allergic patients. These methods range from the final *in vivo* confirmation of allergenicity by double-blinded placebo-controlled food challenges (DBPCFC), over skin prick tests (SPT) to the *in vitro* methods histamine release from basophils and IgE-based techniques such as the RadioAllergoSorbentTest (RAST) and inhibition variants of this technique.

When looking at specific foods and processes, there is a whole spectrum of outcomes ranging from an actual increase in allergenicity by roasting of peanuts, over a constant allergenicity found in fish even after extended periods of cooking to a rapid decline in allergenicity even after mild thermal treatments. The complexity is further increased by patients reacting differently to different components in foods. While it is difficult to identify global rules, it seems to be a general feature that solubility of proteins often decreases after thermal processing, and this poses methodological problems since most analytical approaches are based on proteins in solution. Thus an apparent reduction in allergenicity may in reality be a reduction in solubility or extractability of an allergen. Examples from fish, peanut and other legumes and hazelnuts are discussed.

The formation of qualitative new allergens – the so-called neoallergens – has been proposed, and in particular products of Maillard reactions are in focus. There is still only limited evidence of the clinical relevance of such neo-allergens, but more studies are needed to elucidate this.

* Allergy Clinic, National University Hospital Dept. 7551, Blegdamsvej 9, DK-2100 Copenhagen, Denmark.

Thermal Processing of Food: Potential Health Benefits and Risks
DFG, Deutsche Forschungsgemeinschaft

ISBN: 978-3-527-31909-1

13.1 Introduction

The term allergenicity relies on a proper definition of allergy. In 2001 the European Academy of Allergy and Clinical Immunology (EAACI) proposed a nomenclature [1] which is now being adhered to also by the World Allergy Organisation (WAO). According to this nomenclature food hypersensitivity is a state where a person tolerates less food than a normal person would do. Food hypersensitivity may be divided into reactions where the immune system is involved – called food allergy – and other reactions, called non-allergic food hypersensitivity. The food allergies may be further divided into IgE- and non-IgE-mediated food allergies, and other parties have proposed the division of the former into classic IgE-mediated food allergies, where the food is the primary sensitizer, and cross-reacting foods, where e.g. pollen is believed to be the primary sensitizing agent. Although it can be argued that metabolic diseases such as e.g. lactase deficiency could classify as a non-IgE-mediated food hypersensitivity, I would propose neither to include these conditions nor food poisonings or aversion in the food hypersensitivity group. An overall scheme of the nomenclature is presented in Figure 13.1.

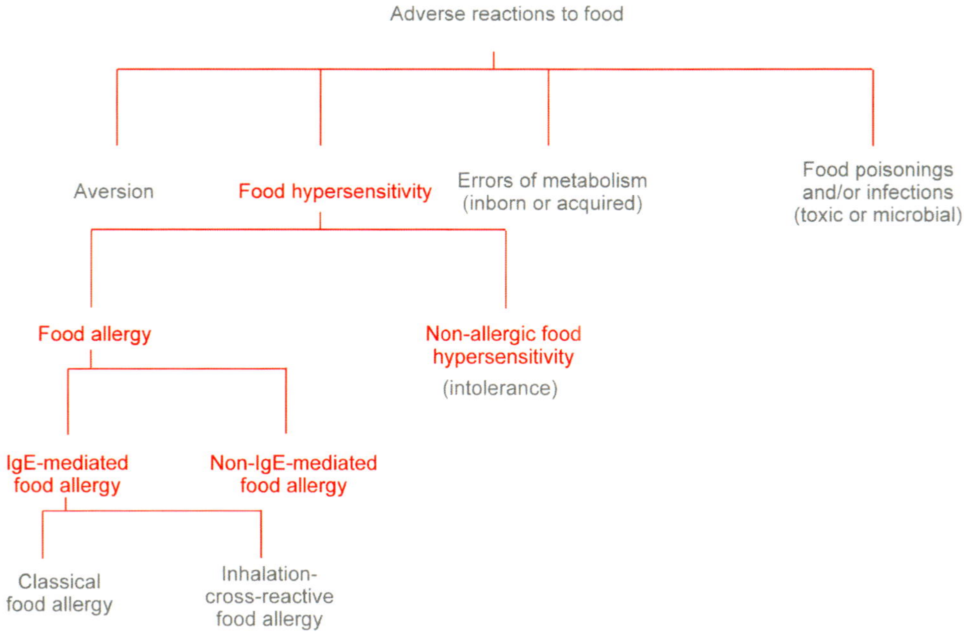

Figure 13.1: Classification of adverse reactions to food. Categories in red characters are endorsed by the European Academy of Allergy and Clinical Immunology.

13.1.1 IgE-mediated Allergies

The present paper will focus on the IgE-mediated food allergies. Although other immunological mechanisms exists [2] such as *celiac disease* where IgG and/or IgA antibodies to gluten are involved [3] or *systemic allergic contact dermatitis* where nickel in the diet elicit T-cell-mediated skin reactions [4], both the frequency and diversity of IgE-mediated food allergies warrant the focus on these diseases. The IgE system is phylogenetically the newest of the immunoglobulins, IgE being found only in mammals and marsupials. To mediate a biological reaction IgE needs to be bound to its high affinity receptors that are found on mast cells in the skin and mucosal surfaces and on the circulating basophil leukocyte. When two or more receptor-bound IgE molecules are cross-linked by binding to an allergen a chain of signal transduction processes is initiated in the effector cell leading to degranulation with release of histamine and other preformed mediators. Furthermore, lipid mediators such as leukotrienes, prostaglandines and platelet aggregating factor (PAF) are synthesized and finally protein synthesis of cytokines and chemokines are induced. These potent inflammatory mediators may cause *acute symptoms* of asthma and hay fever from the airways, urticaria and eczema from the skin, and oral itching, vomiting and diarrhea from gastro-intestinal system. In most severe cases systemic anaphylaxis may occur due to drop in blood pressure. Besides these immediate symptoms that occur within minutes of exposure, making the IgE inflammatory reaction the fastest of the various kinds of hypersensitivities, a *late phase reaction* – lasting for up to 24 hours – often occurs where other inflammatory cells are attracted to the organ by the chemotactic capacity of the mast cell mediators. In particular in asthma these late phase reactions are believed to constitute the most serious part of the disease leading to a more chronic form with remodelling of the lung tissue. The symptomatology of IgE-mediated food allergy may include all of the above reactions, and often more than one organ system are involved [5, 6].

13.1.2 Allergens

By definition an antigen, to which IgE reacts, is called an allergen. Although IgE to low molecular weight substances exist the vast majority of described IgE-specificities are directed against high molecular weight substances, in particular proteins. From what has been mentioned above on IgE-crosslinking, an allergen needs to cross-link at least two IgE-molecules in order to elicit an reaction. Thus, either the allergen needs to be polyvalent in terms of identical epitopes and/or the IgE-immune response needs to be directed against different epitopes on the allergen. Since most allergens are proteins these may be cloned and expressed as recombinants, and more than 1000 allergenic recombinant molecules are described in the literature. These proteins have been described with conventional physical/biochemical methods, but in order to verify that a protein has allergenic properties, the reactions with the immune system need to be ascertained.

While the immune system has the capacity to produce an antibody response to any foreign protein it meets, this production must take place before a reaction can occur. Thus *allergenicity* of a substance in relation to food allergy may be understood in various ways. As with other IgE-mediated diseases the pathogenesis of food allergies can be divided into an afferent – sensitization – phase where the immune system develops an IgE response to one or more constituents of a food, and an efferent – eliciting – phase where a clinical, allergic reaction occurs after ingestion of the food in question. Accordingly the term allergenic may be understood both as the capacity to *sensitize*, i.e. to induce an IgE immune response, and as the capacity to *elicit an allergic reaction* in an individual already sensitized.

13.1.3 Sensitization

Our understanding of the process leading to sensitization of a patient is still very immature, but probably both factors intrinsic (dose, primary to quaternary structure defining what the immune system recognizes as epitopes) and extrinsic (adjuvant effect, host specific factors) to the individual molecule are involved [7]. So far there are still large gaps in our knowledge as to why certain foods and certain proteins in foods become allergenic. There seems to be some crude correlation between the consumption of a food item, and the occurrence of food allergy to that item in certain regions of the world. Among the abundant examples are fish allergy from Scandinavia and coastal North America [8, 9], allergy to crustaceans at the Mexican Golf coast in Louisiana [10], peanuts in the USA in general [11], and *prunoideae* fruits in the Mediterranean region [12]. With the globalization of the food market, it is striking, however, that these differences still exist.

Also there seem to be changes over time. There is an impression that peanut allergy from being primarily a North American problem is spreading to Europe (and perhaps parts of Asia) with the UK as a bridgehead. Both in North America [13, 14] and in the UK [15] the prevalence is reported to increase. A comparison between two Danish birth cohort studies of children born 13 years apart (1985 vs. 1998) gives the impression that egg has taken over from cow's milk as being the most frequent food allergen in infants [16] vs. [17]. Whether this can be ascribed to an abandoned practice of giving a cow's milk based formula during the first night after birth [18] is possible but has not been verified in prospective studies. Other explanations such as changes in patterns of food consumption have been suggested but are hard to verify.

13.1.4 Elicitation

The other meaning of allergenic, i.e. the elicitation of an allergic reaction in an already sensitized individual, has been much more successfully investigated and as described below numerous methods exist for *determination of the biological potency*. Table 13.1 illustrates the hierarchy that exists among these test systems as challenge of human patients is considered as closest to the relevant biological response, i.e. elicitation of an actual allergic response, albeit under controlled and safe circumstances. The next level in the hierarchy is to use the skin as a restricted and localized area for challenge. This system obviously involves the skin mast cells, which must be sensitized by IgE in order to respond to the offending allergen. Leaving the *in vivo* systems, the next step is to use the sensitized basophil granulocyte as a model for the sensitized mast cell present in the relevant organ of the patient. Moving further away from the actual patient, basophil from a non-allergic donor such as cord blood may even be used as a reagent which is then sensitized by IgE derived from an actual patient. A pure system can be obtained by immunochemical assays detecting IgE-allergen binding directly or indirectly by inhibition designs. The above-mentioned human model systems all have their animal experimental counterparts which will only briefly be touched upon. Many of the parameters of the human systems discussed below will also apply to animal models. The basic problem with experimental animals is to actually make them allergic. Although several immunization schemes – often parenteral – are available, which will readily produce an IgE response, it is still not known whether the mere presence of IgE specific to a food allergen gives a good prediction of allergenicity.

Below are discussed some examples of studies on thermal processing and allergy. These have mostly been made on the elicitation phase, i.e. using already sensitized patients or patient sera. At the end of article, I shall briefly discuss the possibilities that thermal processing may create new allergenic properties.

13.2 What Happens with Allergenicity After Thermal Processing?

13.2.1 Fish

The dominating major allergen in codfish is Gad c1, which seems to be extremely stable to thermal treatment. Interestingly the allergen seems to be much less stable to proteolytic processes [19]. Although other molecular weight species have been described as allergens in cod and other fish species, it seems that there is a quite strong correspondence between IgE reactivity to Gad c1 and clinical reactivity to cod [8] and probably other species of fish [20–22]. In a study by Bugajska-Schretter and coworkers a corresponding protein to Gad c1,

Table 13.1: Biological test systems for testing of allergenic potency of foods or food proteins (modified from [38], where the individual test systems are discussed in detail).

In vivo/ in vitro	Target organ/ system	Species	Test system	Examples
In vivo	Entire organism	Human	Challenge of allergic patients	DBPCFC, open challenges
		Experimental animals	Peroral challenge of animals	Anaphylactic response
	Skin	Human	Skin testing of allergic patients	Skin prick tests, intradermal tests
		Experimental animals	Actively or passively sensitized animals	PCA
In vitro	Basophils	Human	Actively or passively sensitized basophils + allergen	Basophil histamine release, cord blood basophil histamine release, CAST
	Basophils or mast cells	Humanized, i.e. transfected with a human IgE receptor	Passively sensitized cells	Mediator release
	Mast cells or basophils	Rodents	Peritoneal mast cells, RBL cell line	Histamine or other mediator release
	IgE	Human	(Inhibition of) IgE-allergen binding in immunochemical assays	RAST or RAST-inhibition
	IgG/IgE	Experimental animal including monoclonal antibodies	Immunochemical assays	ELISAs, Dipsticks etc.

i.e. parvalbumin from carp, has been thoroughly examined [23]. The investigators found that the protein was abundantly distributed in muscle and fibre cells, and by inhibition experiments it was demonstrated that the protein could inhibit for 80–90% of the total IgE binding activity to extracts of cod fish, tuna, and salmon. The parvalbumins to which the carp protein and Gad c1 belong are a family of calcium binding proteins, which play an important part in muscle physiology. This structure is well conserved throughout the animal kingdom, but the homology decreases with decreasing zoological relationship.

In a study where codfish, herring, plaice and mackerel were studied, fresh meat was compared after boiling for 6 min, 1 h and 4 h. The boiled meat was extracted and tested along with the boiling water in histamine release based on patients with clinical allergy to the fish species. In all samples except for the mackerel the allergenic activity could be recovered, indicating that for all practical reasons, thermal processing of fish is neither able to modify the allergenicity nor the solubility of the cod fish allergen Gad c1 or its homologues in other fish species [24].

13.2.2 Peanut and Other Legumes

The changes of peanut allergens during heat treatment are probably the best studied case of thermal processing of foods. By *in vitro* methods it was found that roasting of peanuts increased the biological activity 90 times compared to an extract of raw peanuts [25]. Using a whole peanut extract and purified single allergens Ara h1 and Ara h2, it was demonstrated that heating in the presence of various sugars such as glucose, fructose, xylose and mannose increased the activity of whole peanut extract with a factor of 39–66 [25], strongly suggesting that Maillard reactions are involved. Interestingly only 2–6-fold increase in activity was found for the two major allergens when given the same treatment, and thus other reactions must account for part of the increased activity. It has been suggested that the increase in Ara h1-allergenicity may be caused by a (Maillard reaction induced) irreversible stabilization of the homotrimeric form of the molecule that is known to have increased resistance to digestion [26]. For Ara h2 it has been demonstrated that it has inhibitory effects on trypsin, and that this inhibitory activity is increased in roasted peanut probably due to a partially reduction of Ara h2 [27]. Moreover, the increased IgE-reactivity of Ara h2 with Maillard products has recently been confirmed in an elegant study where the increased reactivity was mapped to specific epitopes by the production of Maillard products of 15-residue peptides spanning the Ara h2 sequence [28].

Many peanut allergic patients display a serological reactivity but only few patients also display a clinical reactivity to other legumes such as soy, peas, and various beans [29, 30]. Interestingly, when testing fresh green peas there was a clinical relevant cross-reactivity with peanut, but even the brief heat-treatment involved in blanching the peas before freezing them as "fresh green peas" was able to destroy the allergen activity in peas, in spite of the strong heat stability of the corresponding cross-reacting peanut allergens (Jensen, personal communication [39]).

The general problem that the extractability of proteins in a food is reduced with increasing processing has been investigated for peanut by Kopper and co-workers [31], who found that treatment with amylase and extraction at extended pH-values (away from the pI of the relevant protein) were indeed able to resolubilize IgE-binding components from a pellet. These findings do not only have methodological consequences but also physiological since amylase from saliva followed by low gastric pH may effectively solubilize immunoreactive proteins from a food.

13.2.3 Hazelnut

In a clinical study comparing raw and roasted (140 °C, 40 min) hazelnuts, 17 patients that reacted clinically in DBPCFC with raw nuts were tested with the roasted nuts [32]. Only five of these reacted and based on skin tests and IgE-binding studies it was estimated that the roasting had reduced the allergenicity approximately 100 times, but based on the overall reaction pattern of the patients in the various tests (SPT, IgE, histamine release) the reduction seems to be of quantitative nature, and it was concluded that while the roasting process had clearly reduced allergenicity it had not made the hazelnuts "safe" from an allergological point of view [32]. Moreover, it should be emphasized that the included patients from Denmark and Switzerland were birch pollen allergics with cross-reactivity to the Bet v1 homologue Cor a 1.04 [33]. Patients from South Europe with food allergy to hazelnuts may react to different – and more heat stable – allergens [34], notably the lipid transfer protein, Cor a 8 [35]. A recent study of the corresponding LTP from apple [36] has confirmed that these LTP allergens are highly resistant to thermal processing.

13.3 Conclusions

It has been debated for many years whether thermal processing may create neo-allergens. Even though sporadic reports on the formation of quite new allergens have occurred [37], there is still no good evidence that qualitatively new epitopes are formed during thermal or other forms of processing of foods. As discussed above, there is amble evidence from the peanut, however, that quantitatively increased allergenic potency may arise from thermal processing. It is interesting that in spite of large research into the Maillard reactions, so far unique "Maillard-epitopes" not present in raw foods have not been unequivocally demonstrated, but this is definitely an area of intense research, and the above statement may be proven wrong few years from now. It is also likely that many other reaction types within food proteins and between food proteins and other constituents may be recovered in the future. This will indeed put a large demand on analytical biochemistry and biophysical chemistry.

Due to the problems with extraction of proteins from processed foods, many of the assays presently used may only give us a part of the whole picture. Thus it is important to emphasize that allergic patients eat foods and not food extracts. This means that the ultimate verification of the allergenicity of a thermally processed food must be made in man. Figure 13.2 outlines the spectrum of tests for allergenicity discussed above. No *in vitro* test has so far been developed that can measure activity in a pellet, so if proteins cannot be brought into solution they are not available for analysis by these tests. To use food allergic patients on the other hand raises a number of problems. First it is neither ethical nor reasonable to use patients as guinea pigs for multiple experiments, and with regards to skin testing with recombinant proteins and processed forms hereof, these reagents are considered diagnostics, which impose demands on GMP- and GCP-procedures for their production and use in patients.

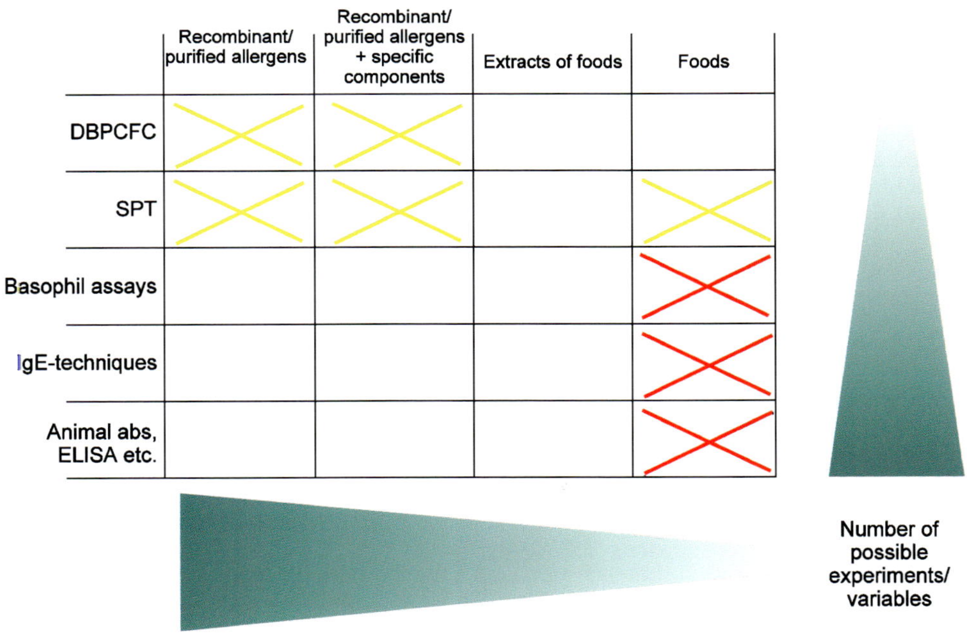

Figure 13.2: Outline of experimental models for investigation of the effect of thermal processing on allergenicity. DBPCFC=Double blind placebo controlled food challenge. SPT=Skin prick test. Red crosses: physically impossible. In order to employ the *in vitro* techniques the test substance must be in solution, and this may also cause problems if treatment of recombinant or purified allergens or extracts results in decreased solubility. Yellow crosses: administrative/economic obstacles due to demands on GMP/GCP.

Thus to further study the influence of thermal processing on food allergenicity an intelligent combination of biochemical, immunochemical and clinical techniques must be applied for future research, and a strong interdisciplinary effort is necessary to shed further light on this complex issue.

References

1. Johansson SG, Hourihane JO, Bousquet J, Bruijnzeel-Koomen C, Dreborg S, Haahtela T, et al. A revised nomenclature for allergy. An EAACI position statement from the EAACI nomenclature task force. *Allergy* 2001 Sep; **56**(9): 813–824.
2. Bruijnzeel Koomen CA, Ortolani C, Aas K, Bindslev-Jensen C, Björksten B, Moneret-Vautrin D, et al. EAACI Position paper: Adverse reactions to food. *Allergy* 1995; **50**: 623–635.
3. Farrell RJ, Kelly CP. Celiac sprue. *N Engl J Med* 2002 Jan 17; **346**: 180–188.
4. Menne T, Vejen N, Sjølin K-E, Maibach HI. Systemic contact dermatitis. *Am J Cont Derm* 1994; **5**: 1–12.
5. Bindslev-Jensen C. Food allergy. *BMJ* 1998 Apr 25; **316**: 1299–1302.
6. Sampson HA. Update on food allergy. *J Allergy Clin Immunol* 2004 May; **113**(5): 805–819.
7. Poulsen LK. In search of a new paradigm: mechanisms of sensitization and elicitation of food allergy. *Allergy* 2005 May; **60**(5): 549–558.
8. Hansen TK, Bindslev-Jensen C, Stahl Skov P, Poulsen LK. Codfish allergy in adults. Specific tests for IgE and histamine release vs double-blind, placebo-controlled challenges. *Clin Exp Allergy* 1996; **26**: 1276–1285.
9. Bernhisel-Broadbent J, Scanlon SM, Sampson HA. Fish hypersensitivity. I. In vitro and oral challenge results in fish-allergic patients. *J Allergy Clin Immunol* 1992 Mar; **89**: 730–737.
10. Daul CB, Morgan JE, Hughes J, Lehrer SB. Provocation-challenge studies in shrimp-sensitive individuals. *J Allergy Clin Immunol* 1988; **81**: 1180–1186.
11. Sampson HA. Epidemiology of food allergy. *Pediatr Allergy Immunol* 1996; **7**(9 Suppl): 42–50.
12. Pastorello EA, Ortolani C, Farioli L, Pravettoni V, Ispano M, Borga A, et al. Allergenic cross-reactivity among peach, apricot, plum, and cherry in patients with oral allergy syndrome: an in vivo and in vitro study. *J Allergy Clin Immunol* 1994 Oct; **94**(4): 699–707.
13. Kagan RS, Joseph L, Dufresne C, Gray-Donald K, Turnbull E, Pierre YS, et al. Prevalence of peanut allergy in primary-school children in Montreal, Canada. *J Allergy Clin Immunol* 2003 Dec; **112**(6): 1223–1228.
14. Sicherer SH, Munoz-Furlong A, Sampson HA. Prevalence of peanut and tree nut allergy in the United States determined by means of a random digit dial telephone survey: a 5-year follow-up study. *J Allergy Clin Immunol* 2003 Dec; **112**(6): 1203–1207.
15. Grundy J, Matthews S, Bateman B, Dean T, Arshad SH. Rising prevalence of allergy to peanut in children: Data from 2 sequential cohorts. *J Allergy Clin Immunol* 2002 Nov; **110**(5): 784–789.
16. Host A, Halken S. A prospective study of cow milk allergy in Danish infants during the first 3 years of life. Clinical course in relation to clinical and immunological type of hypersensitivity reaction. *Allergy* 1990 Nov; **45**: 587–596.
17. Osterballe M, Hansen TK, Mortz C, Host A, Bindslev-Jensen C. The prevalence of food hypersensitivity in an unselected population of children and adults. *Pediatr Allergy Immunol* 2005, **16**: 567–573.

18. Host A. Importance of the first meal on the development of cow's milk allergy and intolerance. *Allergy Proc* 1991 Jul; **12**(4): 227–232.
19. Untersmayr E, Poulsen LK, Platzer MH, Pedersen MH, Boltz-Nitulescu G, Skov PS, et al. The effects of gastric digestion on codfish allergenicity. *J Allergy Clin Immunol* 2005 Feb; **115**(2): 377–382.
20. Hansen TK, Bindslev-Jensen C, Stahl Skov P, Poulsen LK. Codfish allergy in adults. IgE crossreactivity among fish species. *Ann Allergy Asthma Immunol* 1997; **78**: 187–194.
21. Hansen TK, Bindslev-Jensen C, Stahl Skov P, Poulsen LK. Clinical reactivity to fish studied by multiple double-blind placebo-controlled food challenges (DBPCFC) (abstract). *J Allergy Clin Immunol* 1995; **95**: 328.
22. Sten E, Stahl Skov P, Hansen TK, Andersen SB, Torp AM, Bindslev-Jensen U, et al. Cross-reactivity to eel, eelpout and ocean pout in cod-allergic patients. *Allergy* 2004; **59**: 1173–1180.
23. Bugajska-Schretter A, Grote M, Vangelista L, Valent P, Sperr WR, Rumpold H, et al. Purification, biochemical, and immunological characterisation of a major food allergen: different immunoglobulin E recognition of the apo- and calcium-bound forms of carp parvalbumin. *Gut* 2000 May; **46**(5): 661–669.
24. Hansen TK, Stahl Skov P, Poulsen LK, Bindslev-Jensen C. Allergenic activity of processed fish (abstract). *ACI News* 1994; suppl **2**: 445.
25. Maleki SJ, Chung SY, Champagne ET, Raufman JP. The effects of roasting on the allergenic properties of peanut proteins. *J Allergy Clin Immunol* 2000 Oct; **106**(4): 763–768.
26. Maleki SJ, Kopper RA, Shin DS, Park CW, Compadre CM, Sampson H, et al. Structure of the major peanut allergen Ara h 1 may protect IgE-binding epitopes from degradation. *J Immunol* 2000 Jun 1; **164**(11): 5844–5849.
27. Maleki SJ, Viquez O, Jacks T, Dodo H, Champagne ET, Chung SY, et al. The major peanut allergen, Ara h 2, functions as a trypsin inhibitor, and roasting enhances this function. *J Allergy Clin Immunol* 2003 Jul; **112**(1): 190–195.
28. Gruber P, Becker WM, Hofmann T. Influence of the maillard reaction on the allergenicity of rAra h 2, a recombinant major allergen from peanut (Arachis hypogaea), its major epitopes, and peanut agglutinin. *J Agric Food Chem* 2005 Mar 23; **53**(6): 2289–2296.
29. Bernhisel-Broadbent J, Taylor S, Sampson HA. Cross-allergenicity in the legume botanical family in children with food hypersensitivity. II. Laboratory correlates. *J Allergy Clin Immunol* 1989 Nov; **84**: 701–709.
30. Bernhisel-Broadbent J, Sampson HA. Cross-allergenicity in the legume botanical family in children with food hypersensitivity. *J Allergy Clin Immunol* 1989 Feb; **83**: 435–440.
31. Kopper RA, Odum NJ, Sen M, Helm RM, Stanley JS, Burks AW. Peanut protein allergens: the effect of roasting on solubility and allergenicity. *Int Arch Allergy Immunol* 2005 Jan; **136**(1): 16–22.
32. Hansen KS, Ballmer-Weber BK, Luttkopf D, Skov PS, Wuthrich B, Bindslev-Jensen C, et al. Roasted hazelnuts – allergenic activity evaluated by double-blind, placebo-controlled food challenge. *Allergy* 2003 Feb; **58**(2): 132–138.
33. Luttkopf D, Muller U, Skov PS, Ballmer-Weber BK, Wüthrich B, Skamstrup Hansen K, et al. Comparison of four variants of a major allergen in hazelnut (Corylus avellana) Cor a 1.04 with the major hazel pollen allergen Cor a 1.01. *Mol Immunol* 2002 Jan; **38**: 515–525.
34. Schocker F, Luttkopf D, Muller U, Thomas P, Vieths S, Becker WM. IgE binding to unique hazelnut allergens: identification of non pollen-related and heat-stable hazelnut allergens eliciting severe allergic reactions. *Eur J Nutr* 2000 Aug; **39**: 172–180.
35. Pastorello EA, Vieths S, Pravettoni V, Farioli L, Trambaioli C, Fortunato D, et al. Identification of hazelnut major allergens in sensitive patients with positive double-blind, placebo-controlled food challenge results. *J Allergy Clin Immunol* 2002 Mar; **109**: 563–570.

36. Sancho AI, Rigby NM, Zuidmeer L, Asero R, Mistrello G, Amato S, et al. The effect of thermal processing on the IgE reactivity of the non-specific lipid transfer protein from apple, Mal d 3. *Allergy* 2005 Oct; **60**(10): 1262–1268.
37. Maleki SJ. Food processing: effects on allergenicity. *Curr Opin Allergy Clin Immunol* 2004 Jun; **4**(3): 241–245.
38. Poulsen LK. In vivo and in vitro techniques to determine the biological activity of food allergens. *J Chromatogr B Biomed Sci Appl* 2001 May 25; **756**: 41–55.
39. Jensen LB, Andersen M, Skov PS, Poulsen LK, Bindslev-Jensen C. Common serologic reactivities in the legume family do not always reflect clinically allergic reactivities (manuscript submitted for publication).

14 The Acrylamide Minimisation Concept – A Risk Management Tool

Andreas Kliemant * and Angela Göbel **

Abstract

As a reaction towards the threat caused through the discovery of acrylamide in food, the acrylamide minimisation concept was launched in Germany. The basic idea of this concept is the gradual reduction of acrylamide contents in foodstuffs. The pragmatic approach of the acrylamide minimisation concept is based on a voluntary agreement of all partners involved, the governmental organisations as well as the affected industry.

Now, three years after the kick-off of this concept, an evaluation shows its potential. For some food groups a minimisation of the acrylamide contents was already successful. However, further efforts are necessary to perform a comprehensive minimisation in all food groups concerned.

The acrylamide minimisation concept is, in contrast to legal regulations, a soft risk management instrument. It represents a novel approach to cope with situations when a legal limit is not available for instance. The minimisation concept is considered as a smart solution to bring about preventive measures resulting in health protection. As an innovative risk management tool it has the potential to tackle the challenges of consumer protection in the future.

14.1 Introduction

The mitigation of acrylamide formation in foodstuffs is closely related to the German acrylamide minimisation concept. This concept is based on an initiative of the German government to cope with the acrylamide problem. It is operative since August 2002. Immediately after the information about the occurrence of acrylamide in food provided by the Swedish National Food Agency in April 2002 [1] the question arose how to deal with this problem [2, 3]. Acryl-

* Correspondence to: Andreas Kliemant, Unit Prevention, Crisis Management, Rapid Alert Systems, Federal Office of Consumer Protection and Food Safety (BVL), Rochusstr. 65, 53123 Bonn, Germany.

** Unit General Affairs of Food, Food of Non-Animal Origin, Federal Office of Consumer Protection and Food Safety, Berlin, Germany.

Thermal Processing of Food: Potential Health Benefits and Risks
DFG, Deutsche Forschungsgemeinschaft

ISBN: 978-3-527-31909-1

amide has been found carcinogenic and mutagenic in animal tests [4–8] and rated probably carcinogenic and genotoxic for man. With the first discussions it became clear that a legal limit based on adequate risk assessment was not within reach. Sufficient and reliable toxicological data, including those on dietary intake, bioavailability etc. were not available at this moment. The minimisation concept, designed to perform a gradual reduction of acrylamide contents in a wide variety of foodstuffs, appeared an option of considerable promise. The acrylamide minimisation concept was not designed to achieve a minimisation to zero. The basic principle of the concept is the ALARA principle: as low as reasonably achievable. Based on this principle, an effective prevention of the consumer's exposure towards dietary acrylamide could be accomplished during the last three years.

14.2 The Acrylamide Minimisation Concept – The Idea and its Implementation

The information that certain thermally treated foods contain rather high concentrations of acrylamide, provided by the Swedish National Food Agency in April 2002 [1], caused worldwide concern [2, 3]. Until this time acrylamide was well known as a toxic, cancer-causing industrial chemical for technical purposes, primarily used in the preparation of polyacrylamide for plastic materials and also in the processing of drinking water. The new knowledge about acrylamide was a great challenge to the authorities responsible for consumer protection because acrylamide has been found carcinogenic and mutagenic in animal tests [4–8]. The International Agency for Research on Cancer has classified acrylamide as a Group 2A carcinogen [9]. Furthermore a genotoxic mechanism could be verified and attributed to the main metabolite formed in mammals, glycidamide [10]. However, up to now not enough data is available to perform a final risk assessment for man [11]. A potential strategy to handle the potential health risk posed by acrylamide in heated food was to set a limit for acrylamide in food. This appeared, however, to be far from realistic because adequate toxicological data to deduce such a limit were not at hand nor was the technical feasibility given. In addition, the mechanism(s) by which acrylamide is formed during the processing of food was largely unknown and it appeared rather difficult to predict the amount to which the food borne contaminant acrylamide will actually be formed by a given treatment. The formation of acrylamide during the processing as well as during the preparation of food depends on a spectrum of influential factors such as raw materials and processing techniques. Therefore the acrylamide minimisation concept was mainly aimed at developing processing and preparation techniques which would lower the formation of acrylamide in foodstuffs. The intention of this strategy was to influence process parameters with the aim to reach the borders of the technical feasibility without changing the main character of the product. This action followed the ALARA principle: as low as reasonably achievable. The acrylamide minimisation concept represents an instrument which avoids the formation of the food borne contaminant acrylamide as far as possible. This action is expected to accomplish a preventive health protection of consumers, irrespective of currently non sufficient risk assessment for acrylamide.

The acrylamide minimisation concept is voluntarily agreed among the Federal Office of Consumer Protection and Food Safety (BVL), the German federal states, the affected industry and the Federal Ministry of Food, Agriculture and Consumer Protection (BMELV). It is designed as a dynamic strategy to pursue the policy of gradual reduction of acrylamide contents in a wide variety of foodstuffs.

Detailed description of the concept shown in scheme 14.1:

- The BVL sets up a database on acrylamide. In this database 9600 data sets are collected up to now. There is information available on product labelling, producer addresses, batch numbers, best before dates, acrylamide contents, test methods etc. The files are mainly provided by the official food surveillance authorities of the German federal states.

- Foodstuffs concerned are classified into certain food groups. Those foods which make up the ten percent most contaminated products within each group are identified. The lowest of the acrylamide contents of these upper 10 percent is the signal value for this group. If the signal value is higher than 1000 μg/kg, it will be automatically set to 1000 μg/kg.

In addition, individual products from producers with an important market position are monitored as well.

- If acrylamide contents are found above the signal value, the federal states responsible for the food producers affected will be informed by the BVL.

- Food control authorities will contact the respective food producers and start the minimisation dialogue to check whether ingredients or processes could be

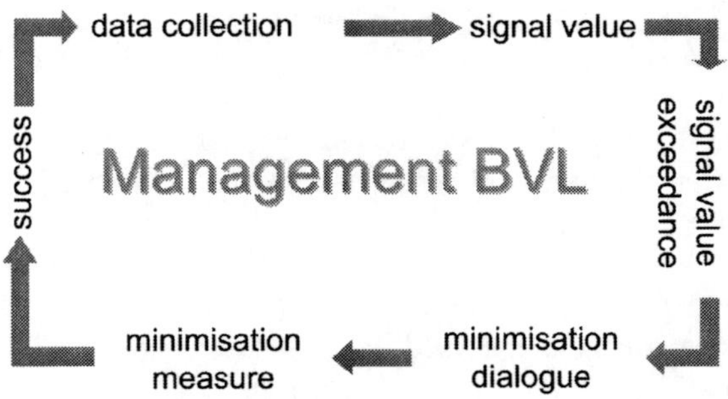

Scheme 14.1: Workflow of the minimisation concept.

changed to minimise acrylamide contents, and which changes this could be. The aim of the minimisation dialogue is to develop corresponding minimisation measures.

- The BVL will update signal values annually. This strategy will balance the effects of seasonal fluctuations in acrylamide contents in foods and of seasonal variations in the food offer.
- Signal values once calculated will not be raised as long as this minimisation concept is pursued, but will be maintained or lowered. This means that acrylamide contents in relevant foods will be continually reduced if the minimisation measures are successful.
- Food with acrylamide contents of more than 1000 µg/kg and from food groups for which no signal values have been set yet will automatically be included in the minimisation dialogue.

The directive to implement or perform minimisation measures within the borders of a nearly unchanged product is of highest importance for the global success of the acrylamide minimisation concept. This action pays attention to consumers' very sensitive reaction on changes of product quality. One solution is to ensure that criteria for quality like crispness of potato crisps are unchanged after a minimisation of acrylamide contents of a product. Another solution is to enlighten the consumer about the necessity to change the product quality. One example is the slogan "golden yellow, not carbonised" to advertise the new colour of chips with low acrylamide contents. Both strategies help to obtain the acceptance of the consumer towards the acrylamide-minimised product.

14.3 The Instruments of the Acrylamide Minimisation Concept

Signal value: The main instrument of the acrylamide minimisation concept is the signal value. The calculation of the signal value is made according to specific definitions. The signal value is based on real data and gives a description of the minimisation situation within the relevant assessment period. Therefore, the signal value is used as a bearing instrument and as an indicator of success. Nevertheless, the validity of the signal value is limited because of deliberate restrictions within the relevant calculation procedure. The signal value has been defined as the lowest acrylamide content in the 10 percent highest contaminated products in a group of food products. For the purpose of the minimisation concept the signal value is subject to certain conditions: it must not exceed 1000 µg/kg, and once it is calculated, it must not be raised, but has to be maintained or reduced. These definitions are necessary to perform a dynamic strategy with the principle of gradual reduction of acrylamide contents. The signal value is the threshold at which a dialogue with the aim to minimise acrylamide contents will start and minimisation measures will be taken. The signal value is a measure of success of minimisation efforts, within the framework of conditions defined for that value under the minimisation concept.

Observer value: The deliberate limitations of the signal value were the reason to create a complementing instrument for the acrylamide minimisation concept, the observer value. The observer value is calculated basically in the same way as the signal value, but without the conditions that it must not be increased and must not exceed 1000 µg/kg, thus providing another instrument to measure the success of minimisation efforts. The observer value acts, however, beyond the framework fixed for the signal value, allowing to assess the situation of acrylamide minimisation even when the signal value is unchanged. For instance, it can show a decrease from 1500 µg/kg to 1050 µg/kg, with the signal value being 1000 µg/kg. This allows a comprehensive description of the real situation of minimisation.

Additional instruments: The degree of success of the concept to minimise acrylamide contents in food products can be additionally evaluated on the basis of classical statistical instruments. Besides the assessment of the development of signal values and observer values for the relevant assessment periods, the development of median values and the frequency distribution of acrylamide contents are assessed, taking into account all acrylamide contents in a food group.

14.4 The Assessment of the Success of the Acrylamide Minimisation Concept: Results of the 4th Calculation of Signal Values

14.4.1 Development of Signal Values and Observer Values

Table 14.1 shows an overview of signal and observer values from the 1st to the 4th calculation. The assessment of the development of signal and observer values is based on the classification into four categories:

- Signal value decreased, 2 product groups
- Signal value unchanged, 1 product group
- Signal value unchanged (with decreased observer value), 5 product groups
- Signal value unchanged (with increased observer value), 5 product groups.

The detailed data on the results of the 4th calculation of signal values presented in Table 14.2 demonstrate the pragmatic approach of this concept. There is no coordination concerning the number of products like in a monitoring programme. This is a result of the deliberate decision to give all suppliers of data a maximum range of decision. This liberty was necessary within a concept vo-

Table 14.1: Signal values and observer values, 1st to 4th calculation.

Food group	Signal (observer) value 1st calculation Sep 17, 2002 acrylamide [µg/kg]	Signal (observer) value 2nd calculation Jan 31, 2003 acrylamide [µg/kg]	Signal (observer) value 3rd calculation Nov 26, 2003 acrylamide [µg/kg]	Signal (observer) value 4th calculation Nov 17, 2004 acrylamide [µg/kg]
Fine bakery ware, short pastry	800	660	575	575 (760) [a]
Breakfast cereals	260	260	200	200 (240) [a]
Coffee, roasted	370	370	370 (520) [a]	370 (420) [a]
Potato crisps	1000 (1500) [a]	1000 (1200) [a]	1000 (1470) [a]	1000 (1029) [a]
Crispbread	610	610	610 (1260) [a]	610 (640) [a]
Chips, prepared	770	570	570	540
Potato fritter, prepared	1000	1000 (1300) [a]	1000 (1080) [a]	1000 (1215) [a]
Gingerbread and bakery ware containing gingerbread	1000	1000 (1370) [a]	1000 (1460) [a]	1000 (1020) [a]
Thin almond biscuits	1000	710	710 (760) [a]	560
Children's biscuits	n.c. [b]	n.c. [b]	360	360
Diabetics' cakes and biscuits	n.c. [b]	n.c. [b]	1000 (1740) [a]	1000 (1010) [a]
Coffee extract	n.c. [b]	n.c. [b]	1000 (1110) [a]	1000 (2380) [a]
Coffee substitute	n.c. [b]	n.c. [b]	1000 (2080) [a]	1000 (2910) [a]

[a] Observer value: calculated basically in the same way as the signal value, but without the condition that it must not be increased and must not exceed 1000 µg/kg.
[b] not calculated.

luntarily agreed among the partners. The wide range between minimum and maximum levels within one food group reflects the wide variety of different products on the market. It also reflects, in particular for products based on potatoes, the change of the composition of the raw material potato during a longer period of time, e.g. during storage. However, differences found between

Table 14.2: Signal values, 4th calculation, valid from November 17, 2004.

Product group	Signal value 4th calculation [µg/kg]	Observer value 4th calculation [µg/kg]	Number of products	Minimum acrylamide [µg/kg]	Median acrylamide [µg/kg]	Maximum acrylamide [µg/kg]
Fine bakery ware, short pastry	575	760	192	5	136	8194
Breakfast cereals	200	240	40	5	40	328
Coffee, roasted	370	420	102	60	277	574
Potato crisps	1000	1029	198	5	395	3483
Crispbread	610	640	86	5	340	1177
Chips, prepared	540	not available	578	5	171	2228
Potato fritter, prepared	1000	1215	26	10	341	2460
Gingerbread and bakery ware containing gingerbread	1000	1020	286	7	251	6891
Thin almond biscuits	560	not available	83	5	181	992
Children's biscuits	360	not available	41	5	73	588
Diabetics' cakes and biscuits	1000	1010	107	5	189	3044
Coffee extract	1000	2380	21	188	852	4948
Coffee substitute	1000	2910	32	248	972	3905

products within one food group are driving minimisation activities, since products with low acrylamide contents demonstrate the possibilities of minimisation.

The difference between the signal value and the associated median value signalises the potential of minimisation within a food group. The food group "roasted coffee" shows a difference of less than 100 µg/kg acrylamide between the signal value and the corresponding median. This allows two interpretations: theoretically, this could reflect either complete absence of minimisation activities or, more plausibly, an only small minimisation potential. The food group "gingerbread" shows a difference of approximately 750 µg/kg acrylamide between the signal value and the corresponding median. This situation allows the interpretation that the minimisation potential in this food group is high. However, a more complex approach regarding the minimisation potential of a food group is the evaluation of the frequency distribution.

The number of food groups observed has been increased during the minimisation concept. This shows the flexibility of this concept. The reason for the extension of food groups was to provide specific data for consumer groups with special dietary requirements and to have in place data regarding another critical food group. Starting with the 3rd calculation of signal values data concerning special products for children and diabetics as well as data on "coffee extract" and "coffee substitute" were collected.

14.4.2 Evaluation of the Frequency Distribution

In addition to the evaluation of development of signal, observer and median values, the evaluation of frequency distribution provides a further tool to comprehensively assess the success of the minimisation concept. This evaluation is based on all acrylamide contents available for a specific food group for the relevant assessment period, thus providing a survey on all acrylamide contents in a given food group. When assessing the frequency distribution, a "shift to the left" reflects an increase in products with acrylamide concentrations below 300 µg/kg or in some product groups below 200 or 100 µg/kg. This shift can be used in combination with the associated signal and maximum values as an indicator for the minimisation potential of the corresponding food group.

The decrease of the signal value and the fact that more than 70% of all tested products have acrylamide contents of less than 300 µg/kg allow to state a minimisation success. Moreover, the characteristics of the frequency distribution, the "shift to the left", can be used in combination with the concomitant existence of high acrylamide levels with more than 1000 µg/kg as an indicator for a very high minimisation potential, as can be seen for the example of chips (Fig. 14.1) and diabetics' cakes and biscuits (Fig. 14.2).

14.4.3 Observation of Individual Products

The observation of individual products is a special tool within the acrylamide minimisation concept. The aim is to provide evidence on the extent of acrylamide minimisation concerning a single product before and after the implementation of minimisation measures. An example for diabetics' cakes and biscuits is shown in Figure 14.3. Thus the observation of single products provides a rationale for the understanding of minimisation processes.

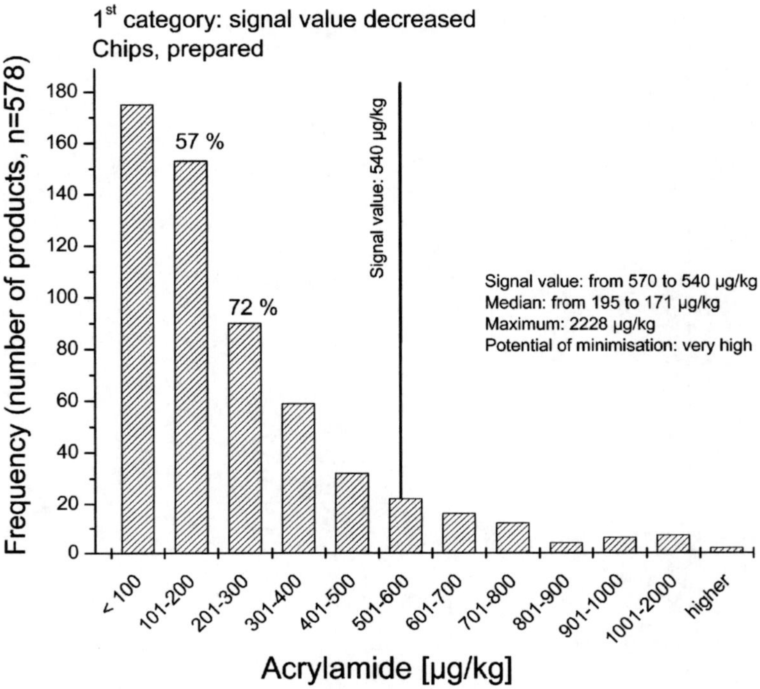

Figure 14.1: Frequency distribution in chips.

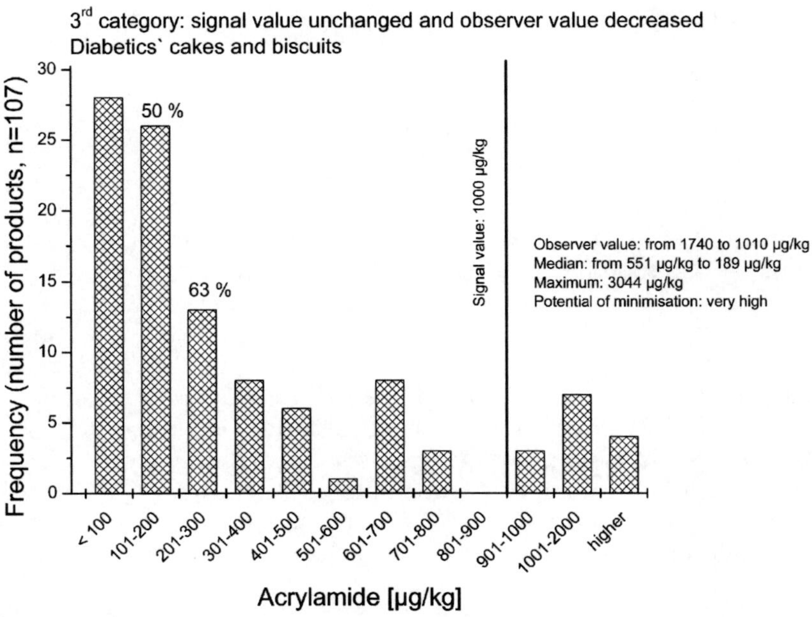

Figure 14.2: Frequency distribution of diabetics' cakes and biscuits.

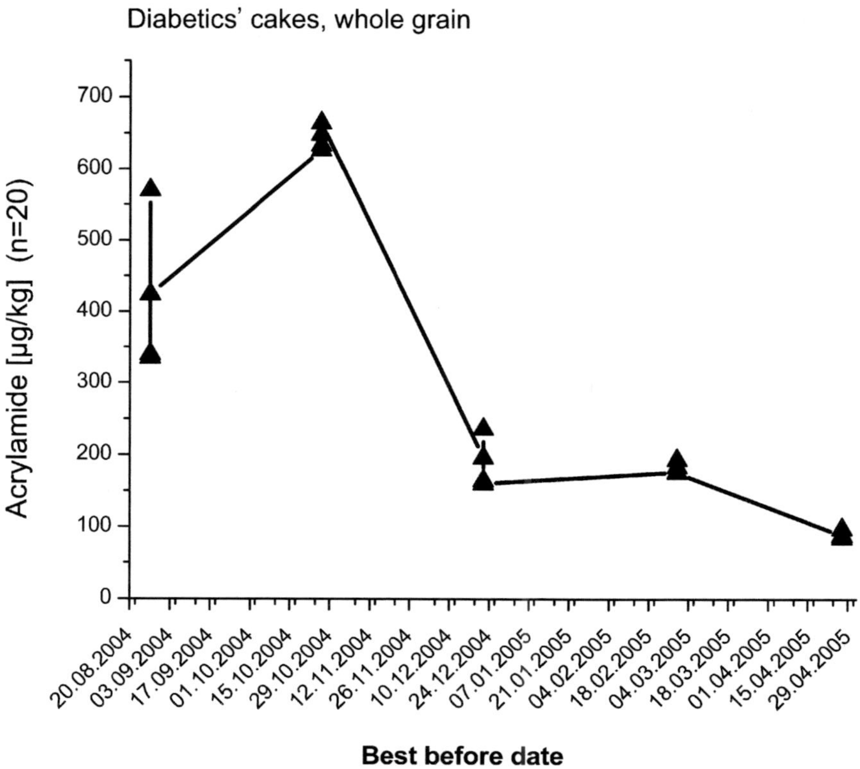

Figure 14.3: Diabetics' cakes, whole grain.

14.5 The Acrylamide Minimisation Concept – Limits and Chances of a Risk Management Tool

An analysis of the limits of this concept needs first to address the fact that the acrylamide minimisation concept up to now represents a national approach to tackle a global problem. In times of globalisation of markets this is a limitation that matters, especially in view of the many products of foreign origin on the domestic market. Unfortunately, those products are not reachable by the German acrylamide minimisation concept. This limitation was recognised in an early stage and an appropriate initiative was started, including the information of authorities of the responsible Member States about products produced by foreign producers exceeding the signal values. Furthermore, all data collected within the framework of the acrylamide minimisation concept were exported to the acrylamide monitoring database of the European Commission, maintained by the Institute for Reference Materials and Measurements. Unfortunately, a European or global solution related to the minimisation concept is not in sight. However, there is a lot of alternative endeavours aiming to minimise acrylamide contents in foodstuffs. For example information was provided by the European Commission on ways to lower the levels of acrylamide formed in food [12]. The Confederation of the Food and Drink Industries of the EU has developed the acrylamide "Toolbox", describing in detail acrylamide formation and providing a set of intervention steps to limit and reduce its formation during manufacturing or processing of food [13]. Furthermore, producers experienced in minimisation through joining the acrylamide minimisation concept transfer their know-how from German to foreign production plants.

Three years after the kick-off of the acrylamide minimisation concept the question arises, whether such a concept has the potential to serve as a general risk management tool. The experience obtained allows the general recommendation that strategies of gradual minimisation should also be applied for other substances than acrylamide. The acrylamide minimisation concept stands for a pragmatic, flexible solution. In particular in cases when a legal limit is not available, this concept can be adapted to arrive at smart solutions for preventive health protection of consumers.

References

1. Swedish National Food Agency (2002) http://www.slv.se
2. World Health Organization (WHO) (2002) FAO/WHO Consultation on the health implications of acrylamide in food. Geneva, 25–27 June 2002. www.who.int
3. European Commission (2002) Opinion of the Scientific Committee on Food on new findings regarding the presence of acrylamide in food. http://europa.eu.int
4. Johnson, K. A.; Gorzinski, S. J.; Bodner, K. M.; Bampbell, R. A.; Wolf, C.H.; Friedman, M. A.; Mast, R. W. (1986) Chronic toxicity and oncogenicity study on acrylamide incorporated in the drinking water of Fischer 344 rats. *Toxicol. Appl. Pharmacol.* **85**, 154–168.
5. Friedman, M. A.; Dulak, L. H.; Stedham, M. A. (1995) A lifetime oncogenicity study in rats with acrylamide. *Fundam. Appl. Toxicol.* **27**, 95–105.
6. Bull, R. J.; Robinson, M.; Laurie, R. D.; Stoner, G. D.; Greisiger, E.; Meier, J. R.; Stober, J. (1984) Carcinogenic effects of acrylamide in Sencar and A/J mice. *Cancer Res.* **44**, 107–111.
7. Bull, R. J.; Robinson, M.; Stober, J. A. (1984) Carcinogenic activity of acrylamide in the skin and lung of Swiss-ICR mice. *Cancer Lett.* **24**, 209–212.
8. Robinson, M.; Bull, R. J.; Knutsen, G. L.; Shields, R. P.; Strober, J. A. (1986) A combined bioassay utilizing both the lung adenoma and skin papilloma protocols. *Environ. Health Perspect.* **68**, 141–145.
9. International Agency for Research on Cancer (IARC) (1994) Acrylamide. *IARC Monogr. Eval. Carcinog. Risks Hum.* **60**, 389–433.
10. Doerge, D. R.; Costa, G. G.; McDaniel, L. P.; Churchwell, M. I.; Twaddle, N. C.; Beland, F. A. (2005) DNA adducts derived from administration of acrylamide and glycidamide to mice and rats. *Mutat. Res.* **580**, 131–141.
11. Joint FAO/WHO Expert Committee on Food Additives (JECFA) (2005) Summary and conclusions of sixty-fourth meeting, Rome, 8–17 February 2005. www.fao.org/es/esn/jecfa/acrylamide_en.stm
12. European Commission, DG Health and Consumer Protection (2003) http://ec.europa.eu/food/food/chemicalsafety/contaminants/acryl_guidance.pdf
13. Confederation of the Food and Drink Industries of the EU (CIAA) (2005) www.ciaa.be

15 The Consequences of Cooking: How the Origin of Cuisine Shaped Human Culture, Ecology, and Biology

Greg Laden *

A strong argument can be made that the last common ancestor of humans and our nearest living relatives (the chimps) was physically and behaviorally similar to modern chimpanzees (*Pan troglodytes*). Therefore, the "story" of human evolution can be thought of as a collection of transitions that can explain the differences between chimps and humans, to the extent that we can clearly characterize each species and thus enumerate or describe differences. My research in many related areas has led me to believe that the most important of these differences have to do with sex, food, and the interaction between the two. There were two main and a few smaller food transitions prior to the invention of agriculture and husbandry: 1) The incorporation of plant underground storage organs as the principle fallback (lean times) food, replacing leaves; and 2) the invention of processing food with fire. While these two transitions are related, it is cooking that had the largest effect with respect to modern human traits. These effects included a few, but not many, changes in the bulk of the gastrointestinal system (chimps and humans are not too different), unprecedented changes in the masticatory system, a cascade of changes in the overall body frame (bones and muscles) both directly and indirectly, and of paramount importance, changes in social, sexual, and cultural behavior without which our species would probably be unrecognizable. The sparks that lit the first controlled fire also created our species.

* Department of Anthropology, University of Minnesota, 395 Hubert H. Humphrey Center, 301 19th Avenue S, Minneapolis MN 55455, USA.

Thermal Processing of Food: Potential Health Benefits and Risks
DFG, Deutsche Forschungsgemeinschaft

ISBN: 978-3-527-31909-1

The Consequences of Cooking:
How the origin of cuisine shaped human culture, ecology, and biology

Greg Laden
Department of Anthropology
University of Minnesota

Collaboration

- Hominid diet
  - Richard Wrangham, Jamie Jones, David Pilbeam, Nancy Lou Conklin-Britian
- Forager Research
  - Colleagues among the Efe and Lese of the Congo
- South African Field Research
  - Sir Francis Thackeray, Lynne Simpson, Peter Beaumont, David Morris, Polke Birkhartz, Henk Stein
- Further Dietary Research
  - Betsy Burr
- Funding and Support
  - National Science Foundation, L.S.B. Leakey Foundation, Lauria Family Trust, Holt Family Trust, MC/Hrdy Fellowship, Transvaal Museum, McGregor Museum

Conclusions

- Humans and other apes *can* eat the same foods, so there is no big story here.
- Humans critically rely on cooked foods. So this is the most important story ever.
- In some ways, cooking allows humans to reach the ultimate goal of all apes, so humans have perfected "apeness."
- However, humans have a social system that is entirely unique and, ultimately, driven by the invention of cooking.

Evolutionary Approach

- What are the human derived features?
- Which derived features are predicted from the mammal pattern? ... the ape pattern?
- Of the remaining features, which are due to thermal processing of food?
- (short answer ... virtually everything of importance is due to the thermal processing of food ...)

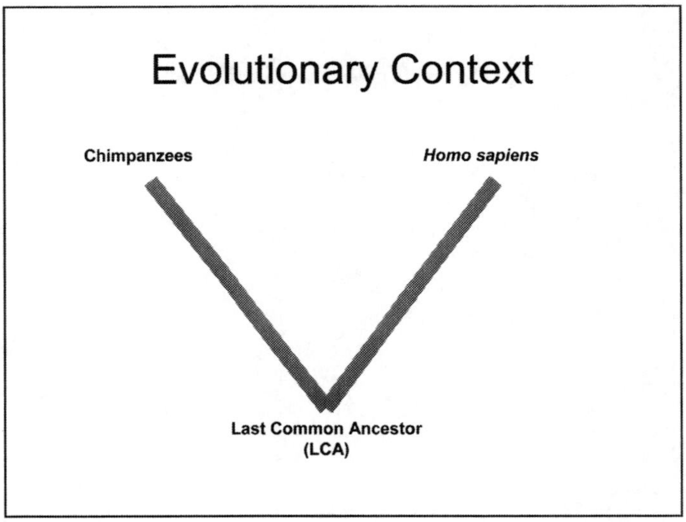
Evolutionary Context
Chimpanzees
Homo sapiens
Last Common Ancestor
(LCA)

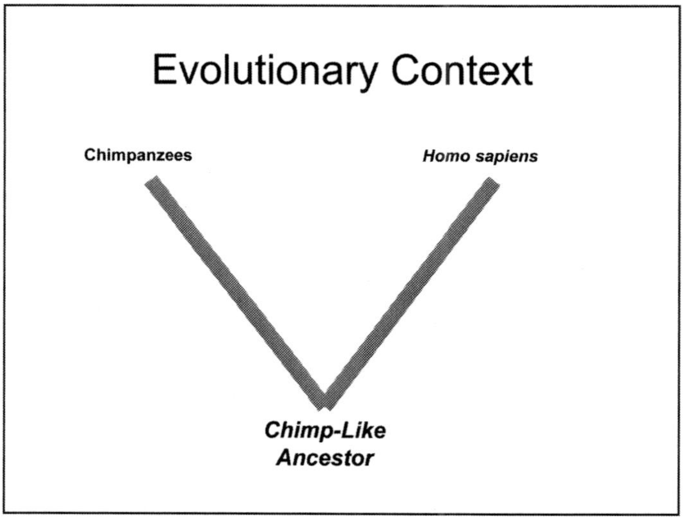
Evolutionary Context
Chimpanzees
Homo sapiens
Chimp-Like Ancestor

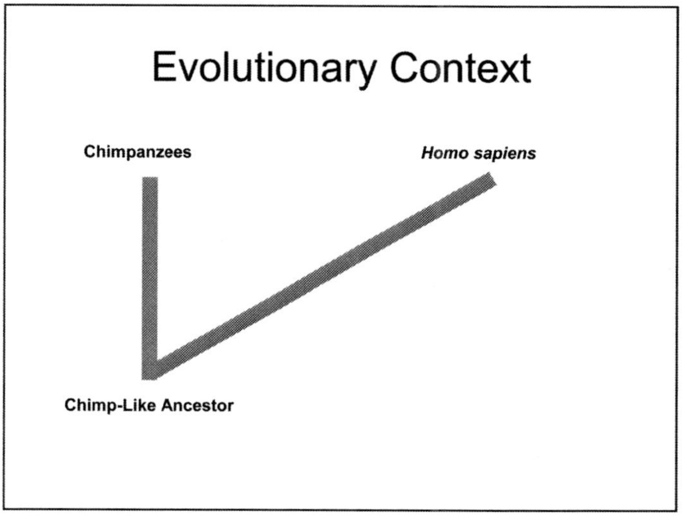
Evolutionary Context
Chimpanzees
Homo sapiens
Chimp-Like Ancestor

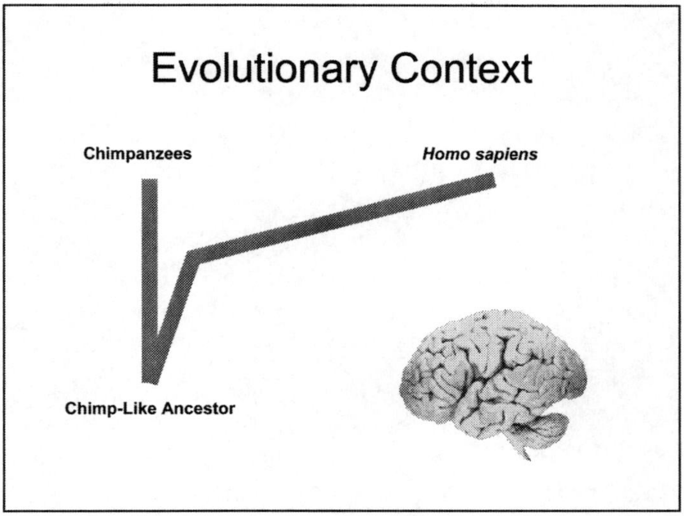
Evolutionary Context
Chimpanzees
Homo sapiens
Chimp-Like Ancestor

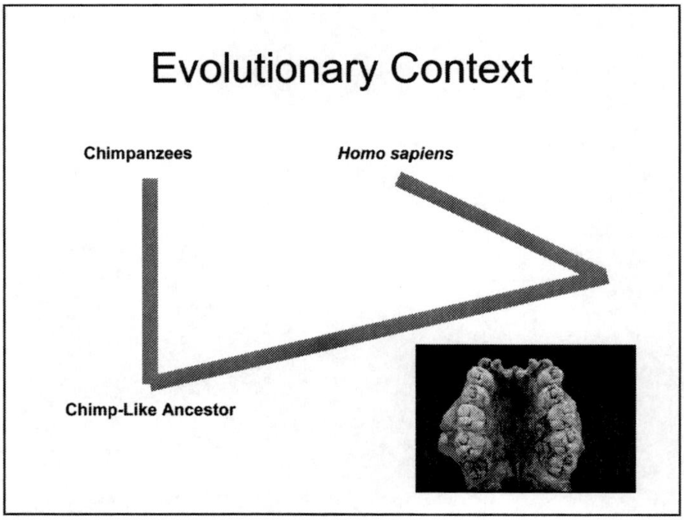
Evolutionary Context
Chimpanzees
Homo sapiens
Chimp-Like Ancestor

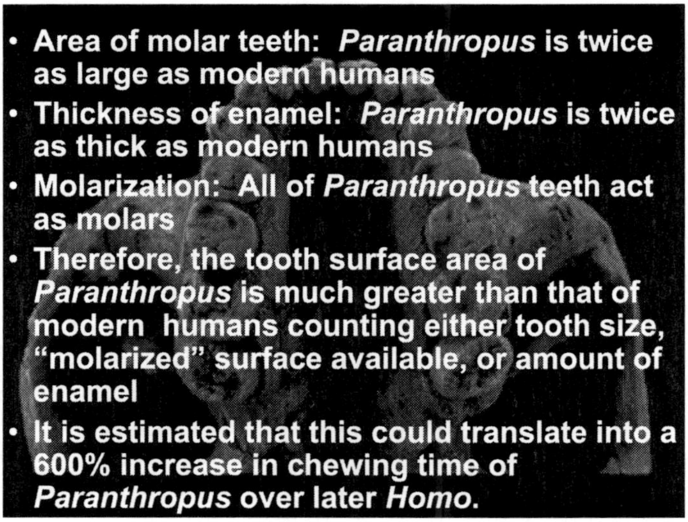
• Area of molar teeth: Paranthropus is twice as large as modern humans
• Thickness of enamel: Paranthropus is twice as thick as modern humans
• Molarization: All of Paranthropus teeth act as molars
• Therefore, the tooth surface area of Paranthropus is much greater than that of modern humans counting either tooth size, "molarized" surface available, or amount of enamel
• It is estimated that this could translate into a 600% increase in chewing time of Paranthropus over later Homo.

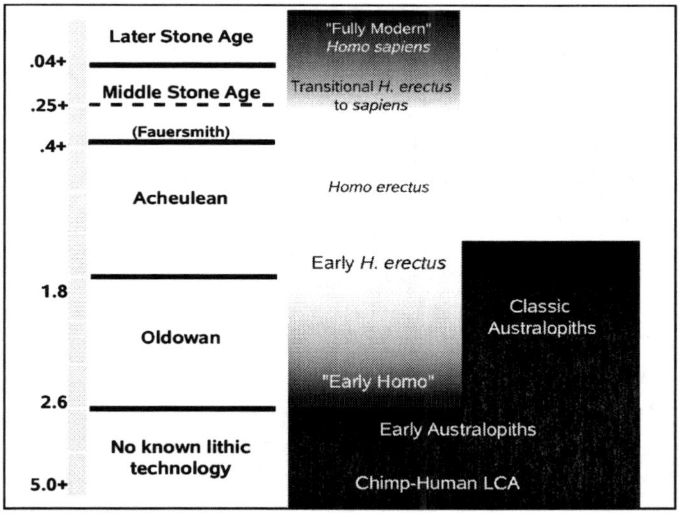
Later Stone Age
.04+
Middle Stone Age
.25+
(Fauersmith)
.4+
Acheulean
1.8
Oldowan
2.6
No known lithic technology
5.0+
"Fully Modern" Homo sapiens
Transitional H. erectus to sapiens
Homo erectus
Early H. erectus
"Early Homo"
Classic Australopiths
Early Australopiths
Chimp-Human LCA

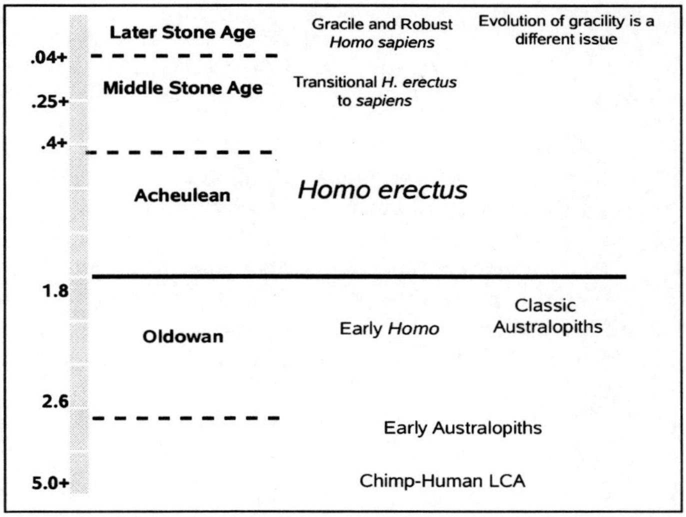
Later Stone Age
Gracile and Robust *Homo sapiens*
Evolution of gracility is a different issue
.04+
Middle Stone Age
Transitional *H. erectus* to *sapiens*
.25+
.4+
Acheulean
Homo erectus
1.8
Oldowan
Early *Homo*
Classic Australopiths
2.6
Early Australopiths
5.0+
Chimp-Human LCA

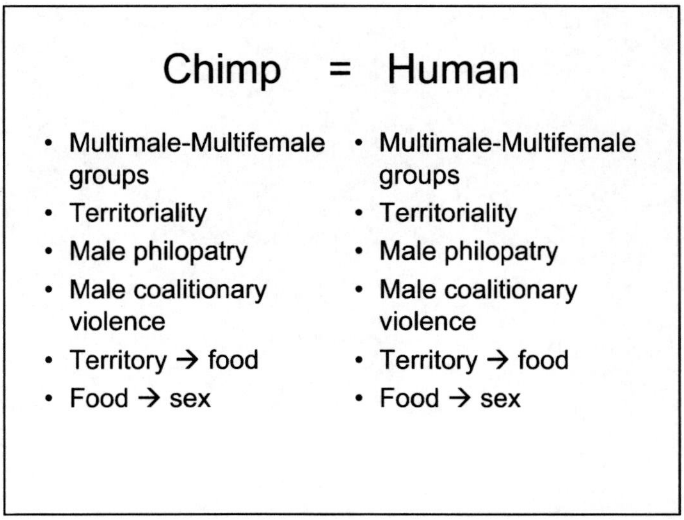
Chimp = Human
• Multimale-Multifemale groups
• Territoriality
• Male philopatry
• Male coalitionary violence
• Territory → food
• Food → sex
• Multimale-Multifemale groups
• Territoriality
• Male philopatry
• Male coalitionary violence
• Territory → food
• Food → sex

Chimp ≠ Human

Chimp	Human
• Promiscuous mating	• Pair bonding
• Continuous simple maximally defendable territories	• Complex undefendable territories
• Male investment in offspring primarily territorial	• Male investment in offspring more diverse
• No inter-territorial interaction	• Complex and critical inter-territorial interaction

Pair bonding

Food sharing

Marriage and food rituals

Ecological and social theory

- Costs and benefits of being solitary
- Costs and benefits of being in a group
- Food quality and distribution
- Female vs. male strategies.

Dietary ecology

- High quality food is rare
- Lower quality food is critically important "fallback" food
- Low quality food requires more "machinery" (larger body size).

Social and dietary factors together

- Grouping (social) behavior arises from conflicting dynamics of competition for food, mates, and protection
- Primates should want to be alone most of the time to minimize feeding competition
- Primates like to be in groups to minimize effects of predation
- Females additionally like to be in groups to avoid infanticidal males
- Ideally "solitary" mammals usually end up as male-female pairs.

Unique derived human traits

- Multi-Female/Multi-Male groups with Exclusive sexual access and "Pair bonding" (Marriage)
- Varying degrees of male Parental Investment
- Territorial in a novel way
- Division of labor in the food quest
- Central place foraging
- Delayed processing of food
- Sharing
- Fixation on social eating.

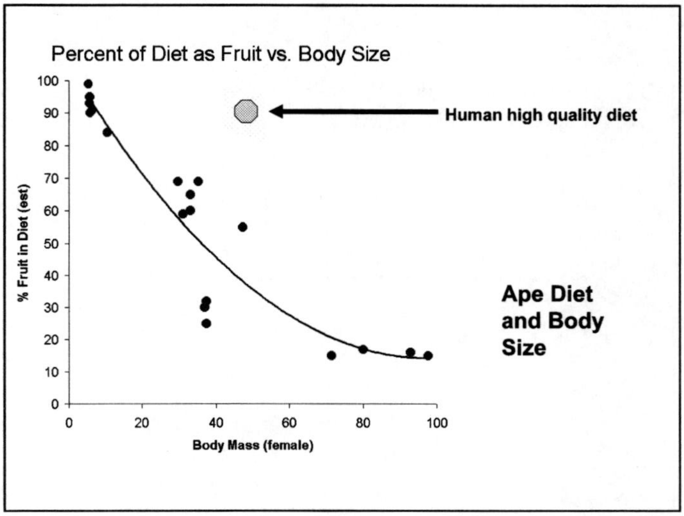
Percent of Diet as Fruit vs. Body Size
Human high quality diet
% Fruit in Diet (est)
Body Mass (female)
Ape Diet and Body Size
0
10
20
30
40
50
60
70
80
90
100
0
20
40
60
80
100

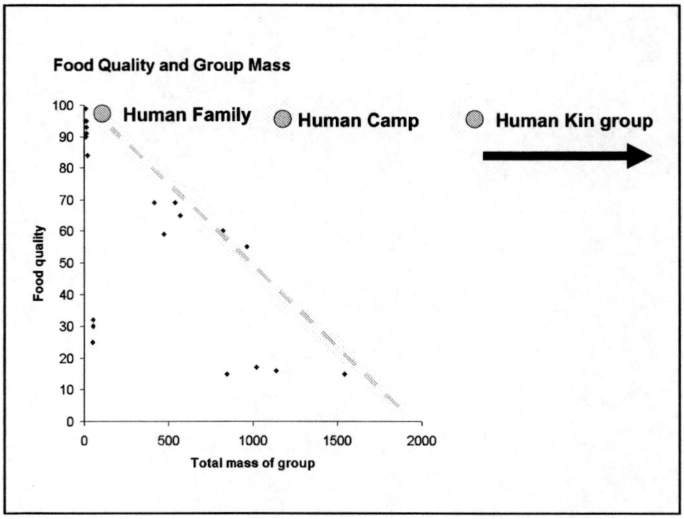
Food Quality and Group Mass
Human Family
Human Camp
Human Kin group
Food quality
Total mass of group
0
10
20
30
40
50
60
70
80
90
100
0
500
1000
1500
2000

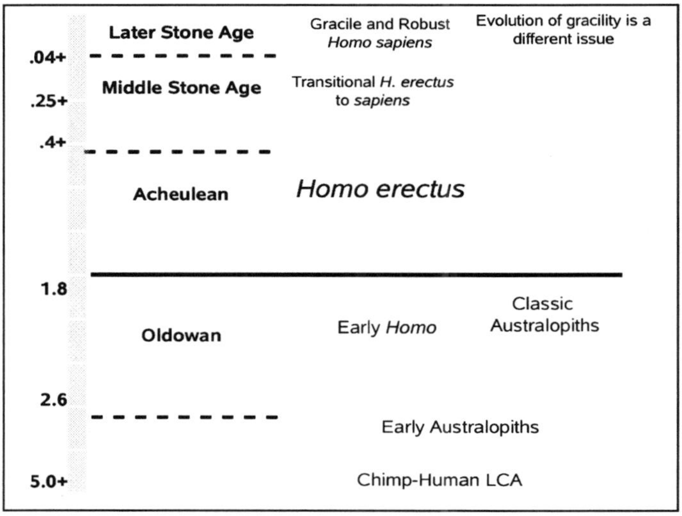
Later Stone Age
Gracile and Robust Homo sapiens
Evolution of gracility is a different issue
.04+
Middle Stone Age
Transitional H. erectus to sapiens
.25+
.4+
Acheulean
Homo erectus
1.8
Oldowan
Early Homo
Classic Australopiths
2.6
Early Australopiths
5.0+
Chimp-Human LCA

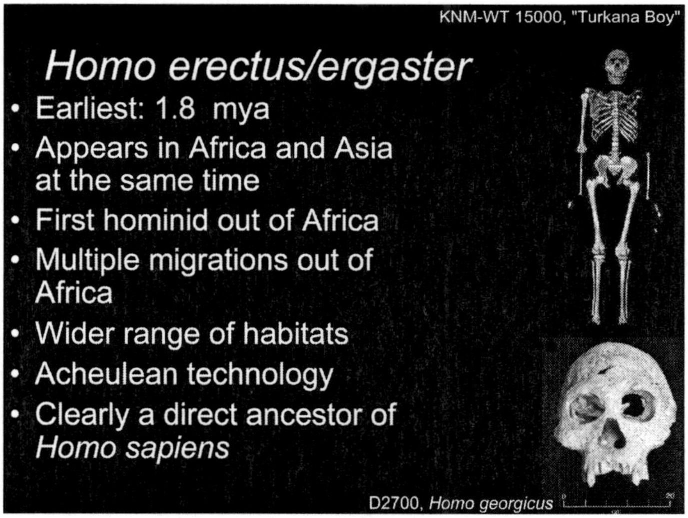
KNM-WT 15000, "Turkana Boy"
Homo erectus/ergaster
• Earliest: 1.8 mya
• Appears in Africa and Asia at the same time
• First hominid out of Africa
• Multiple migrations out of Africa
• Wider range of habitats
• Acheulean technology
• Clearly a direct ancestor of Homo sapiens
D2700, Homo georgicus

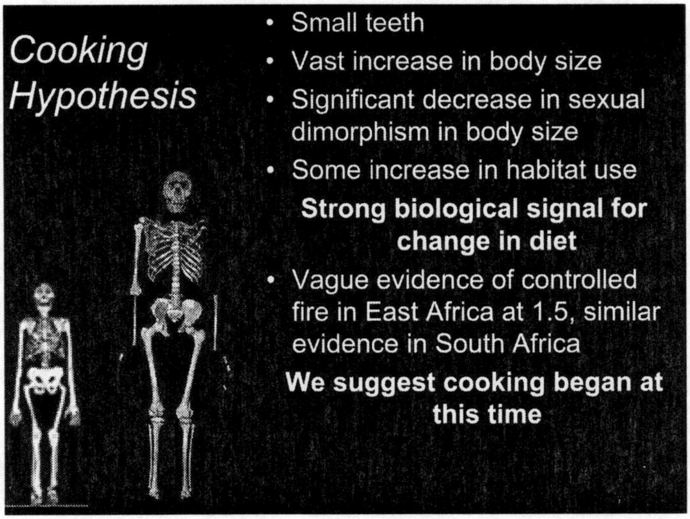
Cooking
Hypothesis
• Small teeth
• Vast increase in body size
• Significant decrease in sexual dimorphism in body size
• Some increase in habitat use
Strong biological signal for change in diet
• Vague evidence of controlled fire in East Africa at 1.5, similar evidence in South Africa
We suggest cooking began at this time

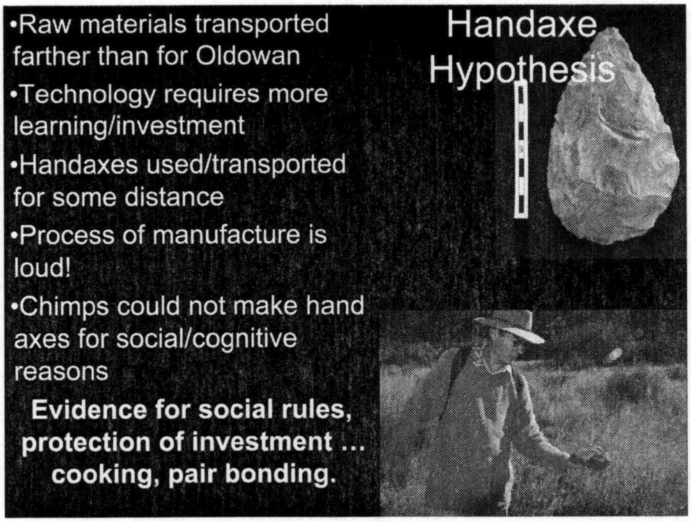
Handaxe
Hypothesis
•Raw materials transported farther than for Oldowan
•Technology requires more learning/investment
•Handaxes used/transported for some distance
•Process of manufacture is loud!
•Chimps could not make hand axes for social/cognitive reasons
Evidence for social rules, protection of investment … cooking, pair bonding.

The consequences from head to toe

- Large brain
  - Symbolic behavior, language
  - Childbirth is potentially deadly
  - Longer juvenile period, more parental investment
- Smaller teeth, fewer teeth
- USO-related toxins, digestion (still to be explored)
- Smaller gut
- Binge eating
- Large body size, ability to vary stature inter-generationally?

The consequences beyond the body

- Large brain
  - Cultural and biological evolution
  - Technology and its consequences
- Pair bonding, marriage
- Extended kinship relations
- Intergroup exchange, gift giving, alliance
- Social and sexual aspects of body image
- Group dynamics surrounding food (preparation and eating)

Further Research Goals

- Testing these ideas in other taxa
- Exploring interaction between symbolic capacities, land use patterns, diet, and technology
- Further explorations of medical consequences

IV Posters

1 Formation, Structural Elucidation, Analysis and Toxicity of Thermal Degradation Products of the *Fusarium* Mycotoxin Nivalenol

Michael Bretz, Anja Knecht, Simon Göckler, and Hans-Ulrich Humpf *

The major class of mycotoxins produced by *Fusarium* moulds are trichothecenes, a large group of sesquiterpenes sharing the same basic chemical structure, a 12,13-epoxytrichothec-9-ene ring system. Their toxic effects range from causing diarrhoea, vomiting and gastro-intestinal inflammation to noncompetitive inhibition of the biosynthesis of proteins in eukaryotic cells. Trichothecenes in general are relatively stable compounds, their degradation is observed only at high temperatures and prolonged heating time.

In order to investigate the stability of the trichothecene nivalenol (NIV) under food processing conditions such as cooking or baking, we performed model heating experiments and screened the residue for degradation products using GC-MS. Heating of nivalenol, especially under mild alkaline conditions, gave a mixture of four compounds (norNIV A, norNIV B, norNIV C and NIV lactone), which were isolated and identified by NMR and MS experiments [1]. Although their formation was also demonstrated in heating experiments with spiked flour samples, only norNIV B was detectable in a screening of several commercially available samples, possibly due to the very low contamination with nivalenol. Furthermore cell culture experiments using immortalized human kidney epithelial (IHKE) cells showed that the four compounds are less cytotoxic (formazan dye cytotoxicity assay) compared to nivalenol. Whereas nivalenol revealed an EC_{50} at 0.9 μmol, all other compounds did not show any significant effect up to 100 μmol [1]. Since the degradation products are less cytotoxic compared to niva-

* Institut für Lebensmittelchemie, Westfälische Wilhelms-Universität Münster, Correns-str. 45, 48149 Münster, Germany.

Thermal Processing of Food: Potential Health Benefits and Risks
DFG, Deutsche Forschungsgemeinschaft

ISBN: 978-3-527-31909-1

lenol we conclude that the formation of the degradation products during heating processes reduces the toxicity of nivalenol contaminated samples. However further toxicity studies are necessary to confirm these results.

References

1. Bretz, M.; Knecht, A.; Göckler, S.; Humpf, H.-U. *Mol. Nutr. Food Res.*, 2005, **49**, 309–316.

2 Influence of Different Heating Processes on the Formation of Nucleotides, Nucleosides and Free Bases in Vegetables

Isabelle Kölling-Speer * and Alfred Montag **

Abstract

Nucleic acids as well as nucleotides, nucleosides and purine and pyrimidine bases are natural components in food. Heating processes change the distribution of the nucleo compounds in comparison to the raw material. Furthermore, it is well-known that the individual components increase the content of uric acid in human serum in a different manner which is important for patients suffering from gout.

Therefore, the individual nucleotides, nucleosides and free bases were analysed in vegetables after cooking, pressure cooking, stewing, and microwaving. After extraction, the nucleo compounds were separated by means of size exclusion chromatography into two fractions. The individual nucleotides as well as the nucleosides and free bases were then determined by RP-HPLC using diode array detection. The obtained results revealed that microwaving seems to be the most suitable heating process for reducing the formation of low molecular nucleo components in vegetables.

2.1 Introduction

Nucleic acids as well as the low molecular compounds such as nucleotides, nucleosides and purine and pyrimidine bases are ubiquitous in food. In fresh unprocessed food the main part of the nucleo compounds – more than 95% – is present as RNA and DNA [1].

During storage and more during different heating processes, e.g. cooking and frying, the proportions of the low molecular nucleo compounds as nucleotides, nucleosides and free bases expand.

---

* Institut für Lebensmittelchemie, Technische Universität Dresden, Bergstr. 66, 01062 Dresden, Germany.

** The data presented in this paper are based on a PhD thesis (Isabelle Kölling-Speer, *Differenzierte Nucleostoffe – Analytik und Veränderungen in Modellsystemen und in Lebensmitteln*, 1993, workgroup of Professor Dr. Alfred Montag †, Universität Hamburg).

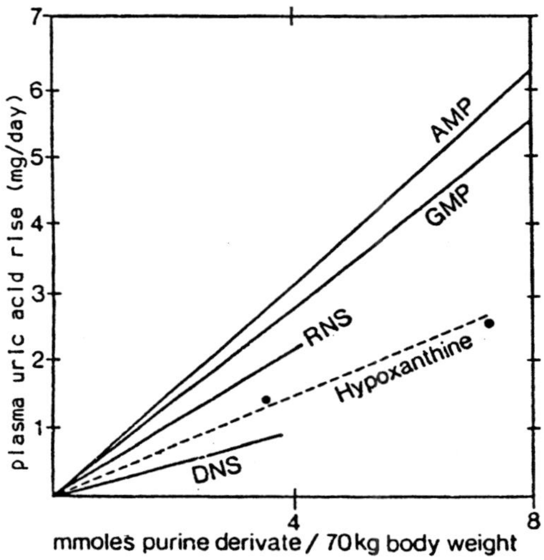

Figure 2.1: Rise of serum uric acid level in comparison to purine load from different sources on molar basis [2].

Furthermore, it is well-known that the individual components increase the serum concentration of uric acid, the end-product of the purine metabolism in man and most animals, in a different manner: the low molecular components mostly exceed a stronger influence than the intact nucleic acids (Fig. 2.1) [2]. Higher uric acid concentrations in serum may cause gout.

Whereas meat and innards are well-known as purine-rich foodstuffs, some vegetables also contain purines in remarkable amounts [3]. Therefore, in this study we investigated the individual nucleotides, nucleosides and free bases in broccoli and leek analysed after usual preparation and cooking in both the vegetables and the water residue.

2.2 Materials and Methods

The broccoli was chopped into clusters and the leek into pieces of 2 cm each. Four different cooking methods were used: normal cooking, pressure cooking, stewing and microwaving. The vegetables and the water residue were each freeze-dried and extracted or dissolved.

The nucleo compounds were separated into two fractions by means of size exclusion chromatography. The individual nucleotides as well as the nucleosides and free bases were then determined by RP-HPLC using diode array detection [4]. Furthermore, the purine and pyrimidine contents were analysed in total after hydrolysis [3, 5].

2.3 Results and Discussion

Although broccoli and leek belong to different botanical families, the same low molecular nucleo compounds were obtained after cooking (Figs. 2.2 and 2.3): besides the nucleotides AMP, GMP, UMP and four cyclic 2':3'-nucleotides, all the expected ribonucleosides as well as deoxyadenosine, deoxyguanosine, and the bases adenine, guanine, and sometimes xanthine, cytosine and uracil were detected (limit of detection: 0.1 µmol/100 g in the boiling water, mostly

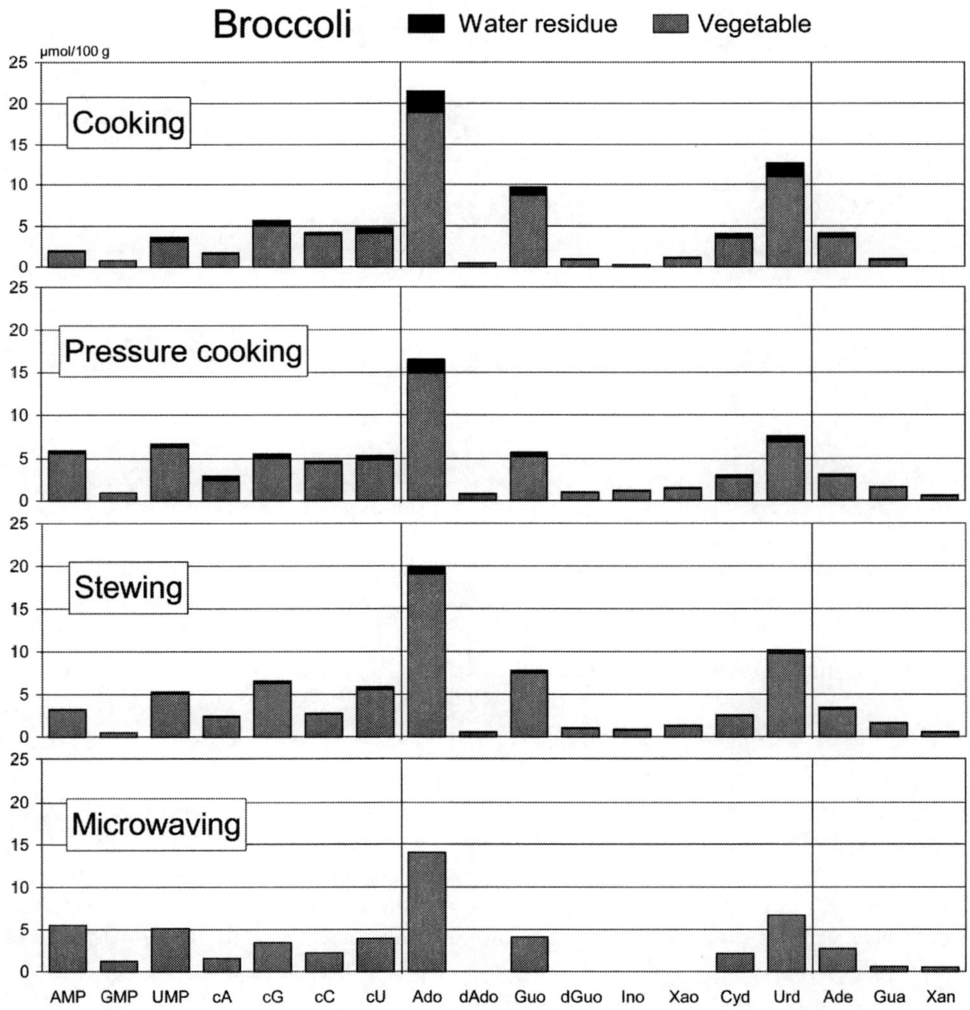

Figure 2.2: Individual amounts of the low molecular nucleo compounds in broccoli after different heating processes.

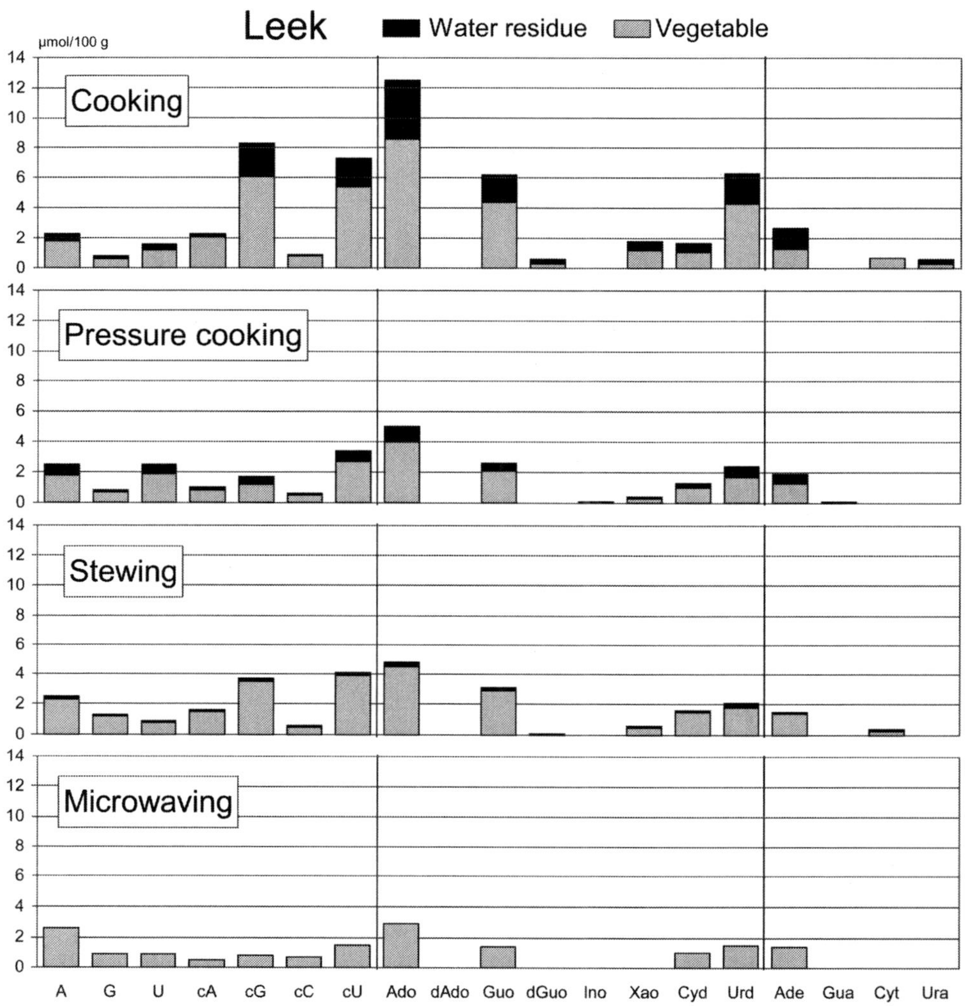

Figure 2.3: Individual amounts of the low molecular nucleo compounds in leek after different heating processes.

0.5 µmol/100 g in the solid matter). The main component is adenosine. Thus, this nucleo compound profile is completely different from that of meat in which the main nucleo compounds are inosine and hypoxanthine [4].

Whereas for broccoli the different cooking procedures influenced the distribution and the amount of the low molecular nucleo compounds only little, there is a great difference between the normal cooking and the gentle microwaving for leek.

Regarding the bases in the low molecular nucleo compounds in relation to their total contents (Tab. 2.1), the differences between the two vegetables also become obvious (Fig. 2.4).

Table 2.1: Total contents of the bases in broccoli and leek after hydrolysis in mmol/kg dry matter.

Base	Broccoli	Leek
Adenine	23.7	14.4
Guanine	14.6	5.1
Hypoxanthine	n.d.	n.d.
Xanthine	n.d.	n.d.
Cytosine	15.6	7.4
Thymine	4.0	3.4
Uracil	17.4	10.0
Dry matter [%]	12.55	10.55

n.d. not detected.

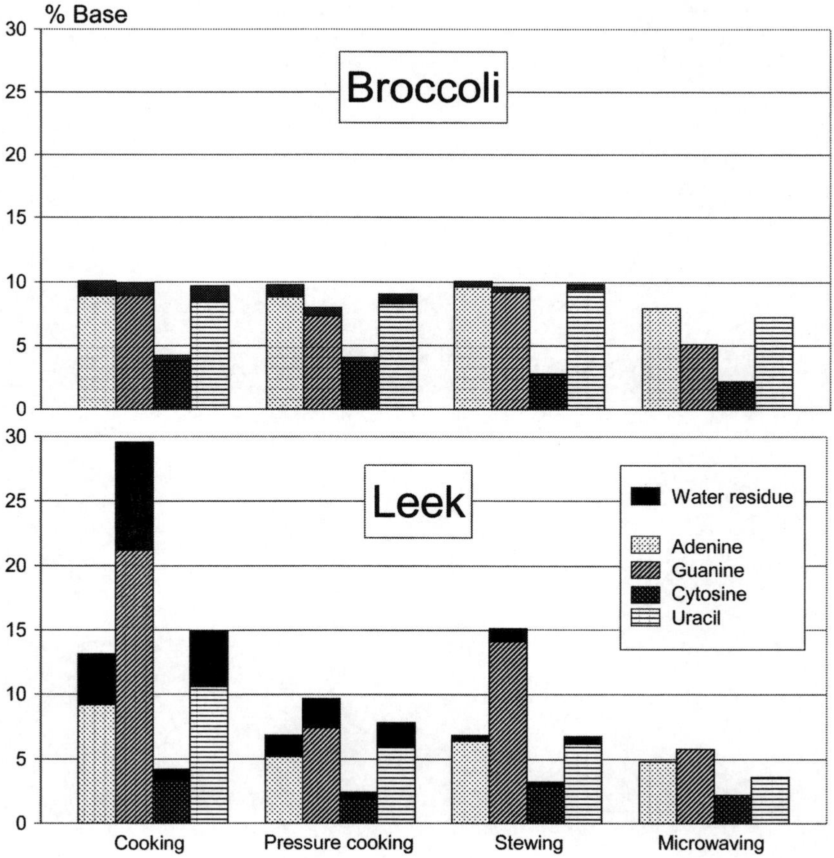

Figure 2.4: Bases in low molecular nucleo compounds in relation to their total contents.

Apparently, the size of the chopped pieces of the leek, the tissue stability, and the wax-coated surface of the broccoli are responsible for these differences.

Because of the smaller portion of the low molecular nucleo compounds, microwaving seems to be the most suitable heating process for vegetables for patients suffering from gout.

References

1. Herbel, W.; Montag, A. (1986) Z. *Lebensm. Unters. Forsch.* **183**, 12–17.
2. Spann, W. K.; Gröbner, W.; Zöllner, N. (1980) *Adv. Exp. Med. Biol.* **122A**, 215–219.
3. Montag, A.; Kölling, I.; Jänicke, S.; Benkmann, R.; Lou, S.N. (1989) Zur Kenntnis des Purinbasengehaltes in Lebensmitteln. *Aktuelle Ernährungsmedizin* **14**, 243–247.
4. Kölling-Speer, I. (1993) PhD thesis, *Differenzierte Nucleostoffe – Analytik und Veränderungen in Modellsystemen und in Lebensmitteln*, Universität Hamburg.
5. Kölling, I.; Montag, A. (1991) Puringehalte in Bieren – Purinbasengehalte alkoholhaltiger und alkoholfreier Biere. *Aktuelle Ernährungsmedizin* **16**, 14–17.

3 Determination of Free 3-Monochloropropane-1,2-Diol in Coffee and Coffee Surrogates

Torsten Kurzrock * and Karl Speer *

Abstract

Free 3-MCPD was analysed in coffee and coffee surrogates. While in roasted coffees and in a mixture of coffee with chicory no free 3-MCPD was detectable (LOD 20 µg/kg), the contents in different coffee surrogates from the German market varied between 40 µg/kg and 800 µg/kg. For sample preparation, a method developed for paper (par. 35, LMBG) was modified. The determination of 3-MCPD was performed by means of GC/MS (SIM-mode) using 3-MCPD-D_5 as the internal standard.

3.1 Introduction

3-Monochloropropane-1,2-diol (3-MCPD) is a toxic chemical substance, and the Scientific Committee on Food recommends a tolerable daily intake (TDI) of 2 µg/kg body weight. 3-MCPD may be formed in foodstuffs by the reaction of chloride with lipids during acidic hydrolysis or during heat treatment at high temperatures [1–3]. So far, a threshold value of 20 µg/kg 3-MCPD was established by the European Union [4] only for hydrolysed plant proteins (HVP).

However, bakery products, especially breads and toasted breads, contain higher 3-MCPD levels. Breitling-Utzmann et al. were able to demonstrate that the concentration of 3-MCPD increases in these foodstuffs with increasing processing temperatures [5]. Therefore, coffee and coffee surrogates may be potential foodstuffs with high 3-MCPD levels because of the high temperatures during the roasting process and the cereal content in coffee surrogates.

A par. 35-LMBG method developed for paper was modified and validated for determination of free 3-MCPD in bakery products [6–8]. This method is also suitable for analysing 3-MCPD in coffee and in coffee surrogates.

* Institut für Lebensmittelchemie, Technische Universität Dresden, Bergstr. 66, 01062 Dresden, Germany.

3.2 Methods

3.2.1 Sample Preparation

Based on the par. 35-LMBG method for paper [6], modified first by Ulbricht [7] and later by Lorenz [8], 3-MCPD was first extracted with a sodium chloride solution. After solid-phase extraction on Chromabond XTR (Macherey & Nagel, Düren, Germany), and evaporation of the solvent, the samples were diluted to 5 mL. Afterwards, 3-MCPD was reacted with 1-(heptafluorobutyryl)imidazole (HFBI) and analysed by means of GC/MS. As internal standard deuterium labelled 3-MCPD (3-MCPD-D_5) was added at the beginning of the sample preparation.

3.2.2 GC/MS

GC/MS was performed by employing an HP 5890 GC equipped with an HP 7673A automatic sampler, an MSD 5970 as detector, and a Gerstel cold injection system. Column: SGE BPX5, 50 m, 0.22 mm ID, 0.25 µm film thickness. Injection volume: 2 µL. Injector program: 60 °C, 12 °C/s to 250 °C, 5 min at 250 °C. Oven temperature: 2 min at 50 °C, 1.5 °C/min to 100 °C, 15 °C/min to 280 °C, 12 min at 280 °C. Ionisation: 70 eV (EI). Detection: SIM mode. 3-MCPD: *m/z*: 253, 289, 453; 3-MCPD-D_5: 257, 294, 456. Dwell time: 50.

3.3 Results and Discussion

3.3.1 3-MCPD in Roasted Coffee

Lorenz' method for the determination of free 3-MCPD in bakery products [8] was adapted for analysing 3-MCPD in coffee and coffee surrogates. The calibration was performed by adding 3-MCPD at four different levels to a roasted coffee. A limit of detection (LOD) of 20 µg/kg and a limit of quantitation (LOQ) of 40 µg/kg were calculated.

In order to study the influence of the lipid content and the influence of the roasting process, an Arabica coffee and a Robusta coffee, each roasted at three different temperatures (light, medium, dark), were analysed. In all coffee samples examined the contents of free 3-MCPD were below our limit of detection (LOD 20 µg/kg). This is in accordance with data published by the European Union [9].

Recently, Doležal et al. [10] also published data for free 3-MCPD. They determined very low concentrations of free 3-MCPD in roasted coffees with contents between 10 µg/kg and 20 µg/kg. In addition, they found up to 400 µg/kg esterified 3-MCPD. However, the transfer of these slightly water-soluble compounds into the beverage needs to be examined before the relevance of these amounts for human nutrition can be discussed.

3.3.2 3-MCPD in Coffee Surrogates

Two insoluble coffee surrogates, four regular soluble samples and four soluble bio-products from the German market as well as one mixture of coffee with chicory were examined. Their ingredients are listed in Table 3.1.

In order to validate the method for this matrix, a standard addition to a sample with low 3-MCPD content was performed at three different levels (Tab. 3.2). The 3-MCPD contents of the analysed coffee surrogates are presented in Table 3.1, in Figures 3.1 and 3.2 chromatograms of samples with high and low 3-MCPD amounts are shown.

In the mixture of coffee and chicory, only levels lower than our limit of quantitation were detectable. In the coffee surrogates, the lowest amounts, between 40 µg/kg and 120 µg/kg, were determined in the insoluble samples, whereas in the soluble coffee surrogates amounts from 80 µg/kg to 760 µg/kg were detected. No significant differences between normal soluble coffee surrogates and bio-products could be observed. Only in the bio-product "K" a very low 3-MCPD level was determined.

Obviously, the 3-MCPD content can be influenced by the roasting conditions and by the content of non-cereal ingredients.

3.3.3 Relevance of 3-MCPD Content in Coffee Surrogates to Human Nutrition

As mentioned above, the European Commission's Scientific Committee on Food [3] has set a tolerable daily intake (TDI) for 3-MCPD of 2 µg/kg body weight. In a worst case – sample F, batch 1 with 760 µg/kg – for an adult weighing 70 kg the level of 140 µg is reached after the consumption of 185 g of this coffee surrogate. To the manufacturer of this sample, who proposes using 5 g/150 mL per cup of the brew, this amount of coffee surrogate corresponds to 37 cups or 5.6 L. For adults, this seems to be less of a problem.

However, in many cases coffee surrogates are consumed by children with a lower body weight. For example, a child with a body weight of 25 kg has a TDI-level of 50 µg/kg. After drinking 7 cups or 1.05 L of this beverage it has consumed 26.6 µg 3-MCPD, this is more than half the TDI-level. After eating other foods with high 3-MCPD amounts like dark toasted bread or cakes [5], the TDI-level may easily be exceeded. In this case, the high 3-MCPD content in a coffee surrogate should not be overlooked.

Table 3.1: Contents of 3-MCPD [µg/kg] in different coffee surrogates.

Sample	Ingredients	Batch 1	Batch 2
Insoluble			
A (Fig. 3.1)	Barley malt	48	43
B	Malt, barley, rye, chicory	117	120
Soluble			
C	Barley, rye, chicory, sugar beet	286	274
D (Fig. 3.2)	Barley, barley malt, chicory, rye	341	344
E	Barley, rye, chicory, malt	373	286
F	Barley, rye, malt	759	550
Soluble bio-products			
G	Barley	377	
H	Malt, barley, chicory	169	
J	Malted barley	483	
K	Rye, chicory, barley, acorn, fig	80	
Mixture with coffee			
L	Coffee, chicory	<20 (LOD)	<40 (LOQ)

Table 3.2: Standard-addition of 3-MCPD to sample A (batch 2).

Sample	Conc. [µg/kg]	Addition [µg/kg]	Target value [µg/kg]	Actual value [µg/kg]	Recovery rate [%]
A 2	43	275	318	307	97
A 2	43	550	593	572	96
A 2	43	825	868	835	96

n: 2; s.d.: <10%.

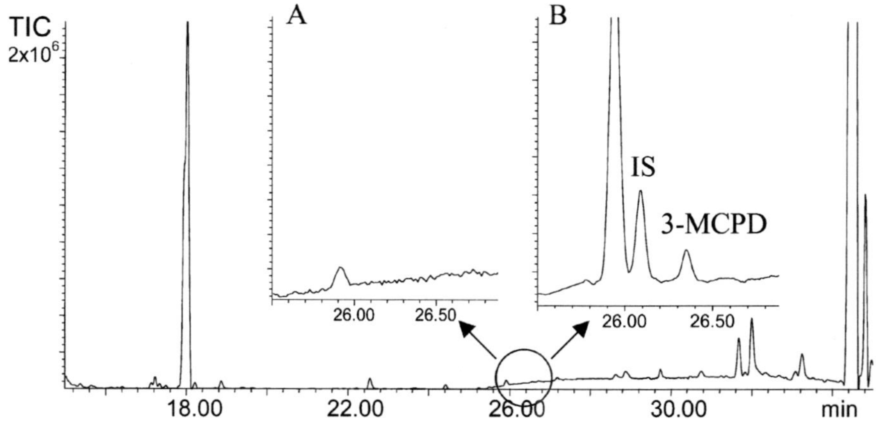

Figure 3.1: Chromatogram (GC/MS) of sample A with low 3-MCPD content; A: TIC×10, B: SIM run, IS: 3-MCPD-D_5.

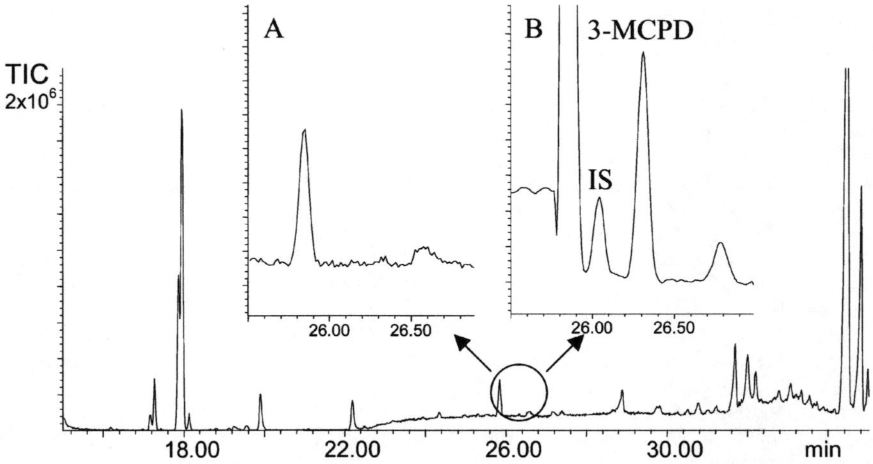

Figure 3.2: Chromatogram (GC/MS) of sample D with high 3-MCPD content; A: TIC×10, B: SIM run, IS: 3-MCPD-D_5.

3.4 Conclusions

The free 3-MCPD contents in roasted coffees are very low with amounts of only up to 20 µg/kg [9, 10]. Contents of esterified 3-MCPD in coffee can be higher [10], but until now, transferring them into the beverage has not been examined. Without these results, a discussion about the influence of esterified 3-MCPD on human health is difficult.

In the coffee surrogates high contents of free 3-MCPD up to 760 µg/kg were determined. The source of the 3-MCPD in coffee surrogates seems to be the cereals from which they are produced.

Furthermore, the wide range of 3-MCPD concentrations in the coffee surrogates demonstrates that it should be possible to decrease the 3-MCPD contents by optimising the composition – by reducing the cereals – and by optimising the roasting process.

Acknowledgments

Our thanks go to Mrs Breitling-Utzmann for supplying us with the 3-MCPD-D_5 standard.

References

1. Collier, P.D.; Cromie, D.D.O.; Davies, A.P. (1991) Mechanism of formation of chloropropanols present in protein hydrolysates. *Journal of the American Oil Chemists* **68**, 785–790.
2. Hamlet, C.G.; Sadd, P.A.; Crews, C.; Velísek, J.; Baxter, D.E. (2002) Occurrence of 3-chloro-propane-1,2-diol (3-MCPD) and related compounds in food: a review. *Food Additives and Contaminants* **19**, 619–631.
3. European Commission SCF/CF/CNTM/OTH/17 (2001) Opinion of the Scientific Committee on Food on 3-Monochloro-propane-1,2-diol (3-MCPD) updating the opinion of 1994 (adopted on 30 May 2001).
4. Commission Regulation (2001) Verordnung (EG) 466/2001 der Kommission vom 08. März 2001 zur Festsetzung der Höchstgehalte für bestimmte Kontaminanten in Lebensmitteln. *Amtsblatt der Europäischen Gemeinschaft* L 77/1.
5. Breitling-Utzmann, C.M.; Köbler, H.; Herbolzheimer, D.; Maier, A. (2003) 3-MCPD-occurrence in bread crust and various food groups as well as formation. *Deutsche Lebensmittel-Rundschau* **99**, 280–285.
6. Method B 80.56-2, Amtliche Sammlung von Untersuchungsverfahren nach § 35 LMBG (2001). Beuth, Berlin.
7. Ulbricht, H. (2003) Landesuntersuchungsanstalt für das Gesundheits- und Veterinärwesen Sachsen, Dresden. *Unpublished information.*
8. Lorenz, K. (2004) Das Vorkommen von 3-Monochlorpropandiol in ausgewählten Lebensmitteln unter Berücksichtigung verschiedener Technologien. *Diplom Thesis*, TU Dresden.
9. EU Directorate – General Health and Consumer Protection (2004) Reports on Tasks for scientific cooperation. Collection and collation of data on levels of 3-monochloropropanediol (3-MCPD) and related substances in foodstuffs.
10. Doležal, M.; Chaloupská, M.; Divinová, V.; Sveisková, B.; Velišek, J. (2005) Occurrence of 3-chlorpropane-1,2-diol and its esters in coffee, *Eur. Food Res. Technol.* **221**, 221–225.

4 Minimisation Concept – A Reaction on the Potential Health Risk of Acrylamide

Andreas Kliemant * and Hartmut Waldner *

A consideration of the acrylamide problem under the angle of risk management in Germany is closely related to the concept of minimisation of acrylamide contents in food products. This minimisation concept is aimed at gradually reducing acrylamide levels in a wide range of food products as a part of preventive health protection of consumers. This concept was agreed among the Federal Office of Consumer Protection and Food Safety (BVL), the authorities of the German federal states, producers and processors and the Federal Ministry of Food, Agriculture and Consumer Protection (BMVEL) and has been pursued for nearly three years now. The concept is based on data compiled by food surveillance authorities of the federal states about acrylamide levels in various foods (9600 sets currently). A calculated signal value is the core of the concept. This signal value is not a legal limit, but just a measure for orientation. Signal values are re-calculated at regular intervals and have been calculated four times so far, at last for 13 groups of food products.

In correspondence with the actual purpose of the minimisation concept, a signal value is subject to certain conditions: it must not exceed 1000 µg/kg, and once it is calculated, it must not be raised, but must be maintained or reduced. If a signal value is exceeded, this will be reported to the food surveillance authority of the federal state where the producers concerned are based. The authority will start a minimisation dialogue with producers with the aim to lower the levels of acrylamide formed in food. Food with acrylamide contents of more than 1000 µg/kg and from food groups for which no signal values have been yet set will automatically be included in this minimisation dialogue. The observer value serves as another measure of success of minimisation efforts, beyond the framework fixed for the signal value under the BVL's minimisation concept. The observer value will allow to assess the situation of acrylamide minimisation even when the signal value is unchanged. For instance, it can show a decrease from 1500 µg/kg to 1050 µg/kg, with the signal value being 1000 µg/kg. This allows to comprehensively describe the real situation of minimisation.

The minimisation dialogue is intended to produce concrete measures for minimisation of acrylamide levels. A number of such measures are currently implemented to good end, including technical changes in production processes and recipes.

* Unit Prevention, Crisis Management, Rapid Alert Systems, BVL, Federal Office of Consumer Protection and Food Safety, Rochusstr. 65, 53123 Bonn, Germany.

The success of the minimisation concept will be measured by the development of signal values and observer values. Signal values and observer values in many product groups have been reduced in the course of the minimisation concept, and current data indicate that we will be able to reduce more signal resp. observer values in the next calculations. In pursuing the minimisation concept we are guided by the principle that there is still much potential for reducing acrylamide levels in foods.

5 Isomerisation of Lycopene Due to Thermal Treatment of Carrot Homogenates: Increased Bioavailability of Total Lycopene and Generation of 5-*cis*-Lycopene in the Human Intestine

Esther Mayer-Miebach *, Diana Behsnilian *, Heike P. Schuchmann **, and Achim Bub ***

Abstract

Thermal induction of lycopene isomerisation and its effect on bioactivity was evaluated using a carrot cultivar with a high lycopene content (*Daucus carota* var. *Nutri Red*). Carrot homogenates (with/without the addition of oil) were heated up to temperatures from 25 °C to 140 °C under nonoxidising conditions and hold for 0.5–5 h. Exposure to temperatures above 100 °C resulted in a significant increase in the 9-*cis*-lycopene isomer and the ratio of all-*trans*- to total *cis*-isomers changed from 90:10 up to 40:60. Carrot homogenates with oil-supplement heated at 130 °C for 0.5 h showed a 9-fold increase in lycopene bioavailability (area under the curve, AUC) as compared to untreated carrot homogenates; AUC from carrots heated at 100 °C for 0.5 h exhibit a 2.5-fold increase. Although no 5-*cis*-lycopene appeared during thermal treatment, this isomer accounted for about 20% of the total lycopene in chylomicrons (AUC).

5.1 Introduction

There is strong evidence that a high intake of a diet rich in lycopene (mainly tomatoes and tomato products) decreases the risks of human cancers and heart diseases [1, 2]. It is still unclear, however, whether human health benefits originate from the consumption of lycopene-rich foods or lycopene itself, and whether lycopene isomers or oxidation products or even accompanying plant-

* Institute of Process Engineering, Federal Research Centre for Nutrition and Food, Haid- und Neu-Str. 9, 76131 Karlsruhe, Germany.

** Institute of Food Process Engineering, Karlsruhe University (TH), Fritz-Haber-Weg 2, 76131 Karlsruhe, Germany.

*** Institute of Nutritional Physiology, Federal Research Centre for Nutrition and Food, Haid- und Neu-Str. 9, 76131 Karlsruhe, Germany.

derived substances are the active compounds [3, 4]. Lycopene, the red colour of tomatoes, is an apolar and acyclic carotenoid completely insoluble in water and only slightly soluble in vegetable oil. Only a minor part is absorbed by humans during consumption of raw vegetables and fruits [5]. Thermal treatment of tomato juices supplemented with 1% of corn oil enhanced lycopene bioavailability [6]. The authors and others [7] propose that *cis*-lycopene may be more bioavailable than all-*trans*-lycopene, although the final proof is still missing. Therefore, the aim of our work was to study thermal processing conditions leading to the formation of *cis*-lycopene and to verify in a second step the *in vivo* bioavailability of various lycopene isomers from product prototypes with low/high *cis*-isomer contents.

5.2 Materials and Methods

Thermal lycopene degradation and isomerisation were investigated using 20 g of homogenated raw carrots (*Daucus carota* var. *Nutri Red*) mixed either with 1% sunflower oil or with 1% water, placed in glass tubes, sealed and heated at 25–140 °C for 0.5–5 hours each (70 °C, 90 °C: 5 h only) using an autoclave equipped with an extern heater. Pressures of up to 3.6 bar were reached. Experiments were replicated threefold.

Homogenates prepared from blanched and deep frozen carrots heated for 0.5 h each at 100 °C and at 130 °C (including 1% oil), or untreated as control samples were prepared for the human intervention study. *In vivo* bioavailability was evaluated in a randomised, double blind, cross-over study with six male volunteers by taking blood samples over a period of 24 h after lycopene ingestion of 30.8 mg per dish. The dishes included identical amounts of vegetable oil. Postprandial lycopene contents were determined in chylomicrons.

5.3 Results

All-*trans*-lycopene contents of homogenised carrots were not affected during heat treatment at temperatures below 70 °C even after heating times up to 5 hours and independent from an oil supplement (Fig. 5.1 a). After heating at 100 °C and 130 °C for 0.5 h, about 90% and 50% of all-*trans*-lycopene remained stable, respectively. Oil supplements did not affect lycopene stability considerably. At temperatures above 100 °C, the total *cis*-isomer content increased to a maximum value of about 440 mg/kg dry mass (d.m.) at 140 °C (heating time

Table 5.1: Lycopene isomer contents after thermal treatment for 0.5 h with/without oil supplement.

	Lycopene isomers [mg/kg$_{w.b.}$]					
	$\Sigma_{Isomers}$	all-*trans*	$\Sigma_{cis\text{-}isomers}$	9-*cis*	13-*cis*	15-*cis*
Raw carrots [1]	120±6	110±6	10±1	1 *	4±1	5±1
Raw carrots with oil [1]	133±12	122±12	11±1	1 *	4±1	6±1
Raw carrots heated at						
100 °C	110±4	96±3	14±1	3±1	2 *	9±4
130 °C	91±0	51±5	40±5	25±4	1 *	14±2
140 °C	98±2	41±2	57±3	36±3	0	21±0
Raw carrots with oil heated at						
100 °C	126±3	99±3	27±1	7±1	2 *	18±3
130 °C	105±12	50±7	55±3	33±1	1 *	21±1
140 °C	105±1	45±1	60±2	40±3	0	22±1

(Mean values; $n=2$, [1] $n=25$; * detection limit).

2 h) (Fig. 5.1 b). 9-*cis*- and 15-*cis*-lycopene were generated during heating with 9-*cis*-lycopene being the predominant isomer. Relative to the untreated raw carrots, contents of 9-*cis*- and 15-*cis*-lycopene increased up to 30-fold and 4-fold, respectively. *Cis*-lycopene contents increased significantly at temperatures above 100 °C even after short heating times (Tab. 5.1).

In contrast, below 100 °C isomerisation only occurred after a 5 h heating time. The ratio of all *trans*- to total *cis*-isomers of lycopene remained unchanged at 100 °C but was modified from 90:10 to 50:50 at 130 °C. We could not detect 5-*cis*-lycopene in any of the heat treated carrot samples.

A 9-fold increase in lycopene bioavailability (area under the curve, AUC) was observed from carrots heated at 130 °C (0.5 h, 1% oil) as compared to untreated carrot homogenates (Fig. 5.2). Bioavailability from carrots heated at 100 °C was 2.5-fold. 9-*cis*-lycopene and 15-*cis*-lycopene, which are generated during heat treatment of carrots, are absorbed and incorporated into human chylomicrons. Although no 5-*cis*-lycopene appeared during thermal treatment, this isomer accounted for about 20% of the total lycopene in chylomicrons (AUC) (Tab. 5.2). Therefore, 5-*cis*-lycopene may be generated in the human gastrointestinal tract or in chylomicrons.

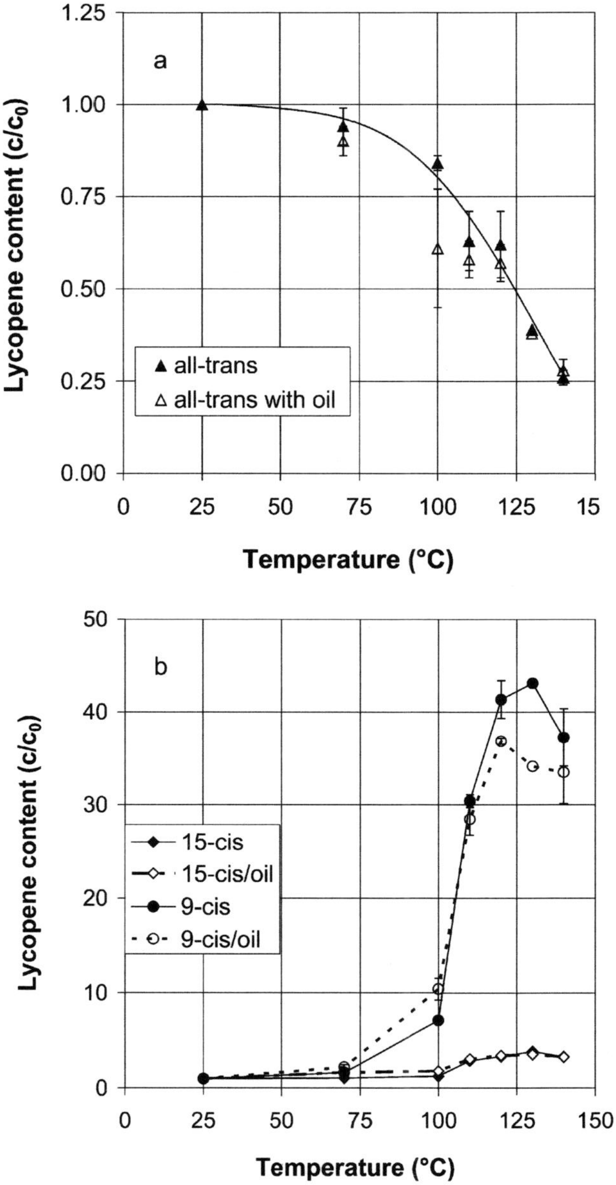

Figure 5.1: Relative lycopene contents in carrot homogenates (relative to the corresponding values in the raw material) after thermal treatment with/without oil supplement; a) all-*trans*-lycopene; b) 9-*cis*, 15-*cis*-lycopene.

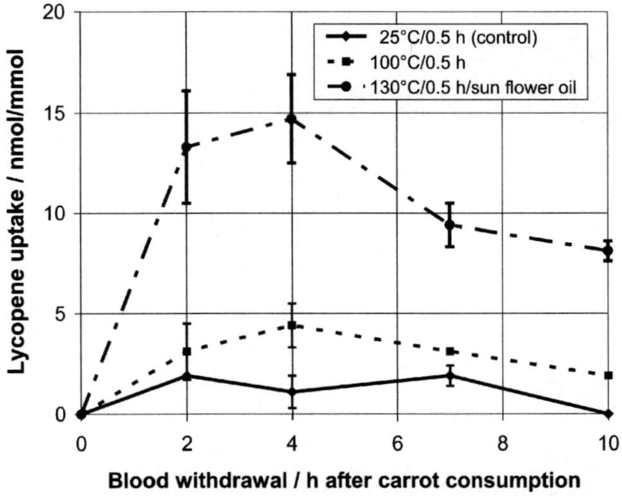

Figure 5.2: Triacylglycerol-standardised lycopene content of postprandial chylomicrons.

Table 5.2: Lycopene isomer contents of postprandial chylomicrons (AUC) from healthy volunteers (n = 6) after carrot ingestion.

	AUC (nmol x h/L)		
	Raw carrots	100 °C	130 °C/oil
all-*trans*-lycopene	11.3±20.8	14.6±4.5	62.0±18.0
5-*cis*-lycopene	–0.6±9.8	5.9±3.5	18.2±10.2
9-*cis*-lycopene	0.1±3.1	0.02±2.8	12.8±6.7
15-*cis*-lycopene	0.7±0.9	8.4±2.5	9.3±2.2
Σ-lycopene	11.5±30.9	28.9±11.4	102±28

5.4 Conclusions

During thermal treatment of carrot homogenates above 100 °C, high contents of various *cis*-lycopene isomers up to 50 mg/kg wet base can be achieved. At the same time thermal processing of carrots improved total lycopene bioavailability by 9-fold while 5-*cis*-lycopene may be generated in the human gastrointestinal tract. Although 5-*cis*-lycopene predominates in chylomicrons and in plasma, it is unlikely that a preferential uptake of 5-*cis*-lycopene exists. The results presented here indicate that the *cis*-lycopene content of carrot homogenates can be modified selectively by applying specific heating conditions. If health benefits are in fact correlated with the uptake of lycopene isomers, selective thermal processing may be used to provide sterilised carrot products with high *cis*-lycopene contents.

Acknowledgments

The work was supported by the German Federal Ministries for Nutrition, Agriculture and Consumer Protection (BMELV) and of Education and Research (BMBF).

References

1. Giovannucci, E. (1999) Tomatoes, tomato-based products, lycopene and cancer: review of the epidemiologic literature. *Journal of the National Cancer Institute,* **91**, 317–331.
2. Arab, L. and S. Steck (2000) Lycopene and cardiovascular disease. *American Journal of Clinical Nutrition,* **71**,1691S–1695S.
3. Paetau, I., Khachik, F., Brown, E. D., Beecher, G. R., Kramer, T. R., Chittams, J. and Clevidence, B. A. (1998) Chronic ingestion of lycopene-rich tomato juice or lycopene supplements significantly increases plasma concentrations of lycopene and related tomato carotenoids in humans. *American Journal of Clinical Nutrition,* **68**, 1187–1195.
4. Boileau, T. W.-M., Liao, Z., Kim, S., Lemeshow, S., Erdman, J. W. and Clinton, S. K. (2003) Prostate carcinogenesis in *N*-methyl-*N*-nitrosourea (NMU)-testosterone-treated rats fed tomato powder, lycopene, or energy-restricted diets. *Journal of the National Cancer Institute,* **95**(21), 1578–1586.
5. West, C. E. and Castenmiller, J. (1998) Quantification of the "SLAMENGHI" factors for carotenoid bioavailability and bioconversion. *International Journal of Vitamins and Nutrition,* **68**, 371–337.
6. Stahl, W. and Sies, H. (1992) Uptake of lycopene and its geometrical isomers is greater from heat-processed than from unprocessed tomato juice in humans. *Human and Clinical Nutrition,* **122**(11), 2161–2166.
7. Boileau, A. C., Merchen, N. R., Wasson, K., Atkinson, C. A. and Erdman, J. W. (1999) *Cis*-lycopene is more bioavailable than *trans* lycopene in vitro and in vivo in lymph-cannulated ferrets. *Journal of Nutrition,* **129**, 1176–1181.

6 Influence of High Hydrostatic Pressure on the Formation of N^{ε}-Carboxymethyllysine and N^{ε}-Carboxyethyllysine in Maillard-type Reactions

Martina Kasper * and Peter Schieberle *

The reaction of reducing carbohydrates with amino acids, known as the Maillard reaction, affects both, positive food attributes, such as taste, aroma and colour, but also negative attributes, like the generation of food borne toxicants (acrylamide) as well as posttranslational protein modifications, known as advanced glycation end products (AGEs). The amounts of these AGEs formed during food processing, however, depend on different parameters like pH value, reaction temperature and time and also high hydrostatic pressure (HHP). However, in particular the influence of the latter parameter on the formation of protein modifications during glycation has scarcely been considered in the literature.

The only available study was recently performed by Schwarzenbolz et al., who showed that pentosidine and pyrraline were differently affected by HHP [1, 2].

The aim of our study was, therefore, to provide new insights into protein modifications, induced by glucose and transient intermediates formed thereof, at the epsilon amino group of lysine when high hydrostatic pressure is applied at higher temperatures. The formation of N^{ε}-carboxymethyllysine (CML) and N^{ε}-carboxyethyllysine (CEL) were in the focus of the study, because both are used as markers to indicate the quality of foods.

For this purpose, model systems containing hippuryllysine and glucose were either reacted at pH 7 under reflux (100 °C) or treated under HHP at the same temperature. The amounts of hippuryl-CML and hippuryl-CEL formed under both conditions were quantified by using a new LC/MS method. Because the results indicated a clear influence of HHP on the concentrations of both AGEs, mechanistic studies using carbon 13-labeled glucose were performed and a new formation pathway of CML will be presented.

References

1. Schwarzenbolz U, Klostermeyer H, Henle T (2000) *Eur Food Res Technol* **211**: 208–210.
2. Schwarzenbolz U, Klostermeyer H, Henle T (2002) *International Congress Series* **1245**: 223–227.

* Chair of Food Chemistry, Department of Chemistry, Technical University of Munich, Lichtenbergstr. 4, 85478 Garching, Germany.

7 Effect of Acrylamide from a Heated Potato Product on the Acrylamide Content in Eggs, Breast Muscle Meat, Liver and Kidney of Hens

Ingrid Halle *, Gerhard Flachowsky *, Marcus Ihling **, Monika Lahrssen-Wiederholt ***, and Horst Klaffke ***

7.1 Introduction

Acrylamide is produced during the thermal processing of food, mainly due to the Maillard reaction of asparagine and reducing sugars. In foods which are baked, roasted, deep fried or similarly prepared, acrylamide levels of over 2000 µg acrylamide per kg have been measured. Potato products in particular have a high potential for accumulating acrylamide. The objective of this study was to determine the carry-over of acrylamide from a heated potato product as part of the daily ratio to eggs, muscles and organs of laying hens.

7.2 Methods

A total of 16 hens were allocated to two groups. Hens were kept individually in a cage battery. The duration time of the experiment was four weeks and commenced when the hens were 42 weeks old. 120 g feed were provided per hen and day for consumption. The diet consisted of 50% potato meal, soy bean meal, barley and premix. The potato meal of the experimental group was baked for 2 hours at 150 degrees. The determined acrylamide content of the experimental diet was 671 ± 32 µg/kg, and of the control group was 125 ± 40 µg/kg. Acrylamide was determined by gas chromatography/mass spectrometry in the diet and samples of hens [1].

* Correspondence to: Gerhard Flachowsky, Institut für Tierernährung, Bundesforschungsanstalt für Landwirtschaft, Bundesallee 50, 38116 Braunschweig, Germany.
** Institut für Ernährungswissenschaften, FSU Jena.
*** Bundesinstitut für Risikobewertung, Berlin.

Table 7.1: Mean concentration of acrylamide per egg [μg/100 g egg].

Group	1st week	2nd week	3rd week	4th week
Control	0.11±0.14	0.39±0.27	0.58±0.10	0.59±0.18
Experimental	1.14±0.47	1.39±0.20	1.64±0.18	1.72±0.05

Table 7.2: Mean concentration of acrylamide in liver, kidney and breast muscle meat [μg/kg].

Samples	Control	Experimental
Liver	5.3 ± 9.2	9.9 ± 4.1
Kidney	3.3 ± 5.8	27.7 ± 15.9
Breast meat	8.7 ± 1.5	15.6 ± 8.3

7.3 Results and Discussion

During the experimental time the acrylamide concentration of eggs in the experimental and control groups was increased (Tab. 7.1). After an experimental period of three weeks, the concentration of acrylamide in the eggs no longer changed. The carry-over coefficients of acrylamide from diet to egg were 0.022/ 0.011 of experimental/control hens.

At the end of the experiment the hens were slaughtered. The highest content of acrylamide was determined in the kidneys, in comparison to contents in breast muscle meat and liver, of the experimental hens (Tab. 7.2).

References

1. Klaffke H. (2004) Vorläufige Methode: Bestimmung von Acrylamid in festen und pastösen Lebensmitteln mittels GC-MS und LC-MS/MS, BfR.

8 PAH in Oil and Tocopherols – Analytical Challenges in Complying with EC Recommendations

Sonja Schittko *, Holger Fritz *, Robert Gatermann *,
and Ansgar Ruthenschrör *

8.1 Introduction

In 2005 the European Commission (EC) defined a maximum level (maximum residue limit, MRL) for benzo(a)pyrene (BaP) in various foods. Oils and fats intended for direct consumption or as an ingredient in food are restricted to a level of up to 2 µg/kg, smoked products up to 5 µg/kg and bivalve molluscs to 10 µg/kg respectively. A lower level of up to 1 µg/kg was required for foods for children [1]. Furthermore the EC laid down a directive with specific directions regarding *the sampling methods and the methods of analysis for the official control of the level of benzo(a)pyrene in foodstuffs* [2]. This requires for example a minimum limit of quantification (LOQ) of 0.9 µg/kg and recovery rates of between 50% and 120% for the analysis of BaP. The regulations were made as a consequence of the collection of PAH data in food where BaP was detected in nearly all analysed samples [3, 4]. It was concluded that 15 genotoxic PAH (so-called SCF PAH) should be monitored to obtain PAH profiles in food and to validate the use of BaP as a marker [5]. This opinion results in new analytical challenges since half of the SCF PAH are not included in the analytical scope of the well known EPA (US Environmental Protection Agency) PAH.

8.2 PAH in Tocopherols

Tocopherols, substances generating the vitamin E effect, are produced from oil seeds. The use of direct firing for seed drying and adsorption from the environment can lead to PAH contamination of some vegetable oils. Refining may reduce the amount of these contaminants in the oils, but during this procedure the tocopherols are extracted from the steam distillates as by-products [6, 7]. Contaminants are enriched in these deodorization distillates and, depending on the type of process, different PAH levels in the tocopherols may result. These

* Eurofins Analytik GmbH, Neuländer Kamp 1, 21079 Hamburg, Germany.

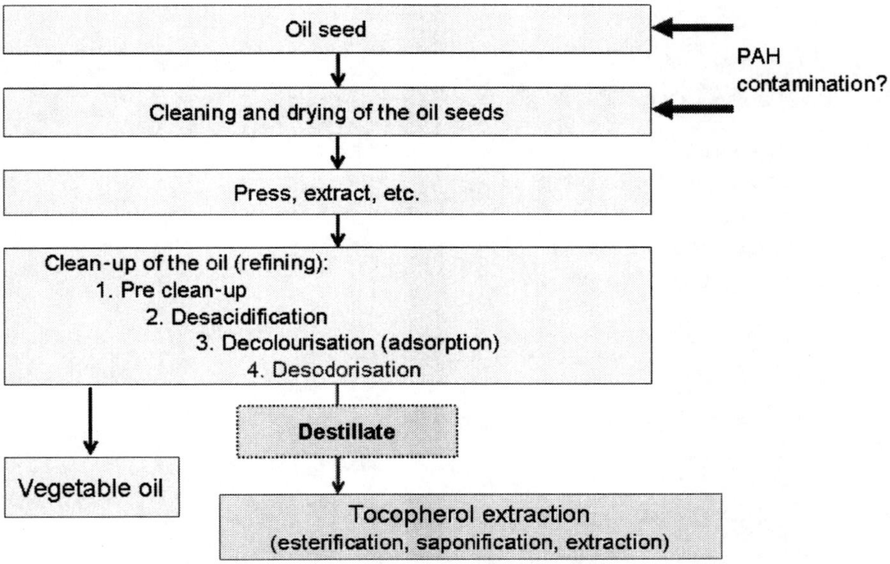

Figure 8.1: Production process of tocopherols and possible sources for PAH contamination.

sample matrices are very complex since the tocopherols are residues originating from plant oils (Fig. 8.1).

8.3 Analytical Procedure

Figure 8.2 compares the PAH analysis of vegetable oils and of tocopherol. While the analysis of vegetable oils simply requires application of gel-permeation chromatography (GPC) before mass spectrometry (GC-MS), the tocopherols need a more complex clean-up with different extraction steps (see Fig. 8.2, [8]).

Chemical ionisation (negative chemical ionisation, NCI) was tested as an alternative detection mode which allows detection of most of the SCF PAH (see Fig. 8.3) and promises higher signal to noise ratios and thereby lower limits of quantification (LOQ) can be expected.

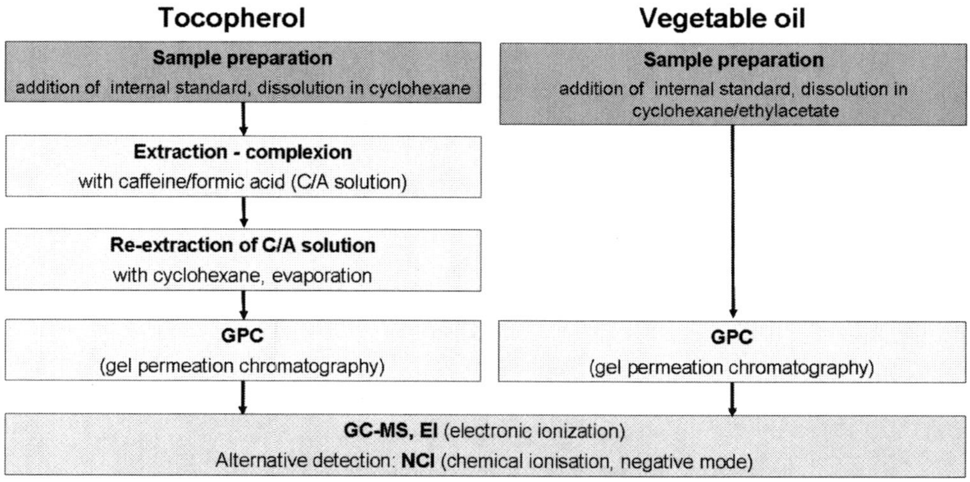

Figure 8.2: Analytical clean-up procedure for oils and tocopherols.

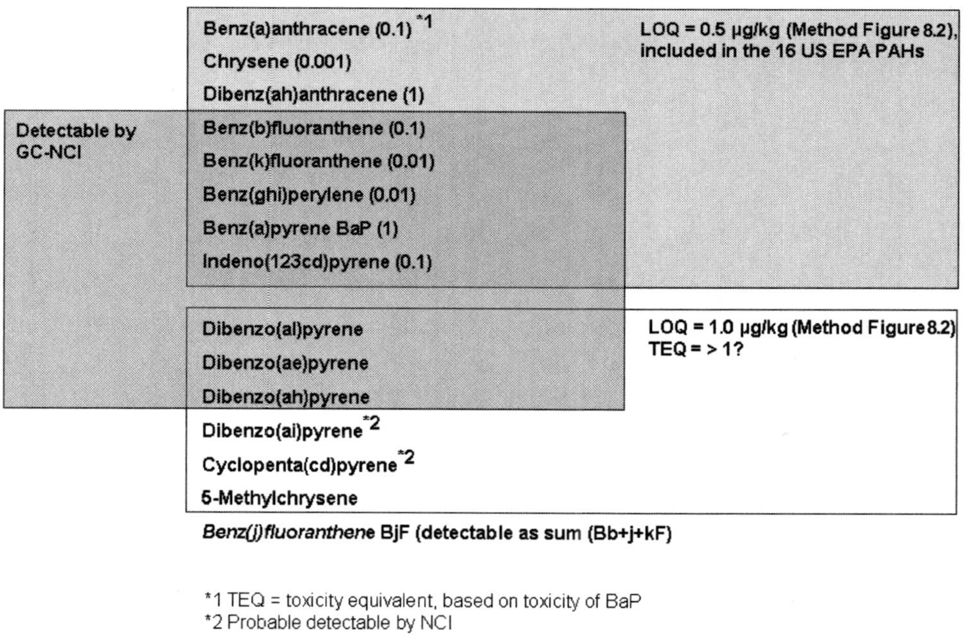

Figure 8.3: The 15 SCF PAH requested by the EC and detectable by method in Figure 8.2, TEQ from [9].

8.4 PAH in Food – Expert Opinion and Recommendation of the EC

Council regulations and directives provide the legal framework for the establishment of MRL's for food contaminants. The member states should assist the commission in their work to achieve the necessary data and information. Based on this an expert opinion of the Scientific Committee in Food (SCF) *concluded* (in 2002) *that benzo(a)pyrene (BAP) may be used as a marker of occurrence and effect of the carcinogenic PAH in food, ...* Following the advice of the SCF the commission considered maximum levels for PAH in foodstuffs and requires therefore PAH data [3]. 8861 samples in total were collected between 1992 and 2003 from the 13 member states and from Norway [4]. Table 8.1 lists the SCF PAH and their occurrence in these samples. The toxicity data which are also presented indicate the genotoxicity and carcinogenicity potential of these contaminants [4, 10].

The collected data indicate that BaP was determined in almost all of the samples which were analysed. However for the vegetable oils many of the samples (85%) contain BaP in the lowest range 0 to 0.5 μg/kg (see also Fig. 8.4) for example. Thus the occurrence of BaP in almost all analysed samples is more of an indication of the well known analytical determination procedures of this leading PAH than it is proof for the presence of this contaminant in foodstuffs. Furthermore the EPA PAH were analysed in almost 22–50% of the samples. In contrast the dibenzo(ax)pyrenes were not analysed during this collecting. The absence of these PAH data is probably due to the lack of analytical experience with these contaminants and not by their actual physical absence.

The concentration ranges of BaP in vegetable oils and oil similar samples are illustrated in Figure 8.4. In most of the samples the greater part of the BaP was in the range up to 0.5 μg/kg. In some samples, especially olive pomace, greater fractions of the total sample set were detected as having higher concen-

Table 8.1: Determination of PAH (shown only the SCF PAH) in samples (source [3]).

	% of analysed samples	Genotoxicity *	Carcinogenicity *
Benz(a)anthracene	38.8	x	x
Benz(b)fluoranthene	43.3	x	x
Benz(k)fluoranthene	24.5	x	x
Benz(ghi)perylene	49.6	x	
Benz(a)pyrene BaP	99.3	x	x
Chrysene	22.1	x	x
Dibenz(ah)anthracene	38.2	x	x
Indeno(123cd)pyrene	47.4	x	x
Cyclopenta(cd)pyrene	0.2	x	x
Benz(b+j+k)fluoranthene	5.8	mixture	mixture

* From IPCS, EHC, 202 [9].

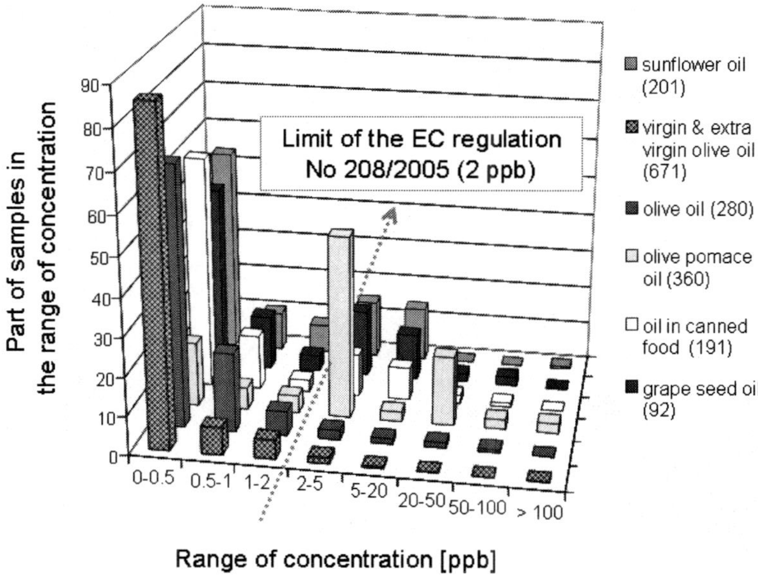

Figure 8.4: Part of samples for the amount of BaP of different oil samples (number of samples for the oil/pomace) [1, 4].

trations. The restriction of the Commission limits these higher contaminated samples with the setting of a MRL of 2 μg/kg [1].

8.5 Conclusion

The collected PAH data confirm the presence of widely varying amounts of BaP in oil samples. The legal regulation of the amount of BaP provides a convenient and reliable way to exclude most highly contaminated samples (Fig. 8.4). The heavier and therefore more toxic PAH are not analysed during the collection of PAH data in food [4]. The latter is probably caused by the lack of analytical experience with these PAH. A greater amount and more reliable data related to the relative toxicities are required in order to obtain PAH profiles in food and in order to verify the reliability of BaP as a marker. In October 2005 the European Commission started the monitoring program to gain the required data related to the SCF PAH. Some of this (see Fig. 8.3) has been analysed as EPA PAH, but most of the data were not included in the analytical procedures until now. With the analytical procedure presented here it is possible to detect the SCF PAH (BjF only detectable as the sum in Bb+j+kF) in oil and even within the complex matrix of Tocopherol up to 1 μg/kg. The necessary lower detection limits can be achieved through the use of GC-NCI which appears to be a fast method when BaP is measured as marker for the heavy PAH.

References

1. Commission Regulation (EC) No 208/2005 of 4 February 2005 amending Regulation No 466/2001 as regards polycyclic hydrocarbons.
2. Commission Directive 2005/10/EC of the 4 February laying down the sampling methods of the analysis for the official control of the level of benzo(a)pyrene in foodstuffs with EEA relevance.
3. Opinion of the Scientific Committee on Food on the risks to human health of Polycyclic Aromatic Hydrocarbons in food, SC/CS/CNTN/PAH/29 Final, 4 December 2002.
4. Report of experts participating in Task 3.2.12, October 2004, Collection of occurrence data on polycyclic aromatic hydrocarbons in food.
5. Commission Recommendation of 4 February 2005 on the further investigations into the levels of polycyclic aromatic hydrocarbons in certain foods (C (2005) 256).
6. Moret, S., Conte, L.S. (2000). Polycyclic aromatic hydrocarbons in edible fats and oils: occurrence and analytical methods. *Journal of chromatography A* **882**, 245–253.
7. Tusan, M., Demirci, M. (2005). Total and individual tocopherols contents of sunflower oil at different steps of refining. *European Food Research Technology* **220**, 251–254.
8. Sagredos, A.N., Sinha-Roy, D. (1979) A Method for rapid determination of polycyclic aromatic hydrocarbons in fats and oils via coffein complexes. *Deutsche Lebensmittel-Rundschau* **11**, 350–352.
9. US Environmental Protection Agency (2004) Human Health Risk Assessment Bulletin – Supplement to RAGS. Available online: www.epa.gov/region4/waste//ots/healthbul.htm, last access: October 28, 2005.
10. IPCS, EHC, 202, International Programme of Chemical Safety, *Environmental Health Criteria* **202**, 1998.

9 Acrylamide in Fried Potato Products – Influence of Process Conditions and Precursor Contents

Knut Franke * and Ernst H. Reimerdes

Deep-fat frying is one of the favored processes to manufacture a broad range of high quality potato products. The process enables a very effective heat transfer with short processing times and provides products with unique flavor and texture properties.

Detection of acrylamide in these products by Törnquist et al. in April 2002 established a new challenge for the potato processing industry and led to many activities with respect to avoid excessive formation of this substance during frying. The level of acrylamide in food products has been lowered distinctly as could be seen in decreased signal values.

However, the ready-to-eat preparation of French fries is often carried out in household fryers using par-fried frozen potato sticks industrially prefabricated. One main part of the acrylamide intake is generated during household frying depending on individual preferences for color (browning) and crispness. All these facts complicate the sustainable reduction of acrylamide intake caused by home-made products.

Otherwise, it is well known that high amounts of free asparagine and reducing sugars as the most effective precursors of acrylamide in potatoes determine the acrylamide formation in these products. To acquire more information about the dependencies of acrylamide formation, precursor contents and frying conditions, four different potato varieties were processed to French fries. Contents of precursors and acrylamide formed during par-frying and ready-to-eat frying with two different durations were measured. Concentrations of reducing sugars, e.g. glucose, and free asparagine were used to estimate the acrylamide contents of fried potatoes after different frying steps applying a multiple regression approach. The correlation coefficient of the regression model was 0.95 indicating a good agreement with the measured acrylamide value by the model calculations.

Therefore, this model can be used to predict acrylamide contents in French fries based on precursor concentrations and frying times. One possible implementation of such a model may be an improved control of household fryers. For this purpose, the content of reducing sugars of the par-fried French fries will be provided for the fryer control by means of the package, for example by barcode. The fryer control will set up proper frying conditions based on the model and will finish the frying process automatically to get high quality French fries with low acrylamide content.

* Deutsches Institut für Lebensmitteltechnik e. V., Prof.-von-Klitzing-Str. 7, 49610 Quakenbrück, Germany.

10 Thermal Stability of Zeaxanthin in Potato Homogenates

Diana Behsnilian *, Esther Mayer-Miebach *, Ping Idda **, and Heike P. Schuchmann **

10.1 Introduction

Zeaxanthin and Lutein belong to the dominant carotenoids found in human blood. They accumulate in the retina and are commonly known as the macular pigment. Epidemiological studies suggest the beneficial role of their dietary intake in the prevention of eye diseases, such as cataract and age-related macular degeneration. Thermal treatment has been proven to induce *cis/trans*-isomerization and to improve the (bio)availability of some carotenoids in different food matrices. However there is little data dealing with the thermal stability of zeaxanthin and lutein contained in fruit and vegetables. The aim of this work was to study the effect of thermal treatment on carotenoid stability and isomerization in a zeaxanthin-containing potato variety.

10.2 Materials and Methods

Raw potato homogenates were used to obtain kinetic data on the thermal stability of zeaxanthin. The homogenates were placed in sample tubes, sealed and exposed to temperatures in the range from 25 °C to 150 °C up to 4 hours. Experiments were repeated three times. Temperature- and pressure-resistant sample tubes were developed for the experiment. Individual carotenoids and some of their isomeric forms were identified and, in addition to total carotenoids, quantified.

* Institute of Process Engineering, Federal Research Centre for Nutrition and Food, Haid- und Neu-Str. 9, 76131 Karlsruhe, Germany.

** Current address: Institute of Food Process Engineering, University of Karlsruhe (TH), 76131 Karlsruhe, Germany.

10.3 Results

The potato variety used for the experiments contained several carotenoids, mainly zeaxanthin (0.2–0.8 mg/100 g), antheraxanthin (0.08–0.35 mg/100 g) and lutein (0.04–0.16 mg/100 g). An increase on the treatment temperature resulted in a decrease of the total carotenoid content. However after 1 h at 70 °C and after 1 h at 150 °C the total carotenoid content in the potato homogenates was respectively 85% and 70% of the initial one. The effect of treatment time was not so determinant regarding carotenoid stability as compared to treatment temperature. Lutein was destroyed yet at 70 °C. On the other hand, a significant fraction of the initial zeaxanthin remained stable even after high-temperature long-time treatments. 9-*cis*-zeaxanthin was formed by heating at 70 °C and the amount of this isomeric form increased with treatment temperature.

Acknowledgments

This work was supported by the German Federal Ministries for Consumer Protection, Nutrition and Agriculture (BMVEL) and of Education and Research (BMBF).

11 Osmotic Treatment as a Pre-Step to Drying and Frying

Diana Behsnilian *, Esther Mayer-Miebach, and Wolf-Dietrich Koller

Abstract

Osmotic treatment as a pre-step to further processing of plant and animal material allows to improve the nutritional and sensory quality of food products. Solid food material is changed by immersion into an aqueous solution of sufficiently high concentration, leading to partial dewatering and simultaneous impregnation with the solution solids. The beneficial effects of an osmotic treatment on processed food products are result of the modification of the composition and/ or structure of the material. Two examples for the positive effects of osmotic treatments as a pre-step to thermal treatments are presented here: (a) reduction of convection drying times of carrot slices with enhanced retention of physiological active compounds and (b) reduction of oil uptake and acrylamide generation when frying potato strips.

11.1 Introduction

Conventional and new food processing techniques are expected to enhance the nutritional, hygienic and sensory quality of food products, improve the processing characteristics of raw materials and semi-finished products, increase the variety of products, and finally take into account the economic and environmental aspects of food processing. Although traditional techniques (e.g. chilling, freezing, convection air drying, pasteurisation, sterilisation) are still the major employed for food processing, progress in their application is still needed, mainly with regard to the loss of nutritional and sensory attributes of the food material throughout processing and storage. Widespread among new processing techniques is the minimisation of thermal stress and the combination of more than one technique, aiming at an optimal retention and enhancement of the nutritional and sensory attributes of processed food [1].

The principle of osmosis as a mean of water removal has been known for a long time; however, application of osmotic treatments to food can be considered among the new or improved techniques, as its main characteristics are that the materials are exposed to minimal thermal stress and that the process is

* Institute of Process Engineering, Federal Research Centre for Nutrition and Food, Haid-und-Neu-Str. 9, 76131 Karlsruhe, Germany.

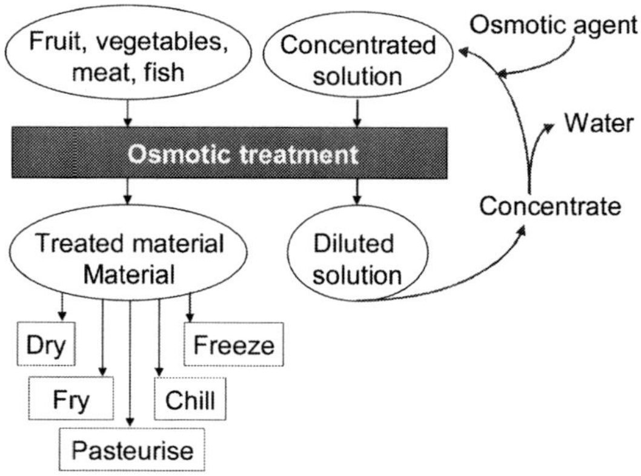

Figure 11.1: Osmotic treatment as a step in the food production chain.

in most cases applied in combination with another processing method (Fig. 11.1). For osmotic treatment, food tissues are immersed in aqueous solutions of sufficiently high concentration at moderate temperatures. A water drain from the tissue into the solution takes place and a solute transfer from the solution into the tissue and a leaching process of the tissue's own solutes are observed. Although much of the initial water content can in this way be removed from the tissue, storage stability of the final product must be assured by another processing technique, as the water content of the osmotically treated product is still high enough to induce spoilage [2].

Partial dewatering of fruit by osmosis in sugar solution before the product is exposed to other processes in order to obtain products of high quality was first suggested in 1966. Since then, osmotic treatments have been studied in combination with air-, vacuum-, freeze- and sun-drying, pasteurising, canning, freezing, frying, addition of preservatives and/or acidification and coating by edible surface layers.

The currently increasing interest in osmotic treatments arises primarily from the need for quality improvement and from economic factors. Quality improvement is related not only to the water removal with minimal thermal stress but also to the impregnated solutes and the modification of the structure. With the correct choice of solutes, and a controlled and equilibrated ratio of water removal and impregnation, it is possible to enhance natural flavour and colour retention in fruit products, hence to avoid additives such as antioxidants; softer textures in partially dehydrated products can be obtained; food ingredients can be designed for particular uses. The economic interest in osmotic treatments focuses on reduced energy consumption for water removal without phase change, as compared to convection drying, and the possible reduction of the refrigeration load by partial concentration prior to freezing of fruit or vegetables.

Two examples for the positive effects of osmotic treatments are shown here: as a pre-step to convection drying and as a pre-step to frying.

11.2 Materials and Methods

Warm air drying: Osmotic treatment was applied to carrot slices (3 mm) as a pre-step to air drying. Osmotic treatment was performed at 25 °C during 60 min using three aqueous solutions, all with a_W = 0.91: 15% sodium chloride, 62% sucrose and a combined 7% sodium chloride/34% sucrose solution. For the control samples with no osmotic treatment, blanching at 95 °C, 1 min was applied. Convection drying was performed in pilot scale (Heindl, Mainburg, Germany) at 70 °C [3]. Dried slices were stored up to four months under nitrogen atmosphere in light-tight bags at 25 °C. Investigated quality parameters: vitamin C [4] and carotenoid retention [5].

Frying: Osmotic treatment of potato strips was performed at 40 °C during 30 min in a 10% sodium chloride aqueous solution. For the control samples with no osmotic treatment, the strips were introduced in water at 40 °C for 30 min. After osmotic treatment or soaking in water the potato sticks (200 g portions) were fried at 175 °C for 7 min in sunflower oil in a catering fryer. Investigated quality parameters: oil uptake (Soxhlet method, diethyl-ether) and acrylamide content [6].

11.3 Results

11.3.1 Warm Air Drying

A main part of the water present in the fresh material is removed during the osmotic treatment at 25 °C, e.g. after 60 min osmotic pre-treatment in a sodium-chloride-sucrose solution the water content (X) is reduced from 8.9 $kg/kg_{d.m.}$ to 1.7 $kg/kg_{d.m.}$.

After osmotic pre-treatment the thermal stress for achieving products with X=0.15 is radically reduced: using a ternary sucrose-sodium chloride aqueous solution air drying times can be reduced up to 50% as compared to the drying of material not pre-treated (Fig. 11.2).

As both sucrose and sodium chloride have been proposed as retarding or inhibiting the oxidation of ascorbic acid by different mechanisms [7, 8], an enhanced retention of this phytochemical can be expected when applying an osmotic pre-treatment in sucrose or sodium chloride solution. During osmotic treatment a slight leaching of the Vitamin C was observed (Fig. 11.3). However, after convection drying, osmotic pre-treated carrots showed a much higher Vitamin C retention compared to blanched carrots. According to McLaughlin and Magee Vitamin C degrades exponentially during drying [9]. The shorter drying times of pre-treated carrots could partly explain the better retention. However,

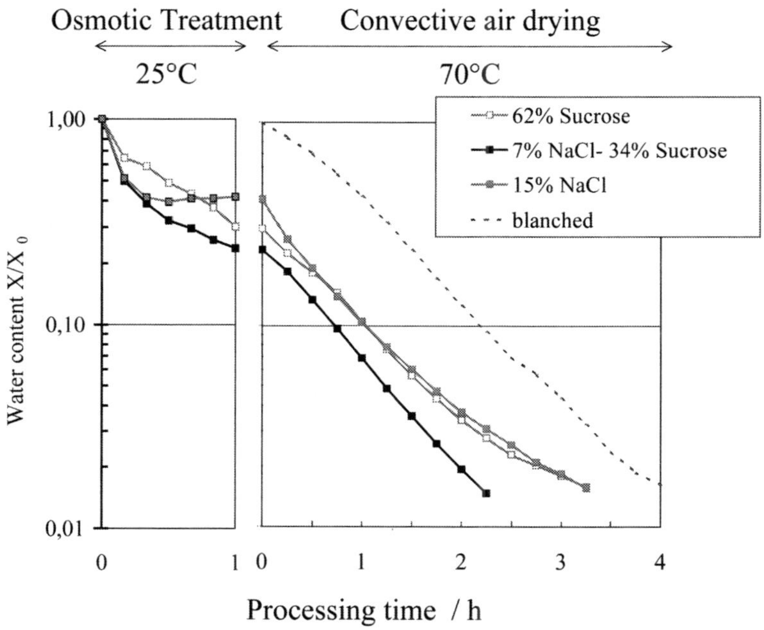

Figure 11.2: Kinetic of the water removal from: (a) carrot slices during osmotic pre-treatment in three different solutions and further convection drying; (b) blanched carrot slices during convective drying.

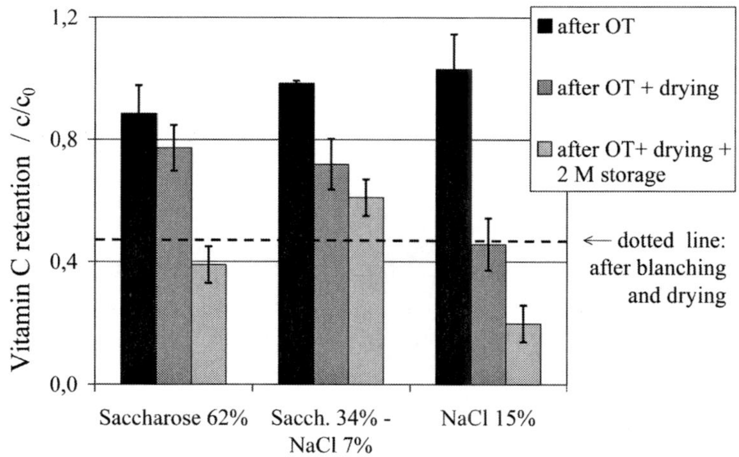

Figure 11.3: Vitamin C retention during processing and storage of carrot slices.

a protective effect from sucrose is also confirmed: carrots pre-treated in sucrose or in sodium chloride solutions need the same drying times to reach the desired product, however almost 80% of the initial Vitamin C is retained when treating in sucrose and only 50% when treating in sodium chloride. Furthermore pre-

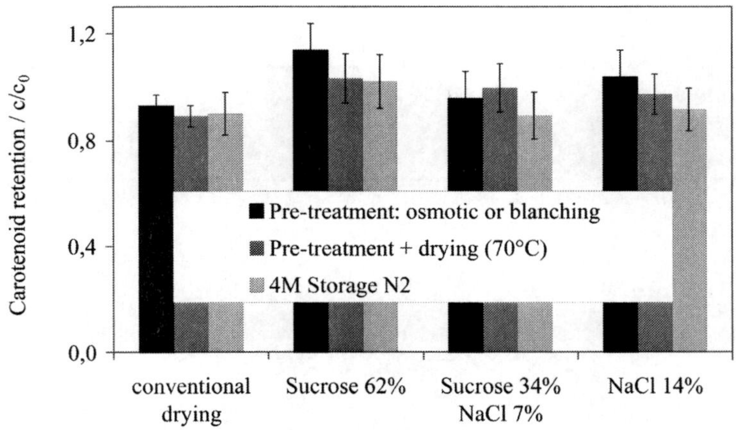

Figure 11.4: Carotenoid retention during processing and storage of carrot slices.

treating in a ternary solution leads to an excellent retention, even after two months storage.

Concerning carotenoid retention, the osmotic pre-treatment had no significant effect. Only with the pre-treatment in sucrose solution, a slight enhancement of the carotenoid retention has been observed (Fig. 11.4).

11.3.2 Frying

All techniques applied to reduce fat uptake during deep-frying deal with the reduction of moisture loss and/or modification of the surface upon frying. According to Krokida et al. [10] and Bunger et al. [11] it is possible to obtain products with a clearly reduced fat content when an osmotic pre-treatment is incorporated in the production line of potato chips. Compared to chips produced without the pre-treatment, fat content can be reduced up to 30% applying an osmotic treatment in 20% sodium chloride solution and up to 60% in 40% sucrose solution. However regarding the possible increased acrylamide generation after an impregnation in sucrose solution, the use of sucrose is not recommendable.

Our results show that chips produced including an osmotic step contained less fat than those obtained without OT (Tab. 11.1).

During frying the fat not only remains at the surface but also penetrates the potato strips through the capillars and channels connected to the surface. During the osmotic treatment together with the water removal and uptake of the solution solutes, the surface of the strips is modified: the cells of the outer layers shrink and up to a certain extent the tissue collapses, giving rise to a compacted surface. The penetration of the fat in the osmotic pre-treated potato strips is in this way hindered and the fat uptake is reduced, as confirmed by observing the microstructure of fried potatoes (SEM) (Fig. 11.5).

Table 11.1: Water loss and solid gain during OT and during frying of pre-treated and control potato strips.

kg/100 kg raw potato strips	OT	Frying	
		OT	Control
Water loss	25	45	47
Solid gain *	4,7	2	5

* Solid gain: NaCl after osmotic treatment; fat after frying.

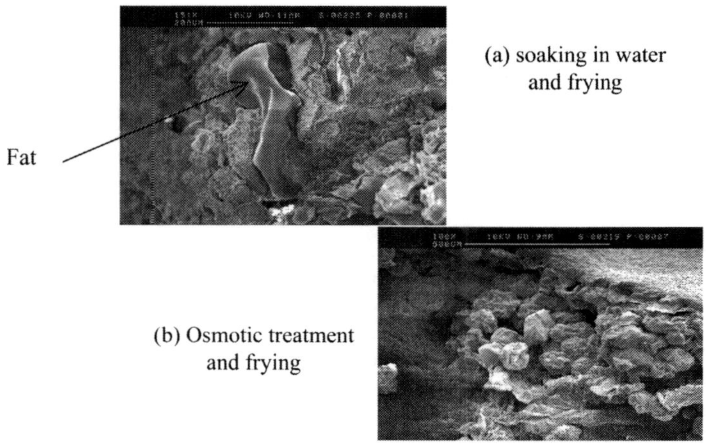

Figure 11.5: Scanning electron microscopy pictures of fried potato strips with a previous (a) soaking in water (control) and (b) osmotic treatment. Penetrated fat can be observed in the control, but not in the osmotic pre-treated fried potatoes.

Furthermore the acrylamide content in OT-chips was approx. 60% less than in the control chips. A slight leaching of reducing sugars takes places during the osmotic treatment, but also during the soaking in water. The modification of the surface of the potato strips (compactness) plays a significant role in the reduction of acrylamide generation: the surface in contact with the frying fat is reduced, reducing the thermal stress inside the potato strip during frying and thus reducing the acrylamide generation.

11.4 Conclusions and Outlook

The beneficial effects of an osmotic treatment on dried carrot slices and fried potato chips has been shown. The drying time of carrot slices was reduced and the retention of physiological active compounds enhanced. For the production of potato chips the fat uptake and the acrylamide generation were reduced.

Reduced processing times at higher temperatures could also reduce processing costs. A financial study of the complete process including recycling of the osmotic solution is needed to confirm this point.

Although not presented here, frying times can be shortened applying an osmotic pre-treatment. This optimisation of the frying time of osmotic pre-treated potato strips, too, could lead to a further reduction of the fat uptake and the acrylamide generation.

Acknowledgments

Part of this work was supported by the German Federal Ministries for Nutrition, Agriculture and Consumer Protection (BMELV) and of Education and Research (BMBF).

References

1. Spieß, W.E.L.; Behsnilian, D. (1998) Osmotic treatments in food processing – current state and future needs. In: Drying '98; C.B. Akritidis, D. Marinos-Kouris, G.D. Saravakos (Eds.), Ziti Editions, Thessaloniki, Vol A, 47–56.
2. Behsnilian, D. (2005) Osmotische Vorbehandlung. In: Lebensmittelverfahrenstechnik: Rohstoffe, Prozesse, Produkte. Schuchmann, H.P. & Schuchmann H. (Eds.), Wiley-VCH, Weinheim, 311–317.
3. Regier, M.; Mayer-Miebach, E.; Behsnilian, D.; Neff, E.; Schuchmann, H.P. (2005) Influence of drying and storage of lycopene-rich carrots on the carotenoid content. *Drying Tech.*, **23**, 989–998.
4. Bognar, A.; Dawood, H.G. (2000) Simple in-line postcolumn oxidation and derivatization for the simultaneous analysis of ascorbic and dehydroascorbic acids in foods. *J. Chromatogr. Sci.*, **38**, 162–168.
5. Mayer-Miebach, E.; Spieß, W.E.L. (2003) Influence of cold storage and blanching on the carotenoid content of Kintoki carrots. *J. Fd. Eng.*, **65**, 211–213.
6. Koller, W.-D. (2005) Eine Methode zur Bestimmung des Bildungspotentials für Acrylamid von Rohware und eine einfache Methode zur Bestimmung des Acrylamidgehaltes von Lebensmitteln mittels GC/MS. In: BfEL Jahresbericht 2004, BfEL (Ed.), 206.
7. Hsieh, Y.P.; Harris, N.D. (1993) Effect of sucrose on oxygen uptake of ascorbic acid in a closed aqueous system. *J. Agric. Fd. Chem.*, **41**, 259–262.
8. Harel, S. (1994) Oxidation of ascorbic acid and metal ions as affected by NaCl. *J. Agric. Fd. Chem.*, **42**, 2404–2406.
9. McLaughlin, C.P.; Magee, T.R.A. (1998) The effect of shrinkage during drying of potato spheres and the effect of drying temperature on vitamin C retention. *Fd. & Bioprod. Processing*, **76**, 138–142.
10. Krokida, M.K.; Oreopoulou, V.; Maroulis, Z.B.; Marinos-Kouris, D. (2001) Effect of osmotic dehydration pretreatment on quality of french fries. *J. Fd. Eng.*, **49**, 339–345.
11. Bunger, A.; Moyano, P.; Rioseco, V. (2003) NaCl soaking treatment for improving the quality of french-fried potatoes. *Food Res. Inter.*, **36**, 161–166.

12 Influence of Maillard Reaction Products on the Inflammatory Cellular Response of Macrophages

Sonja Muscat * and Monika Pischetsrieder

Maillard reaction products (MRPs) are formed through a non-enzymatical reaction between reducing sugars and proteins or amino acids. This so called Maillard reaction takes place in heated foods like e.g. bakery products. Furthermore, so called "advanced glycation end products" (AGEs) are formed through analogical reaction steps in the human body. Highly increased AGE-levels can be detected in patients with diabetes mellitus (increased glucose level) or renal failure (diminished excretion). It's well known that AGE can induce an inflammatory cellular response. For signal transduction, a receptor-dependent induction of oxidative stress, e.g. via RAGE (receptor for advanced glycation end products) is suggested. So far however little is known about the question if food-derived MRPs can directly trigger inflammatory processes in the intestinal or if they can lead to an inflammatory state in the human body after resorption. In this regard, macrophages which are important mediators of inflammatory responses are of particular interest. To investigate the pro-inflammatory properties of MRPs, macrophages were stimulated with a reaction mixture of lysine and ribose, which was heated for 30 min at 120 °C. As an early event in the inflammatory signal transduction the activation of the transcriptional factor Nuclear factor-κB (Nf-κB) was immunochemically detected. It could be shown that stimulation of macrophages with the reaction mixture leads to a significant, 18-fold Nf-κB activation compared to the control. The observed effect was due to MRPs, because stimulation with the unheated reaction mixture did not lead to a Nf-κB activation. Furthermore the influence of coffee, a food with high MRPs content, on the inflammatory response of macrophages was investigated. Whereas an extract of raw coffee beans did not lead to Nf-κB activation, freshly brewed coffee leads to a 13-fold Nf-κB activation. The cellular mechanism of the MRPs-induced Nf-κB activation was investigated by using RAGE-transfected and untransfected human embryonic kidney cells (HEK) for the stimulation experiments. The reaction mixture could induce a 2-fold Nf-κB activation in the RAGE-transfected HEK-cells. However, stimulation of the untransfected cells, which do not express the receptor, leads also to a 2-fold Nf-κB activation. The data indicate that MRPs can induce an inflammatory response in macrophages. Since Nf-κB activation was not RAGE-dependent, a second mechanism of cell activation by MRPs must exist.

* Institute for Pharmacy and Food Chemistry, Henriette Schmidt-Burkhardt Chair of Food Chemistry, Schuhstr. 19, 91052 Erlangen, Germany.

13 Dosimetry of Acrylamide and Glycidamide Binding to Proteins in Human Blood

Daniel Bertow *, Matthias Baum *, and Gerhard Eisenbrand *

Abstract

The carcinogen acrylamide (AA) is an α-, β-unsaturated carbonyl compound. AA and its metabolite glycidamide (2,3-epoxypropanamide, GA) react with nucleophilic centers in biomolecules like DNA, as well as sulfhydryl and amino groups in proteins. GA, which binds covalently to DNA, is directly genotoxic and is therefore regarded as the ultimate carcinogen of AA.

AA and GA are inactivated by conjugation with glutathione (GSH), but reactions with other "noncritical" targets like hemoglobin and other blood proteins might also contribute to their inactivation.

Human blood, the first target compartment following absorption through the gut was used as a model system to investigate concentration-dependent formation of adducts between the N-terminal valine in Hb and AA/GA. Blood samples from volunteers were incubated with AA or GA (0.3–10.0 μM) for 1–6 h. Adducts formed were isolated using the N-alkyl Edman method as pentafluorophenylthiohydantoin (PFPTH, Fig. 13.2) derivatives and quantified by GC/MS.

In order to evaluate the extent of AA effectively bound to erythrocytes and plasma proteins we investigated the distribution of radioactivity into blood compartments after incubation with ^{14}C-radiolabeled AA.

It is intended to compare Hb adduct formation with concentration-dependent induction of genotoxic effects in white blood cells to approach the goal of establishing potential correlations between biomarkers of exposure (Hb adducts) and biomarkers of effect (DNA damage) under conditions simulating as close as possible AA exposure from the diet.

In blood from non-smoking probands we found an average background valine adduct level in the range of 20–45 pmol/g Hb with AA and in the range of 6–40 pmol/g Hb for GA. This corresponds well with results of further studies. Adduct levels of acrylamide with N-terminal valine in Hb were significantly increased after incubation with 0.3 μM AA (4 h), respectively 3 μM (1 h) and beyond. GA-valine adducts became significant after 4 h incubation with 3 μM GA. GA reacts more slowly with hemoglobin as compared to AA.

About 35–45% of the label was found in the erythrocyte fraction and about 60% was found in plasma after incubation with 30 μM AA. Hemoglobin con-

* University of Kaiserslautern, Erwin-Schroedinger-Str. 52, 67663 Kaiserslautern, Germany.

tained about 10% of the radio label. In plasma 10–20% was bound to proteins (acetone precipitation). These results show that substantial amounts of AA are bound to Hb and plasma proteins after 1 h. In total 25–35% radioactivity are bound to proteins or cell fragments.

13.1 Introduction

Acrylamide (AA) is carcinogenic in animal experiments. AA and its epoxide metabolite glycidamide (2,3-epoxypropanamide, GA) react with nucleophilic sites in biomolecules like DNA as well as sulfhydryl and amino groups in proteins. GA binds covalently to DNA causing genotoxic effects and is therefore suspected to be responsible for the carcinogenic potency of AA exposure [1]. In human blood lymphocytes genotoxicity is detectable at 10 μM as lowest concentration, as evidenced by the comet assay (Baum et al., Contribution III-6). Adduct formation between GA and DNA have been characterized, mainly with N-7 of guanine [2, 3]. The extent of conversion of AA to GA was estimated from the hemoglobin adduct data. The ratio of AUCs for GA to AA (AUC_{GA}/AUC_{AA}) was estimated to be about 0.3 in humans [4, 5]. Applying mercapturic acids as biomarkers of exposure a ratio of GAMA/AAMA of 0.1 was found, which suggests that AA conversion to glycidamide in humans is rather low [6, 7].

In adult humans AA half-lives were found between two and seven hours, which is sufficient for distribution among all tissues [5, 8]. The electrophilic double bond in AA and the electrophilic epoxide ring in GA can react with nucleophilic centers in biomolecules. Adduct formation of AA and GA with thiol and amino groups of hemoglobin is well suited to examine long term internal

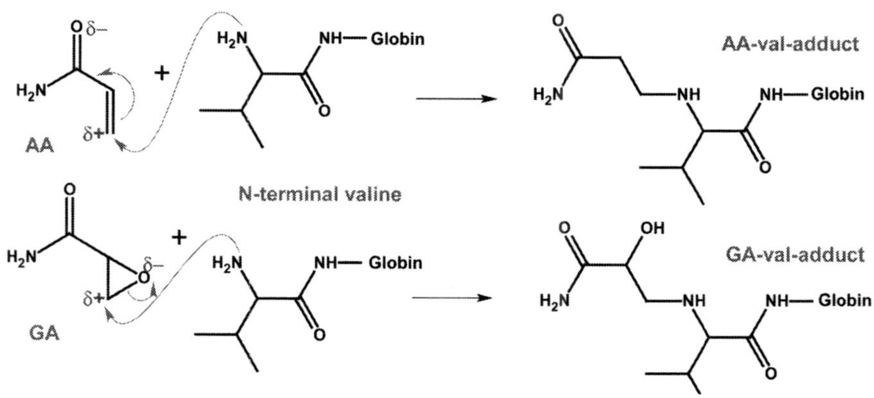

Figure 13.1: Adduct formation of acrylamide and glycidamide with N-terminal valine.

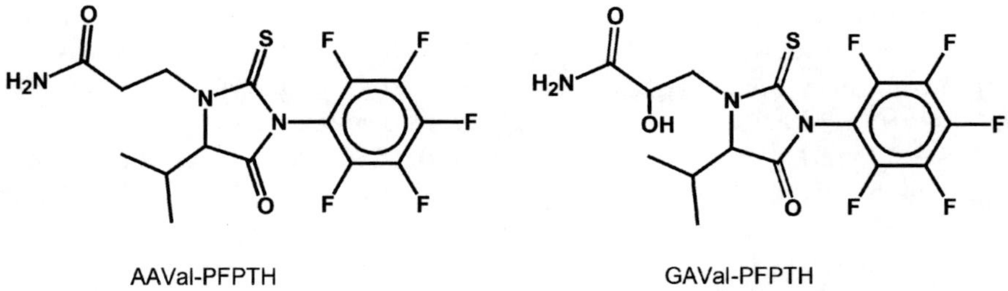

Figure 13.2: Chemical structure of pentafluorophenylthiohydantoins of acrylamide and glycidamide.

exposure with AA and GA since red blood cells have a life span around three months [9, 10]. Adducts formed by reaction between AA or GA with the α-NH_2 group of N-terminal valine are considered to be a useful biomarker of exposure (Fig. 13.1) [11–13]. Such a biomarker can be applied to detect human exposure via intake of food which is besides tobacco smoke an important human exposure source. Background adduct levels of AA around 40 pmol/g hemoglobin lead finally to the conclusion that processed food under certain conditions represents a source of AA [14]. Human exposure via this route is estimated as 1 μg/kg bw by FAO/WHO [15]. In studies investigating background levels of AA, adducts with N-terminal valine in the range of 10–40 pmol/g hemoglobin were reported in non-smokers. Adduct background levels of GA were equivalent [5, 16, 17].

Concentration-dependent reaction kinetics of AA and GA adduct formation with hemoglobin in human blood and DNA damage in the nuclei of leucocytes can be examined in parallel in order to arrive at a protein marker-guided dosimetry of genotoxic effects [18]. For risk assessment it is of particular interest to verify whether the reaction of AA and its metabolite GA with "noncritical" blood components like Hb or GSH is favoured in the low concentration range.

13.2 Material and Methods

13.2.1 Adduct Formation of AA/GA with N-terminal Valine in Hemoglobin

Blood samples (2 mL) were incubated with acrylamide and glycidamide (0.3, 3 and 10 μM) in 20 μL of a phosphate buffer solution (pH 7.4). Adduct background levels were also determined. All samples were stored at 37 °C during incubation time. Further sample preparation was performed according to Schettgen et al. [17] with slight modifications.

Acetonization was performed by addition of 100 μL freshly prepared 1% sulfuric acid in acetone. After two hours samples were neutralized with 150 μL of a freshly prepared 0.1 M $NaHCO_3$ solution. 1 mL toluene was added and samples were washed with 2 mL bidistilled water twice. Samples were dried under nitrogen, resolved in 100 μL toluene, treated in an ultrasonic bath for 5 min, centrifuged at 3500 g for 10 min and analysed by GC/MS.

Gaschromatographic separation: Agilent 6890 gaschromatograph in combination with an Agilent 5973 inert mass spectrometer; Agilent 7683 autosampler; Agilent 7673 PTV injector, operating in pulsed splitless mode; DB-XLB capillary column (30 m, 0.25 μm i.d., 0.25 μm film, Agilent); carrier gas: helium 5.0 at constant flow 1.6 mL/min; initial column temperature of 90 °C, 1 min, raised at a rate of 25 °C/min to 120 °C, increased at a rate of 10 °C/min to 240 °C and finally at a rate of 25 °C/min up to 310 °C where it remained for 5 min.

Mass spectrometric detection: Agilent 5973 inert mass spectrometer, chemical ionisation (CI) mode with 45 vol% of methane as reactant gas; Single Ion Monitoring (SIM) mode.

13.2.2 Distribution of ^{14}C-AA in Human Blood (Fig. 13.3)

Human blood samples (1.5 mL) were incubated with ^{14}C-labelled acrylamide (^{14}C-AA, 8300 Bq) in 5 μL ethanol (final concentration 30 μM). Incubations were performed in Eppendorf tubes (2 mL), 37 °C, 4 h under slight shaking in a thermomixer. Weighing of all fractions followed every fractionation step to compensate differences in sample processing.

Separation of erythrocytes and plasma was done by centrifugation (100 g, 10 min), plasma was frozen. Erythrocytes were washed with 500 μL physiological sodium chloride solution and centrifuged (100 g, 10 min). Hemolysis was achieved by addition of 100 μL bidistilled water, vortex treatment and storage at –20 °C (at least 30 minutes). An aliquot of lysed erythrocytes (100 μL) was separated for hemoglobin isolation by adding 700 μL of a solution of 50 mM HCl in 2-propanol. After vortexing and centrifugation (3500 g, 8 min) the sepa-

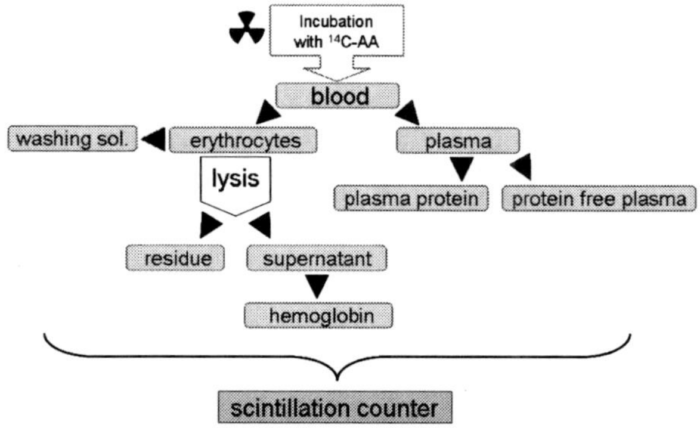

Figure 13.3: Sample preparation scheme for distribution of ^{14}C-AA in human blood.

rated supernatant was spiked with 700 μL ethyl acetate (–20 °C) and stored for 15 min in a freezer. After centrifugation (3500 g, 8 min) precipitated hemoglobin was washed with 100 μL ethyl acetate, solved in 500 μL of a physiological sodium chloride solution, vortexed and incubated with protease (40 μL) for 2 h at 37 °C under strong shaking in a thermomixer.

To thawed plasma samples aliquots (100 μL) 400 μL cold acetone was added to precipitate proteins. After centrifugation (100 g, 4 min) protein precipitate was washed with 200 μL cold acetone. Protein was immediately resolved in 400 μL physiological sodium chloride solution by vortexing, addition of protease (10 μL) and storage for 1/2 h at 37 °C under strong shaking in a thermomixer.

Lysis of whole blood was done by adding 300 μL bidistilled water and 100 μL sodium dodecyl sulfate solution (SDS) while lysis of erythrocytes was done by adding 100 μL of bidistilled water, freezing, thawing and vortexing.

Radioactivity was measured using a Beckman LS 1701 scintillation counter. 100 μL of each fraction (radioactive and blank samples) were mixed with 10 mL scintillation cocktail Rotiszint® eco plus.

13.3 Results and Discussion

In studies measuring occupational exposure, adducts formed with N-terminal valine in hemoglobin from erythrocytes have been applied as sensitive biomarkers for the quantification of internal AA or GA doses in exposed persons. We investigated background adduct levels of AA and GA with N-terminal valine. In blood from non-smoking probands, we found an average background valine adduct level in the range of 20–45 pmol/g Hb with AA and in the range of 6–40 pmol/g Hb for GA. This corresponds well with results of further studies [5, 11, 16, 17, 19–21].

Table 13.1 summarizes data of background levels of AA and GA taken from human studies reflecting the wide range of adduct formation with N-terminal valine in hemoglobin.

Vesper et al. recently published data of background levels of AA and GA which exceeds data previously published [21]. This difference might be explained by the application of peptides as internal standard to consider efficiency of the modified Edman degradation.

Incubation experiments of human blood with AA and GA were performed in order to examine the concentration–time dependence of adduct formation with N-terminal valine (Fig. 13.4). Adduct levels of AA with N-terminal valine in Hb were significantly increased after incubation with 0.3 μM AA (4 h), respectively 3 μM (1 h) and beyond. GA-valine adduct formation became significant after 4 h incubation with 3 μM GA. Adduct formation appeared linear over the measured concentration range. GA reacts more slowly with hemoglobin in whole blood as compared to AA.

Table 13.1: Data of background levels of AA and GA taken from human studies.

Publication	Number of participants	AA-val [pmol/g Hb] ∅±SD	GA-val [pmol/g Hb] ∅±SD
[19]	5 NS, 5 S	<10 (detected in one S)	n.d.
[11]	2 NS, 2 S	41 (NS); 81 (S)	3 (NS); 7 (S)
[16]	5 NS	27±6 (NS)	26±6 (NS)
[17]	13 NS, 16 S	19±7 (NS); 80±48 (S)	17±7 (NS); 53±30 (S)
[20]	296 NS, 99 S	~12±6 (NS); ~50±45 (S)	n.q.
This study	12 NS	35±5 (NS)	16±12 (NS)
[21]	96	129	97

NS: non-smoker; S: smoker; n.d.: not detected; n.q.: not quantified.

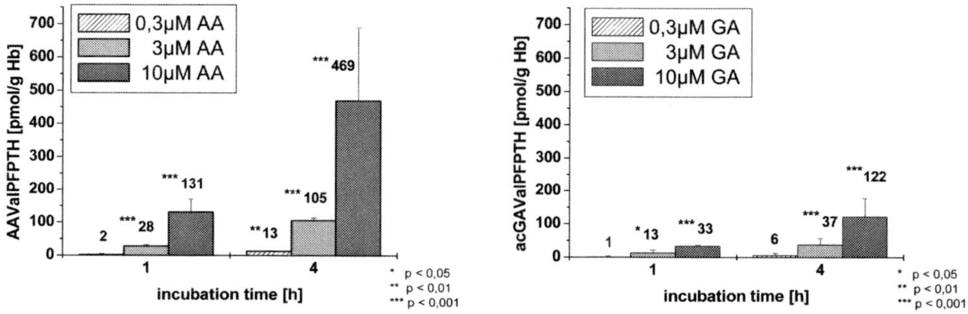

Figure 13.4: Dosimetry of AA and GA with N-terminal valine in Hb.

About 35–45% of the label was found in the erythrocyte fraction and about 60% was found in plasma after incubation with 30 µM AA, 4 h. Hemoglobin contained about 10% of the radio label. In plasma 10–20% was bound to proteins (acetone precipitation). Xie et al. showed in a low dose experiment using accelerator mass spectrometry (AMS) that adduct levels with serum albumin exceed these of hemoglobin [22]. Our results indicate that substantial amounts of AA are bound to Hb and plasma proteins after 1 h (Fig. 13.5). In total 25–35% radioactivity are bound to proteins or cell fragments. Whether the soluble radioactivity found in plasma can be ascribed to unreacted AA and/or to AA reaction products with low molecular weight blood compounds like glutathione needs further investigation.

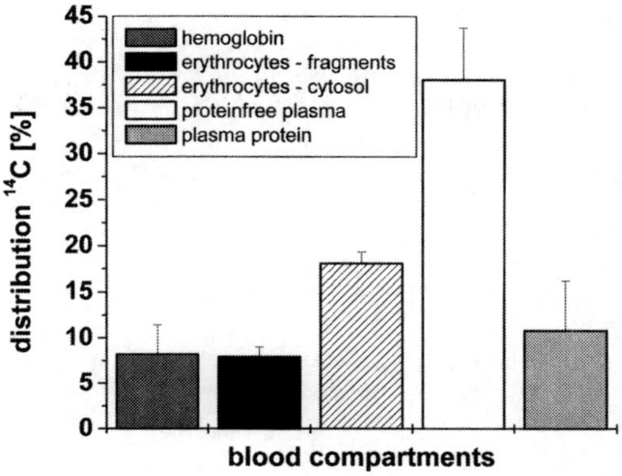

Figure 13.5: Distribution of ^{14}C-AA in human blood after 4 h incubation.

Acknowledgments

This work was supported by grants of AIF 108, Bundesministerium für Wirtschaft und Arbeit via Forschungskreis der Ernährungsindustrie and Bund für Lebensmittelrecht und Lebensmittelkunde e.V. (BLL).

References

1. Friedman, M., *J. Agric. Food Chem.*, 2003, **28**: 4504.
2. Solomon, J.J., et al., Direct alkylation of 2'-deoxynucleosides and DNA following in vitro reaction with acrylamide. *Cancer Res.*, 1985, **45**(8): 3465–3470.
3. Gamboa da Costa, G., et al., DNA adduct formation from acrylamide via conversion to glycidamide in adult and neonatal mice. *Chem. Res. Toxicol.*, 2003, **16**(10): 1328–1337.
4. Calleman, C.J., The metabolism and pharmacokinetics of acrylamide: implications for mechanisms of toxicity and human risk estimation. *Drug Metab. Rev.*, 1996, **28**(4): 527–590.
5. Fennell, T.R., et al., Metabolism and hemoglobin adduct formation of acrylamide in humans. *Toxicol. Sci.*, 2005, **85**(1): 447–459.
6. Boettcher, M.I., et al., Excretion of mercapturic acids of acrylamide and glycidamide in human urine after single oral administration of deuterium-labelled acrylamide. *Arch. Toxicol.*, 2006, **80**(2): 55–61.
7. Fuhr, U., et al., Toxicokinetics of acrylamide in humans after ingestion of a defined dose in a test meal to improve risk assessment for acrylamide carcinogenicity. *Cancer Epidemiol. Biomarkers Prev.*, 2006, **15**(2): 266–271.
8. Kirman, C.R., et al., A physiologically based pharmacokinetic model for acrylamide and its metabolite, glycidamide, in the rat. *J. Toxicol. Environ. Health A*, 2003, **66**(3): 253–274.
9. Watson, W.P. and A. Mutti, Role of biomarkers in monitoring exposures to chemicals: present position, future prospects. *Biomarkers*, 2004, **9**(3): 211–242.

10. Dybing, E., et al., Human exposure and internal dose assessments of acrylamide in food. *Food Chem. Toxicol.*, 2005, **43**(3): 365–410.
11. Perez, H. L., et al., Simultaneous analysis of hemoglobin adducts of acrylamide and glycidamide by gas chromatography-mass spectrometry. *Anal. Biochem.*, 1999, **274**(1): 59–68.
12. Paulsson, B., J. Grawe, and M. Tornqvist, Hemoglobin adducts and micronucleus frequencies in mouse and rat after acrylamide or N-methylolacrylamide treatment. *Mutat. Res.*, 2002, **516**(1/2): 101–111.
13. Tornqvist, M. and L. Ehrenberg, Estimation of cancer risk caused by environmental chemicals based on in vivo dose measurement. *J. Environ. Pathol. Toxicol. Oncol.*, 2001, **20**(4): 263–271.
14. Tareke, E., et al., Analysis of acrylamide, a carcinogen formed in heated foodstuffs. *J. Agric. Food Chem.*, 2002, **50**(17): 4998–5006.
15. FAO/WHO, Joint FAO/WHO expert committee on food additives. Sixty-fourth meeting, Rome, Italy, 7–17 February 2005.
16. Paulsson, B., et al., Hemoglobin adducts from glycidamide: acetonization of hydrophilic groups for reproducible gas chromatography/tandem mass spectrometric analysis. *Rapid Commun. Mass Spectrom.*, 2003, **17**(16): 1859–1865.
17. Schettgen, T., et al., Determination of haemoglobin adducts of acrylamide and glycidamide in smoking and non-smoking persons of the general population. *Int. J. Hyg. Environ. Health*, 2004, **207**(6): 531–539.
18. Baum, M., et al., Acrylamide and glycidamide: genotoxic effects in V79-cells and human blood. *Mutat. Res.*, 2005, **580**(1/2): 61–69.
19. Bergmark, E., et al., Determination of hemoglobin adducts in humans occupationally exposed to acrylamide. *Toxicol. Appl. Pharmacol.*, 1993, **120**(1): 45–54.
20. Bader, M., et al., Querschnittsstudie zur ernährungs- und tabakrauchbedingten Belastung mit Acrylamid. *Deutsches Ärzteblatt*, 2005, **102**(39): 2640–2643.
21. Vesper, H. W., et al., Automated method for measuring globin adducts of acrylamide and glycidamide at optimized Edman reaction conditions. *Rapid Commun. Mass Spectrom.*, 2006, **20**(6): 959–964.
22. Xie, Q., et al., Adduction of biomacromolecules with acrylamide (AA) in mice at environmental dose levels studied by accelerator mass spectrometry. *Toxicol. Lett.*, 2006, **163**(2): 101–108.

14 DNA Strand Breaking Capacity of Acrylamide and Glycidamide in Mammalian Cells

Doris Marko *, Zeina Tjaden, Florian Fueller, and Nicole Puppel

Since the first synthesis of acrylamide (AA) in 1949, this compound has gained importance in many application fields from wastewater treatment, adhesive development, to the intensive use in laboratory gels, to name some examples. However, the image of an useful but toxic synthetic chemical has been turned completely upside down by the unexpected finding that AA is also formed in thermally processed food. The extent of AA formation strongly depends on the heating conditions and the concentration of respective precursors. The daily intake of AA by food was estimated for Germany as roughly 0.6 μg/kg bodyweight [1]. Depending on nutritional habits severalfold higher intake levels are possible. In longterm studies in rodents AA was found to be carcinogenic and has been classified as a probable carcinogen in humans. However, the mechanism of carcinogenicity is not fully understood. AA has been shown to be clastogenic and mutagenic *in vivo*. In rodents, after interperitoneal injection of AA the formation of a depurinating DNA-adduct, N7-(2-carbamoyl-2-hydroxyethyl)-guanine, has been reported. However, AA itself reacts quite slowly with DNA. Within the cells, AA can react with intracellular glutathione (GSH), but can also be oxidized to the more reactive glycidamide (GA) by cytochrome P450-dependent monooxygenase (CYP2E1). GA is known to be clastogenic and mutagenic *in vitro* and *in vivo*.

We compared the DNA damaging potency of acrylamide (AA) and its metabolite glycidamide (GA) in cell systems differing with respect to species origin and CYP2E1 expression (V79, Caco-2, primary rat hepatocytes) using the comet assay. After 24-h incubation only at the highest concentration of AA (6 mM) a slight but significant increase in DNA damage was observed in V79 and Caco-2 cells. In primary rat hepatocytes, however, expressing substantial amounts of CYP2E1, no induction of DNA strand breaks was found. In contrast, direct treatment with the AA metabolite GA resulted in a significant increase in DNA damage in V79 cells and primary rat hepatocytes at concentrations $\geq$100 μM (24 h). Caco-2 cells were found to be slightly less sensitive, exhibiting an increase in DNA strand breaks at concentrations $\geq$300 μM GA. These data confirm the higher genotoxic potential of GA compared to AA. On the other hand the data indicate that high expression of CYP2E1 per se is not necessarily associated with increased genotoxicity of AA. We therefore addressed the question whether the intracellular glutathione (GSH) level might be a critical deter-

* Institute of Applied Biosciences, Division of Food Toxicology, University of Karlsruhe, Kaiserstr. 12, 76131 Karlsruhe, Germany.

minant for the genotoxicity of AA in cells with different CYP2E1 status. Co-incubation of rat hepatocytes and V79 cells with DL-buthionine-[S,R]-sulfoxime and 1 mM AA resulted in a depletion of intracellular GSH as well as in a significant induction of DNA strand breaks. However, at higher concentrations (≥1.25 mM) a strong increase in cytotoxicity, resulting in a severe loss of viability, was observed. In summary, the DNA strand breaking effect of AA appeared not to be directly correlated with the CYP2E1 status of the cells. Depletion of GSH is associated with an increase in AA genotoxicity but seems also to lead to a substantial enhancement of cytotoxicity [2].

Acknowledgments

The study was funded by the Bundesinstitut für Risikobewertung, Berlin.

References

1. Madle, S., Broschinski, L., Mosbach-Schulz, O., Schönig, G., Schulte, A., Zur aktuellen Risikobewertung von Acrylamid in Lebensmitteln. *Bundesgesundheitsblatt – Gesundheitsforschung – Gesundheitsschutz,* 2003, **5**: 405–415.
2. Puppel, N., Tjaden Z., Füller, F., Marho, D., DNA strand breaking capacity of acrylamide and glycidamide in mammalian cells. *Mutant. Res.*, 2005, **580**(1–2): 71–80.

15 2-Dodecylcyclobutanone, the Radiolytic Product of Palmitic Acid, is Genotoxic in the Comet Assay and Induces Chromosomal Aberrations in Preneoplastic Cells Using 24-Color-FISH

Nadine Knoll *, Anja Weise **, Michael Glei *, Uwe Claussen **, Brigitte Marian ***, and Beatrice L. Pool-Zobel *

The irradiation of fat-containing food for preservation leads to alterations in fatty acids, resulting in the formation of 2-alkylcyclobutanones (2ACB). These contaminants do not occur naturally in non-irradiated foods. The most abundant 2-dodecylcyclobutanone (2dDCB) is derived from palmitic acid. Results of previous *in vitro* studies with primary human colon cells and *in vivo* experiments with rats fed with 2-alkylcyclobutanones indicated that these radiolytic derivatives may be genotoxic and enhance the progression of colon tumors at high concentrations. The underlying mechanisms of these effects, however, are not clearly understood. Aim of the present study was to further investigate the genotoxic potential of lower concentrations of 2dDCB in human colon adenoma cells (LT97) and to analyse whether 2dDCB also induces chromosomal abnormalities. LT97 adenoma cells in suspension culture were exposed to 2dDCB for 0–120 min (150–2097 μM) and adherent cells were exposed for 6 h (30–150 μM) followed by an additional cell cycle. Trypan blue exclusion, comet assay, and 24-color-Fluorescence-*In-Situ*-Hybridisation (FISH) were the methods used to determine cytotoxicity, DNA strand breaks and chromosomal abnormalities, respectively. 2dDCB was cytotoxic (EC_{50}: 589 μM, 90 min treatment) and genotoxic at sub-cytotoxic doses (300 μM, 30 min treatment). The long-term incubation of LT97 adenoma cells (30–150 μM, 6 h) with 2dDCB resulted in dose-dependent clastogenic effects. In particular, chromosomal translocations or deletions were observed in those chromosomes, which harbour tumour-relevant genes. These findings clearly demonstrate that 2dDCB has a marked genotoxic potential in human colon cells, which were derived from preneoplastic colon adenoma. Mutations in adenoma cells may potentiate the transition from adenoma to carcinoma. The compound may be regarded as a possible risk factor for progression of colon carcinogenesis.

* Department of Nutritional Toxicology, Institute for Nutritional Sciences, Friedrich Schiller University, Dornburger Str. 25, 07743 Jena, Germany.
** Institute of Human Genetics and Anthropology, Friedrich Schiller University, Kollegiengasse 10, 07743 Jena, Germany.
*** Institute for Cancer Research, University of Vienna, Austria.

V Participants and Contributors of the Symposium

Prof. Dr. Hans-Jürgen Altmann
Federal Institute for Risk Assessment
Thielallee 88–92
14195 Berlin

Dr. Matthias Baum
Division of Food Chemistry
and Environmental Toxicology
University of Kaiserslautern
67663 Kaiserslautern

Quim. Farm. Diana Behsnilian
Institute of Process Engineering
Federal Research Centre
for Nutrition and Food
Haid-und-Neu-Str. 9
76131 Karlsruhe

Dr. Alexis Cockroft
Food Standards Agency
Process Contaminants Unit
125 Kongsway
London WC2B 6NH
UK

Dr. Ruthild Creutziger
BASF Aktiengesellschaft
Abt. GUP/CA, Gebäude Z470
67056 Ludwigshafen

Prof. Dr. Erik Dybing
Norwegian Institute of Public Health
PO Box 4404 Nydalen
0403 Oslo
Norway

Prof. Dr. Manfred Edelhäuser
Ministerium für Ernährung
und Ländlichen Raum
Kernerplatz 10
70182 Stuttgart

Prof. Dr. Gerhard Eisenbrand
Chair of DFG-Senate Commission
on Food Safety
Division of Food Chemistry
and Environmental Toxicology
University of Kaiserslautern
67663 Kaiserslautern

Prof. Dr. Karl-Heinz Engel
Lehrstuhl für Allgemeine
Lebensmitteltechnologie
TU München
Am Forum 2
85350 Freising-Weihenstephan

Barbara Engeli
Bundesamt für Gesundheit
Stauffacherstr. 101
8004 Zürich
Switzerland

Thermal Processing of Food: Potential Health Benefits and Risks
DFG, Deutsche Forschungsgemeinschaft

ISBN: 978-3-527-31909-1

Prof. Dr. Gerhard Flachowsky
Institut für Tierernährung
Bundesforschungsanstalt
für Landwirtschaft
Bundesallee 50
38116 Braunschweig

Dr. Henrik Frandsen
Danish Institute for Food
and Veterinary Research
Mørkhøj Bygade 19
2860 Søborg
Denmark

Dr. Knut Franke
Deutsches Institut
für Lebensmitteltechnik e. V.
Prof.-von-Klitzing-Str. 7
49610 Quakenbrück

Dr. Gerd Fricke
Bundesministerium Ernährung,
Landwirtschaft und Verbraucherschutz
Rochusstr. 1
53123 Bonn

Dr. Angela Göbel
Bundesamt für Verbraucherschutz
und Lebensmittelsicherheit
Postfach 480447
12254 Berlin

Prof. Dr. Werner Grunow
Federal Institute for Risk Assessment
Thielallee 88–92
14195 Berlin

Dr. Sabine Guth
DFG-Senate Commission
on Food Safety
Scientific Office
Division of Food Chemistry
and Environmental Toxicology
University of Kaiserslautern
67663 Kaiserslautern

Dr. Norbert Haase
Federal Research Centre
for Nutrition and Food
Schuetzenberg 12
32756 Detmold

Dr. Michael Habermeyer
DFG-Senate Commission
on Food Safety
Scientific Office
Division of Food Chemistry
and Environmental Toxicology
University of Kaiserslautern
67663 Kaiserslautern

Prof. Dr. Andrea Hartwig
Institut für Lebensmitteltechnologie
und Lebensmittelchemie
TU Berlin
Gustav-Meyer-Allee 25
13355 Berlin

Dr. Katrin Hoenicke
Eurofins Analytik GmbH
Wiertz-Eggert-Jörissen
Stenzelring 14 b
21107 Hamburg

Prof. Dr. Thomas Hofmann
Institut für Lebensmittelchemie
Universität Münster
Corrensstr. 45
48149 Münster

Maike Hühne
Tchibo GmbH
Süderstr. 293
20537 Hamburg

Prof. Dr. Hans-Ulrich Humpf
Institut für Lebensmittelchemie
Universität Münster
Corrensstr. 45
48149 Münster

Prof. Dr. Hans-Georg Joost
Deutsches Institut
für Ernährungsforschung
Arthur-Scheunert-Allee 114–116
14558 Nuthetal

Prof. Dr. Fred Kadlubar, Director
Division of Pharmacogenomics
and Molecular Epidemiology
National Center
for Toxicological Research
3900 NCTR Road
Jefferson, AR 72079
USA

Dr. Hennicke Kamp
BASF Aktiengesellschaft
Experimental Toxicology and Ecology
67056 Ludwigshafen

Martina Kasper
TU München
Lichtenbergstr. 4
85748 Garching

Dr. Monika Kemény
DFG-Senate Commission
on Food Safety
Scientific Office
Division of Food Chemistry
and Environmental Toxicology
University of Kaiserslautern
67663 Kaiserslautern

Dr. Andreas Kliemant
Bundesamt für Verbraucherschutz
und Lebensmittelsicherheit
Rochusstraße 65
53123 Bonn

Dipl. Troph. Nadine Knoll
Department of Nutritional Toxicology
Institute for Nutrition
Friedrich Schiller University
Dornburger Str. 25
07743 Jena

Prof. Dr. Dietrich Knorr
Berlin University of Technology
Dept. of Food Biotechnology &
Food Process Engineering
Königin-Luise-Str. 22
14195 Berlin

Prof. Dr. Ib Knudsen
Chief Adviser in Food Safety
and Toxicology
Danish Institute for Food
and Veterinary Research
Mørkhøj Bygade 19
2860 Søborg
Denmark

Dr. Gunhild Kozianowski
Südzucker AG Mannheim/Ochsenfurt
Wormser-Str. 11
67283 Obrigheim/Pfalz

Dr. Mariella Kuilman
DSM Food Specialities
PO Box 1
2600 MA Delft
The Netherlands

Prof. Dr. Greg Laden
Department of Anthropology
University of Minnesota
395 Hubert H. Humphrey Center
301 19th Avenue S
Minneapolis MN 55455
USA

Prof. Dr. Doris Marko
Universität Karlsruhe
Institut für Angewandte
Biowissenschaften
Abteilung für Lebensmitteltoxikologie
Fritz-Haber-Weg 2
76128 Karlsruhe

Prof. Dr. Reinhard Matissek
Lebensmittelchemisches Institut (LCI)
des Bundesverbandes der Deutschen
Süßwarenindustrie
Adamsstr. 52–54
51063 Köln

Dr. Esther Mayer-Miebach
Bundesforschungsanstalt
für Ernährung und Lebensmittel
(BFEL)
Institut für Verfahrenstechnik
Haid-und-Neu-Str. 9
76131 Karlsruhe

Sonja Muscat
Institute for Pharmacy
and Food Chemistry
Henriette Schmidt-Burkhard Chair
of Food Chemistry
Schuhstr. 19
91052 Erlangen

Marino Petracco
Università ca Foscari
Via Rio Spinoleto, 38/2
34149 Trieste
Italy

Prof. Dr. Lars Poulsen
Allergy Clinic
National University Hospital
Department 7551
Blegdamsvej 9
2100 Copenhagen
Denmark

Walther Quasigroch
Bundesministerium Ernährung,
Landwirtschaft und Verbraucherschutz
Rochusstr. 1
53123 Bonn

Prof. Dr. Dr. Gerhard Rechkemmer
Lehrstuhl für Biofunktionalität
der Lebensmittel
Technische Universität München
Hochfeldweg 1
85350 Freising

Dr. Jack Reuvers
DSM Food Specialities
PO Box 1
2600 MA Delft
The Netherlands

Dr. Elke Richling
Lehrstuhl für Lebensmittelchemie
Universität Würzburg
Am Hubland
97074 Würzburg

Prof. Dr. Peter Schieberle
Lehrstuhl für Lebensmittelchemie
Technische Universität München
Lichtenbergstr. 4
85748 Garching

Dr. Josef Schlatter
Swiss Federal Office of Public Health
Stauffacherstr. 101
8004 Zürich
Switzerland

Gabriele Scholz
Nestlé Research Centre Lausanne
Vers-Chez-Les-Blanc
1000 Lausanne 26
Switzerland

Prof. Dr. Dr. Dieter Schrenk
Division of Food Chemistry
and Environmental Toxicology
University of Kaiserslautern
67663 Kaiserslautern

Dr. Gerrit Speijers
RIVM
Antonie van Leeuwenhoeklaan 9
3720 BA, Bilthoven
The Netherlands

Dr. Richard Stadler
Nestlé Product Technology Centre
1350 Orbe
Switzerland

Prof. Dr. Pablo Steinberg
Chair of Nutritional Toxicology
University of Potsdam
Arthur-Scheunert-Allee 114–116
14558 Nuthetal

Dr. Heike Strelen
Deutsche Forschungsgemeinschaft
(DFG)
Kennedyallee 40
53175 Bonn

Prof. Dr. Robert J. Turesky
Wadsworth Center
NYS Dept. of Health
Empire Plaza
Albany, NY 12201
USA

Prof. Dr. Stefan Vieths
Paul-Ehrlich-Institut
Abteilung Allergologie
Paul-Ehrlich-Str. 51–59
63225 Langen

Dr. Doris Wolf
DFG-Senate Commission
on Food Safety
Scientific Office
Division of Food Chemistry
and Environmental Toxicology
University of Kaiserslautern
67663 Kaiserslautern